AF834452

IRMA Lectures in Mathematics and Theoretical Physics 2

Edited by Vladimir G. Turaev

Institut de Recherche Mathématique Avancée
Université Louis Pasteur et CNRS
7 rue René Descartes
67084 Strasbourg Cedex
France

IRMA Lectures in Mathematics and Theoretical Physics

1 Deformation Quantization, *Gilles Halbout* (Ed.)

Locally Compact Quantum Groups and Groupoids

Proceedings of the meeting
of theoretical physicists and mathematicians
Strasbourg, February 21–23, 2002

Rencontre entre physiciens théoriciens et mathématiciens
Strasbourg, 21–23 février 2002

Editor

Leonid Vainerman

Walter de Gruyter · Berlin · New York 2003

Editor
Leonid Vainerman
Département de Mathématiques et Mechanique, Université de Caen, Campus II
Boulevard de Marechal Juin, B. P. 5186, 14032 Caen Cedex, France
e-mail: Leonid.Vainerman@math.unicaen.fr

Series Editor
Vladimir G. Turaev
Institut de Recherche Mathématique Avancée (IRMA), Université Louis Pasteur – C.N.R.S.,
7, rue René Descartes, 67084 Strasbourg Cedex, France, e-mail: turaev@math.u-strasbg.fr

Mathematics Subject Classification:
17B37, 22D25, 46Lxx

Key words:
quantum group, quantum groupoid, Hopf algebra, von Neumann algebra, duality

⊗ Printed on acid-free paper which falls within the guidelines of the ANSI
to ensure permanence and durability.

Library of Congress Cataloging-in-Publication Data

Locally compact quantum groups and groupoids : proceedings of the meeting of theoretical physicists and mathematicians, Strasbourg, February 21–23, 2002 / editor, Leonid Vainerman.
p. cm. – (IRMA lectures in mathematics and theoretical physics ; 2)
ISBN 3 11 017690 4
1. Quantum groups – Congresses. 2. Quantum groupoids – Congresses. 3. Locally compact groups – Congresses. 4. Mathematical physics – Congresses. I. Vainerman, Leonid. II. Series.
QC20.7.G76 L63 2002
530.14'3–dc21
2002191190

ISBN 3-11-017690-4

Bibliographic information published by Die Deutsche Bibliothek

Die Deutsche Bibliothek lists this publication in the Deutsche Nationalbibliografie;
detailed bibliographic data is available in the Internet at <http://dnb.ddb.de>.

© Copyright 2003 by Walter de Gruyter GmbH & Co. KG, 10785 Berlin.
All rights reserved, including those of translation into foreign languages. No part of this book may be reproduced in any form or by any means, electronic or mechanical, including photocopy, recording, or any information storage and retrieval system, without permission in writing from the publisher.
Printed in Germany.
Cover design: I. Zimmermann, Freiburg.
Depicted on the cover is the Strasbourg Cathedral.
Typeset using the authors' $\TeX$ files: I. Zimmermann, Freiburg.
Printing and binding: Hubert & Co. GmbH & Co. KG, Göttingen.

Preface of the Series Editor

This volume of IRMA Lectures in Mathematics and Theoretical Physics contains the proceedings of the workshop "Quantum Groups, Hopf Algebras and their Applications" held in Strasbourg in February 2002. The workshop was hosted by IRMA (Institute of Advanced Mathematical Research) in the framework of a longstanding wide-range program of meetings between mathematicians and theoretical physicists. This program was initially called "Cooperative Research Program" and was introduced by Jean Frenkel and Georges Reeb in 1965. Since then, these meetings between mathematicians and physicists have taken place at IRMA on the average twice a year. They are sponsored by CNRS (National Center of Scientific Research, France) and IRMA.

The proceedings of a number of these meetings have appeared as IRMA preprints, but were never published. The proceedings of the previous (68th) meeting "Deformation Quantization" appeared as the first volume of IRMA Lectures in Mathematics and Theoretical Physics. The 69-th meeting, whose proceedings constitute this volume, was organized by Leonid Vainerman and myself

The papers published in this volume concern the theory of quantum groups and quantum groupoids. The book should be useful to specialists in this area and related areas, as well as to students of quantum groups.

Préface de l'éditeur de la collection

Ce deuxième volume de "IRMA Lectures in Mathematics and Theoretical Physics" présente les actes du colloque "Groupes quantiques, algèbres de Hopf et leurs applications" qui s'est tenu à l'IRMA (Strasbourg) en février 2002. Le colloque s'est déroulé dans le cadre du programme général de rencontres entre physiciens théoriciens et mathématiciens. Ce programme intitulé initialement "Recherche Coopérative sur Programme" (RCP) a été créé en 1965 sur l'initiative de Jean Frenkel et Georges Reeb avec l'aide de Jean Leray et de Pierre Lelong. Depuis 1965 les rencontres entre physiciens et mathématiciens se déroulent à l'IRMA en moyenne deux fois par an. Ces rencontres sont soutenues financièrement par le CNRS et l'IRMA.

Les actes de plusieurs de ces rencontres avaient donné lieu aux prépublications de l'IRMA sans pour autant être publiés. Les actes de la rencontre précédente (68-ème) sur le thème "Deformation Quantization" sont parus dans le premier volume de la présente collection "IRMA Lectures in Mathematics and Theoretical Physics". La 69-ème rencontre "Groupes quantiques, algèbres de Hopf et leurs applications" – dont les actes constituent ce volume – a été organisée par Leonid Vainerman et moi-même.

Les articles de ce volume traitent de la théorie des groupes quantiques et des groupoïdes quantiques. Ce livre sera utile aux mathématiciens et physiciens travaillant sur ce sujet ainsi qu'à ceux qui étudient la théorie des groupes quantiques.

Strasbourg, novembre 2002 *Vladimir Turaev*

Table of Contents

Introduction of the editor

Leonid Vainerman

Département de Mathématiques et Méchanique
Université de Caen, Campus II – Boulevard de Maréchal Juin
B.P. 5186, 14032 Caen Cedex, France
email: `leonid.vainerman@math.unicaen.fr`

This volume contains seven papers written by participants of the 69 th meeting of theoretical physicists and mathematicians held in Strasbourg (February 21–23, 2002). One of the main topics discussed there was "Locally compact quantum groups and groupoids" which is the title of the volume. The purpose of this introduction is to recall some motivations and ideas from which the above topic emerged and to present the above mentioned papers.

1 Locally compact quantum groups

1.1 Kac algebras

The initial motivation to introduce objects which are more general than usual locally compact groups was to extend classical results of harmonic analysis on these groups, including the Fourier transform theory and the Pontryagin duality. It is well known that the above theories work perfectly in the framework of abelian locally compact groups. If G is such a group, then the role of exponents is played by the unitary continuous characters of G, and the set $\hat{G}$ of all such characters is again an abelian locally compact group – the dual group of G. The Fourier transform maps functions on G to functions on $\hat{G}$, and the Pontrjagin duality claims that the dual of $\hat{G}$ is isomorphic to G.

If a locally compact group G is not abelian, the set of its characters is too small, and to extend the harmonic analysis and duality in a reasonable way, one should consider instead the set $\hat{G}$ of (classes of) its unitary irreducible representations and also their matrix coefficients. For compact groups, this point of view leads to the widely known Peter–Weyl theory; the duality theory for such groups was done by T. Tannaka [75] and M. G. Krein [42]. A new feature of this duality is that $\hat{G}$ does not carry a structure of a group, but can be equipped with some quite different structure (block-algebra or Krein algebra [30]); however, starting with such a structure, the initial compact group can be reconstructed. Such a non-symmetric duality was later established by W. F. Stinespring [69] for unimodular groups, and by P. Eymard [28] and T. Tatsuuma [76] for general locally compact groups.

In 1961 G. I. Kac [33], [34] proposed a completely new idea, which allowed to restore the symmetry of the duality for unimodular, not necessarily abelian, groups. Namely, he introduced a category of objects (he called them *ring groups*), containing both unimodular groups and their duals, and constructed the Fourier transform and duality within this category. His duality extended those of Pontryagin, Tannaka–Krein and Stinespring.

In algebraic terms, one can think of a ring group as of a Hopf *-algebra with an involutive antipode S (i.e., $S^2 = \mathrm{id}$). In topological terms, its algebra A is a von Neumann algebra, and the comultiplication $\Delta\colon A \to A \otimes A$ and the antipode $S\colon A \to A$ are von Neumann algebra maps. On the contrary, its counit is not a well defined von Neumann algebra map, that is why it is not present in the definition of a ring group. Instead, A is required to be equipped with a faithful normal trace φ compatible with Δ and S and playing the role of a Haar measure. Without discussing here this definition in detail, let us show two standard examples of ring groups related to an ordinary unimodular group G with a Haar measure μ.

Example 1.1. $A = L^\infty(G,\mu)$, $\Delta\colon f(g) \mapsto f(gh)$, $S\colon f(g) \mapsto f(g^{-1})$, $\varphi(f) = \int_G f(g)d\mu(g)$, where $g, h \in G$, $f(\cdot) \in L^\infty(G,\mu)$.

Example 1.2. $A = \mathcal{L}(G)$ – the von Neumann algebra generated by left translations L_g or by left convolutions $L_f = \int_G f(g)L_g d\mu(g)$ with continuous functions $f(\cdot) \in L^1(G,\mu)$ $\Delta\colon L_g \mapsto L_g \otimes L_g$, $S\colon L_g \mapsto L_{g^{-1}}$, $\varphi(f) = f(e)$, where $g \in G$, e is the unit of G.

G.I. Kac showed that for any commutative (resp., co-commutative) ring group $\mathcal{G}$, i.e., such that the algebra A is commutative (resp., $\sigma \circ \Delta = \Delta$, where $\sigma : a \otimes b \mapsto b \otimes a$ is the usual flip in $A \otimes A$), there is a unimodular group G such that $\mathcal{G}$ is isomorphic to the ring group of Example 1.1 (resp., 1.2) related to G. Thus, the category of unimodular groups (resp., their duals) is embedded into the category of ring groups.

The theory of ring groups used, as a technical tool, I. Segal's theory of traces on von Neumann algebras, which is a non-commutative extension of the classical theory of measure and integral. In [36], [37], [38] G. I. Kac and V. G. Paljutkin gave concrete examples of non-trivial, i.e., non-commutative and non-co-commutative, ring groups, which were neither ordinary groups nor their duals. As it was mentioned by V. G. Drinfeld [13], the Kac–Paljutkin examples were the first concrete examples of quantum groups.

The theory was completed in the early '70s, when the Tomita–Takesaki theory and the foundations of the theory of weights on operator algebras became available – our reference to these topics is [70]. Namely, G. I. Kac and L. Vainerman [39], on the one hand, and M. Enock and J.-M. Schwartz [21], on the other hand, extended the category of ring groups in order to cover all locally compact groups (certain results in this direction were obtained also by M. Takesaki [72], [73]). They allowed φ and $\varphi \circ S$ to be different weights on A playing respectively the role of a left and a right Haar measure (for ring groups $\varphi = \varphi \circ S$ was a trace), gave appropriate axioms and extended the construction of the dual.

To emphasize the importance of the pioneering work of G. I. Kac, M. Enock and J.-M. Schwartz called these more general objects *Kac algebras*. Locally compact groups and their duals were embedded in this category respectively as commutative (see [72]) and co-commutative (see [81]) Kac algebras, the corresponding duality covered all versions of duality for such groups. The standard reference to the Kac algebra theory is [22]. C^*-algebraic Kac algebras have been discussed in [63], [24] (see also [82]).

1.2 From Kac algebras to locally compact quantum groups

The discovery of quantum groups by V. G. Drinfeld [13] was accompanied by the arrival of new important examples of Hopf $*$-algebras, obtained by deformation either of universal enveloping algebras of Lie algebras [13], [31], or of function algebras on Lie groups [92], [93], [68]. Their operator algebra versions did not fit into the Kac algebra theory, because the antipodes were neither involutive, nor even bounded maps. This provided a strong motivation to construct a more general theory, which would be as elegant as that of Kac algebras but would also cover these new examples. The first steps in this direction were made in [92], [94], where S. L. Woronowicz constructed the theory of compact quantum groups and developed for them the Peter–Weyl theory and the Tannaka–Krein duality. Moreover, he managed to deduce the existence of a Haar measure from his set of axioms rather than assume it, as was the case in the Kac algebra theory (and, as we will see below, in some of its extensions). The last feature holds also for discrete quantum groups – see [64], [16], [15], [87].

Remark 1.3. 1) The Haar theorem for compact C^*-algebraic ring groups has been proven by V. G. Paljutkin [63] (see also [82]).

2) In [11], the Peter–Weyl theory was constructed for much more general objects than compact quantum groups, for which the comultiplication is not necessarily an algebra map.

In the case of non-compact and non-discrete quantum groups, an in-depth prior analysis of concrete examples was necessary. It was not so difficult to construct such examples in terms of generators of certain Hopf $*$-algebras and commutation relations between them. It was much harder to represent these generators as (typically, unbounded) operators acting on a Hilbert space and to give a meaning to the relations of commutation between these operators. Finally, it was even more difficult to associate an operator algebra with the above system of operators and commutation relations and to construct comultiplication, antipode and invariant weights as applications related to this algebra. There is no general approach to these highly nontrivial problems, and one must design specific methods in each specific case [95]–[98], [64], [1], [90].

There are other examples of operator algebraic quantum groups which are easier to construct. For example, given a non-commutative locally compact group G, one can replace the comultiplication Δ of the co-commutative Kac algebra described in Example 1.2 with the new comultiplication of the form $\Delta_\Omega(\cdot) = \Omega\Delta(\cdot)\Omega^{-1}$, where Ω is an element from $\mathcal{L}(G) \otimes \mathcal{L}(G)$ such that Δ_Ω remains co-associative. This construction (called *twisting*) was developed on a purely algebraic level by V.G. Drinfeld [14] and on an operator algebraic level in [23], [83] and [55], where numerous concrete examples were obtained as well. Note that a, in a sense dual, construction has been proposed by M. Rieffel [65].

The other construction has been developed in [35]. Given two finite groups, G_1 and G_2, viewed respectively as a co-commutative ring group $(\mathcal{L}(G_1), \Delta_1)$ (see Example 1.2) and a commutative ring group $(L^\infty(G_2), \Delta_2)$ (see Example 1.1), let us try to find a ring group (A, Δ) which makes the sequence

$$(L^\infty(G_2), \Delta_2) \to (A, \Delta) \to (\mathcal{L}(G_1), \Delta_1) \tag{1}$$

exact. G. I. Kac explained that: 1) (A, Δ) exists if and only if G_1 and G_2 are subgroups of a group G such that $G_1 \cap G_2 = \{e\}$ and $G = G_1 G_2$. Equivalently, G_1 and G_2 must act on each other (as on sets), and these actions must be compatible. 2) To get all possible (A, Δ) (they are called *extensions* of $(L^\infty(G_2), \Delta_2)$ by $(\mathcal{L}(G_1), \Delta_1)$), one must find all possible 2-cocycles for the above mentioned actions, compatible in certain sense. Under these conditions, [35] gives the explicit construction of (A, Δ) (the cocycle bicrossed product construction). The famous Kac–Paljutkin examples of non-trivial ring groups [36], [37], [38] are exactly of this type. Later on, both algebraic and analytic aspects of this construction were intensively studied by S. Majid [50]–[53] who gave also a number of examples of operator algebraic quantum groups, some of them were not Kac algebras. Very recently, the theory of extensions of the form (1), with locally compact G_1 and G_2, has been developed in [80].

An important step in the generalization of the Kac algebra theory was the theory of *multiplicative unitaries*. Already W. F. Stinespring [69] mentioned an important role in the construction of the dual for a unimodular nonabelian group G played by the unitary

$$W_G(\xi)(g, h) = \xi(g, g^{-1}h) \tag{2}$$

acting on $L^2(G, \mu) \otimes L^2(G, \mu)$. G. I. Kac, in order to construct his duality for ring groups, introduced in this more general context a similar unitary

$$W^*(\Lambda(a) \otimes \Lambda(b)) = (\Lambda \otimes \Lambda)(\Delta(b)(a \otimes 1)), \tag{3}$$

where $a, b \in \mathfrak{N}_\varphi := \{x \in A : \varphi(x^*x) < \infty\}$, Λ is the GNS-mapping for φ [70]. Moreover, he was the first to point out that W verifies the *Pentagonal relation*:

$$W_{12} W_{13} W_{23} = W_{23} W_{12} \tag{4}$$

and to show that all the information about the ring group could be encoded in W.

On the contrary, S. Baaj and G. Skandalis [2] took a unitary verifying (4) (they called it a multiplicative unitary), as the starting point of their theory. They have constructed two Hopf C^*-algebras in duality out of a given multiplicative unitary, under certain regularity conditions, and gave a number of important constructions of C^*-algebraic quantum groups in this framework (including the bicrossed product construction). The investigation of the above mentioned regularity conditions and alternative *manageability* conditions [96] is one of the most important topics in the theory of multiplicative unitaries [1], [3], [96], [5]. Note that several examples of C^*-algebraic quantum groups, more general than Kac algebras, were given in [2], [67].

T. Masuda and Y. Nakagami proposed an extension of the Kac algebra theory by requiring the antipode S to have a polar decomposition consisting of a unitary part and a generator of one-parameter group of automorphisms of a von Neumann algebra A. The idea of such a polar decomposition of S is due to E. Kirchberg (unpublished). The Kac algebra case is exactly the situation when S equals its unitary part and for that reason is involutive and bounded. A certain disadvantage of this approach was the necessity for some quite complicated axioms which disappears in the Kac algebra case. A joint work by T. Masuda, Y. Nakagami and S. L. Woronowicz on the C^*-algebra version of this theory is still in progress.

To sum up, one can say that trying to extend the Kac algebra theory in order to cover important concrete examples of quantum groups, one faces a mixture of algebraic and analytic problems. That is why it was important to design a purely algebraic framework, where the main algebraic features of the future theory would be present. It was done by A. Van Daele in [88], [89] and in his joint work with J. Kustermans [46], where the notion of a multiplier Hopf *-algebra with positive integrals was proposed and a natural duality was constructed. As for analytic aspects of the story, by the end of the '90s the theory of weights on C^*-algebras had been further developed, mainly by J. Kustermans, and after that the theory of locally compact quantum groups was proposed by J. Kustermans and S. Vaes [43]–[45].

A locally compact quantum group is a collection $\mathcal{G} = (A, \Delta, \varphi, \psi)$, where A is either a C^*- or a von Neumann algebra equipped with a co-associative comultiplication $\Delta\colon A \to A \otimes A$ and two faithful semi-finite normal weights φ and ψ - right and left Haar measures. The antipode is not explicitly present in this definition, but can be constructed from the above data, as well as its polar decomposition, using the multiplicative unitary, canonically associated with $\mathcal{G}$ by means of the formula (3). Kac algebras, compact and discrete quantum groups are special cases of a locally compact quantum group, but what is even more interesting, all important concrete examples of operator algebraic quantum groups fit into this framework. One can find an exposition of this theory in [47] and [79]. In the present book, more information on locally compact quantum groups can be found in the Preliminaries of the article by S. Vaes and L. Vainerman. To simplify the notations, in what follows we denote a locally compact quantum group by (A, Δ); usually we deal with the case when A is a von Neumann algebra and $\Delta\colon A \to A \otimes A$ is a normal monomorphism of von

Neumann algebras. Let us present now the three papers on locally compact quantum groups contained in this volume. We start with a paper by J. Kustermans and E. Koelink devoted to a concrete example of a locally compact quantum group, related to $\mathrm{SU}_q(1,1)$. As a Hopf *-algebra, $\mathrm{SU}_q(1,1)$ is one of the three real forms of $\mathrm{SL}_q(2,\mathbb{C})$, the two others being $\mathrm{SU}_q(2)$ and $\mathrm{SL}_q(2,\mathbb{R})$. Remark that the quantum group $\mathrm{SU}_q(2)$ and its dual are well understood on the operator algebra level [92]–[94], [68]; such an understanding of $\mathrm{SL}_q(2,\mathbb{R})$ is still an open problem.

Concerning $\mathrm{SU}_q(1,1)$, in 1991 S. L. Woronowicz showed that this object cannot exist as a C^*-Hopf algebra, and this result was a source of pessimism for several years. Then L. Korogodsky explained that it was reasonable to deform rather the normalizer $\widetilde{\mathrm{SU}}(1,1)$ of $\mathrm{SU}(1,1)$ in $\mathrm{SL}_q(2,\mathbb{C})$ than $\mathrm{SU}_q(1,1)$ itself. The paper of J. Kustermans and E. Koelink gives a clear overview of the highly nontrivial construction of quantum $\widetilde{\mathrm{SU}}(1,1)$ and its dual as locally compact quantum groups and their theory of representations. The main tool they use is the explicit analysis of eigenfunctions of certain unbounded operators in terms of special functions of q-hypergeometric type. The paper also contains historical remarks and shows the contribution of other specialists.

The paper by A. Van Daele is a survey of the theory of algebraic quantum groups (multiplier Hopf *-algebras with positive integrals) and their relations with locally compact quantum groups. As was mentioned above, this theory provided one of the main motivations for the development of locally compact quantum groups by J. Kustermans and S. Vaes and showed almost all algebraic features of the latter. On the other hand, it is much easier technically, even if much attention is attached to the links with the corresponding operator algebraic results. The category of algebraic quantum groups contains the categories of compact and discrete quantum groups (but not all the ordinary locally compact groups), is self-dual and closed under several constructions, such as, for example, the Drinfeld double. An important tool used in the paper is the Fourier transform. Thus, algebraic quantum groups provide a good and relatively simple model for studying more general objects. So the paper will be of interest both for students and experts.

The paper by S. Vaes and L. Vainerman is devoted to extensions of *Lie groups* of the form (1). In this case, instead of the condition $G = G_1 G_2$, one should require $G_1 G_2$ to be an open dense subset of G, as in [3]. Then, for the corresponding *Lie algebras* we have $g = g_1 \oplus g_2$ – the direct sum of vector spaces. So, to construct examples of locally compact quantum groups, one can start with such a decomposition of Lie algebras and try to construct a corresponding pair of groups (G_1, G_2). But this problem proves to be not so easy to resolve (typically, one must deal with non-connected Lie groups), and often it has no solution at all. In the paper the case of complex and real Lie groups G_1 and G_2 of low dimensions is studied in detail. In particular, a complete classification of the corresponding locally compact quantum groups with two or three generators is obtained, and all the ingredients of their structure are computed, as well as their infinitesimal objects (Hopf *-algebras and Lie bialgebras).

2 From quantum groups to quantum groupoids

2.1 Actions of locally compact quantum groups and subfactors

Classical groups are interesting first of all as groups of transformations, acting on certain spaces. Similarly, one can define a (left) action of a locally compact quantum group (A, Δ) on a von Neumann algebra N (which plays the role of a "quantum space") as a normal monomorphism $\alpha\colon N \to A \otimes N$ of von Neumann algebras such that $(\mathrm{id} \otimes \alpha)\alpha = (\Delta \otimes \mathrm{id})\alpha$. Now the *fixed point subalgebra* can be defined as

$$N^\alpha := \{x \in N : \alpha(x) = 1 \otimes x\},$$

and the *crossed product* $A \ltimes N$ as the von Neumann algebra generated by $\alpha(N)$ and $\hat{A} \otimes \mathbb{C}$, where $\hat{A}$ is the von Neumann algebra of the dual. An action is said to be *outer* if $(N^\alpha)' \cap N = \mathbb{C}$. For the motivations and details see [77], [79], [80] and Preliminaries of the article by S. Vaes and L. Vainerman in this volume. There is a series of nice results on such actions that extend classical results on actions of locally compact groups on von Neumann algebras [77], [79], but here we focus our attention on the links with subfactors.

Starting with a given inclusion $N_0 \subset N_1$ of von Neumann algebras and performing step by step the well known *basic construction* of V. Jones, one can obtain the *Jones' tower* of von Neumann algebras [32]:

$$N_0 \subset N_1 \subset N_2 \subset N_3 \subset \cdots .$$

Recall that the initial inclusion is said to be *irreducible*, if $N_0' \cap N_1 = \mathbb{C}$ (in this case all the N_i $(i = 0, 1, 2, \dots)$ are *factors*, i.e., have trivial centers), and *of depth* 2, if the triple of relative commutants

$$N_0' \cap N_1 \subset N_0' \cap N_2 \subset N_0' \cap N_3$$

is again the basic construction.

Example 2.1. Given an outer action α of a locally compact group G on a factor N_1, the inclusion $N_0 = N_1^\alpha \subset N_1$ is irreducible and of depth 2, and N_2 is isomorphic to $G \ltimes N_1$.

M. Enock and R. Nest [20], [17] showed that, conversely, for any irreducible depth 2 subfactor $N_0 \subset N_1$ satisfying a natural regularity condition, the von Neumann algebra $A = N_1' \cap N_3$ can be given the structure of a locally compact quantum group (A, Δ) with an outer action α of the commutant $(A, \Delta)'$ on N_1, such that $N_0 = N_1^\alpha$ and the triples $N_0 \subset N_1 \subset N_2$ and $\mathbb{C} \otimes N_1^\alpha \subset \alpha(N_1) \subset A' \ltimes N_1$ are isomorphic (in fact, this result was precised by S. Vaes [77]).

Remark 2.2. The idea that outer actions of Kac algebras are closely related to the structure of irreducible depth 2 subfactors, is due to A. Ocneanu (see, for example, Postface in [22]). Finite index irreducible depth 2 subfactors of type II_1 were characterized in terms of outer actions of finite-dimensional Kac algebras by R. Longo [48], W. Szymanski [71] and M. C. David [12].

This beautiful result motivates the following natural hypothesis: if we drop the irreducibility condition keeping however the depth 2 condition for a subfactor, this situation should be characterized in terms of an outer action of some more general structure than a locally compact quantum group. And this is a way to approach the notion of a locally compact quantum groupoid.

Indeed, already finite index depth 2 subfactors of type II_1 reveal the purely algebraic aspect of the story. It is shown in [59] that in this case the above mentioned result is still true, up to notations, if one gives the finite-dimensional algebra $A = N_0' \cap N_2$ a structure of a *weak C^*-Hopf algebra* (introduced in [7], [6]) acting outerly on N_1. Like a finite-dimensional Kac algebra, a weak C^*-Hopf algebra is a finite-dimensional C^*-algebra A equipped with a co-associative comultiplication, an antipode and a counit. The main difference between them is that this comultiplication is not necessarily a unital map and the counit is not necessarily a homomorphism of algebras $A \to \mathbb{C}$. This implies the existence of a canonical C^*-subalgebra R of A, called counital or *base subalgebra*, playing a fundamental role within this structure. For a weak C^*-Hopf algebra coming from subfactors we have $R = N_0' \cap N_1$; clearly, $R = \mathbb{C}$ if and only if the subfactor is irreducible. One can show that the dual vector space for a weak C^*-Hopf algebra carries the structure of the same type, i.e., this notion is self-dual.

Like in Examples 1.1 and 1.2, the algebra of functions and the groupoid algebra of a usual finite groupoid give respectively standard examples of a commutative and cocommutative weak C^*-Hopf algebra [58], [61] which justifies the usage of the term "quantum groupoid". Moreover, the notion of the base subalgebra naturally extends the function algebra on the set of units of a usual groupoid. For examples of non-trivial (i.e., non-commutative and non-cocommutative) quantum groupoids see [7], [57], [58], [59], [61], [26].

Initially, weak C^*-Hopf algebras were introduced in [7] as symmetries of certain models in algebraic quantum field theory. Another source of interest in them is their representation category, which is flexible enough to describe all rigid monoidal C^*-categories with finitely many classes of simple objects (in general, in this representation category a unit object is not a counit because the latter is not a representation, and the tensor product differs from the usual tensor product of vector spaces) [8], [57], [60], [61]. So, quantum dimensions of irreducible representations need not to be integer, and these categories have interesting applications in low-dimensional topology [57], [61]. A survey of the theory of finite quantum groupoids and their applications can be found in [61].

2.2 Multiplicative partial isometries and pseudo-multiplicative unitaries

As noted above, multiplicative unitaries are of fundamental importance in the theory of locally compact quantum group. So, it would be natural to define and to study similar objects also for quantum groupoids. Since for any weak C^*-Hopf algebra there exists a positive linear form on its C^*-algebra A playing the role of a Haar measure [6], one can define an operator W by (3). Now W is not in general a unitary, but just a partial isometry verifying the Pentagonal equation (4) [9], [86].

Like in the case of quantum groups, the inverse problem is more subtle, and in order to resolve it one should impose some regularity conditions on a given partial isometry. Namely, J. M. Vallin showed in [86] that any regular multiplicative partial isometry generates two quantum groupoids in duality, which extends the above mentioned result of S. Baaj and G. Skandalis on multiplicative unitaries. In the paper published in this book, J. M. Vallin continues the study of the structure of regular multiplicative partial isometries acting on a finite-dimensional Hilbert space, in the spirit of [4], where finite-dimensional multiplicative unitaries were studied in detail.

First, it is shown that, after an amplification and reduction, any regular multiplicative partial isometry is isomorphic to an *irreducible* one, i.e., verifying a certain quite strong condition. The latter condition allows to prove *quantum Markov properties*; for instance, the existence of a faithful positive linear form on the involutive algebra generated by the two quantum groupoids associated to the partial isometry (the Weyl algebra) that extends both normalized Haar measures of these quantum groupoids. In its turn, this implies that any regular multiplicative partial isometry is a composition of two very simple partial isometries. Finally, it is shown that a regular multiplicative partial isometry is completely determined by the two quantum groupoids associated and by the spaces of its fixed and cofixed vectors, and a complete characterization of quantum groupoids in duality acting on the same Hilbert space in the irreducible situation is obtained.

The notion of a locally compact quantum groupoid is much less transparent in the infinite-dimensional case, which corresponds to the infinite index depth 2 inclusions of von Neumann algebras (in fact, the development of this theory is still in progress). The reason is that in this case complicated analytical aspects play a significant role, as well as the presence of nontrivial base von Neumann algebra. In particular, instead of usual tensor products of Hilbert spaces and von Neumann algebras one should inevitably use the relative tensor product of Hilbert spaces and the "fiber" product of von Neumann algebras over a base algebra. In the finite-dimensional case these new notions reduce respectively to a subspace of the usual tensor product of Hilbert spaces and to a reduced subalgebra of the usual tensor product of von Neumann algebras. For the definitions and explanations see the paper by M. Enock on infinite-dimensional locally compact quantum groupoids published in this volume, which also outlines the nearest prospects for this field.

To approach the notion of a locally compact quantum groupoid, it is necessary first to figure out, what kind of objects can be associated with an ordinary locally compact groupoid in the spirit of Examples 1.1 and 1.2 and the formula (2). It was explained in [84], [86] that one gets this way two *Hopf bimodules* in duality – commutative and co-commutative, and a *pseudo-multiplicative unitary*. The same objects were associated with depth 2 inclusions of von Neumann algebras in [25], [18]; moreover, in both cases one can even equip the Hopf bimodules with antipodes having polar decompositions. For the definitions and explanations see the survey by M. Enock. We only remark that both these structures are defined over a base von Neumann algebra, and that in the finite-dimensional case they reduce respectively to a weak C^*-Hopf algebra and to a multiplicative partial isometry.

Like in the theory of locally compact quantum groups, it is crucial to understand exact relations between these two "faces" of a locally compact quantum groupoid. It was shown in [25], [18] that, given a pseudo-multiplicative unitary, one can construct in a natural way two Hopf bimodules in duality (as we mentioned above, in the cases related to a usual locally compact groupoid and to depth 2 inclusions of von Neumann algebras, one can even equip these objects with antipodes having polar decompositions). The work by F. Lesieur on a converse result is still in progress. Finally, in [19], the theory of quantum groupoids of compact type is developed, following the strategy of [2].

2.3 On purely algebraic quantum groupoids

Until now we have discussed quantum groups and groupoids only in the framework of operator algebras. As for purely algebraic quantum groupoids, there are several versions of them, designed from various motivations. Let us mention first the notion of a *weak Hopf algebra* [6] extending substantially the one of a Hopf algebra. Like in the C^*-case, the main difference between them is that the comultiplication of a weak Hopf algebra A is not necessarily a unital map and the counit is not necessarily a homomorphism of algebras $A \to k$ (k is the ground field), and this implies the existence of a base subalgebra R of A, which is automatically separable (if R is commutative, we get a notion of a *face algebra* [29]). The theory of these objects in the finite-dimensional case nicely extends that of Hopf algebras [6], [56], [91], [8], [57], [61]. Their representation categories cover all rigid monoidal categories with finitely many classes of simple objects, even in the case of a commutative base subalgebra [29], [62]. So, they can be used as an appropriate tool for the study of such categories [27], [57]. Dropping the antipode in a weak Hopf algebra we get a *weak bialgebra* whose representation category is monoidal, but not necessarily rigid.

The notion of a weak Hopf algebra (resp., weak bialgebra) is a partial case of that of a *Hopf algebroid* (resp., *bialgebroid*) in the sense of [49] and [99] – see [26] (resp., [66]). The definition of the latter two structures was motivated by the analogy with a usual (semi)groupoid, their base algebra naturally extends the function algebra on the set of its units. On the other hand, the notion of a bialgebroid is equivalent to that of a $\times_R$-bialgebra introduced earlier by M. Takeuchi [74] (here also, R denotes a base algebra) – see [10]. It was shown in [66] that a $\times_R$-bialgebra with a separable base is a weak bialgebra. For all the above mentioned objects, their representation category is monoidal.

Brief discussion of some other versions of quantum groupoids can be found in [61].

Now we are ready to present the two remaining papers of this volume. P. Schauenburg discusses a construction that allows to replace the base algebra R in any $\times_R$-bialgebra A with a Morita-equivalent algebra S (i.e., having equivalent representation category) in order to obtain a $\times_S$-bialgebra whose representation category is equivalent to that of A as monoidal categories. He gives a spectacular illustration: for a concrete example of a weak Hopf algebra from [60], [61] this Morita base change reduces the dimension of A from 122 to 24 without affecting the monoidal category of representations (the base algebra changes from $\mathbb{C} \oplus M_2(\mathbb{C})$ to $\mathbb{C} \oplus \mathbb{C}$).

The starting point for the paper by K. Szlachányi is a *balanced depth* 2 *extension of algebras* $N \subset M$ which is a purely algebraic generalization of the notion of finite index depth 2 von Neumann subfactors – see the definition in the text. For such an extension, the endomorphism ring $A = \mathrm{End}_N M_N$ carries a bialgebroid structure (its base R is the relative commutant of N in M) equipped with the canonical action on M, whose subalgebra of A-invariants is N [40]. This generalizes the above mentioned result of [59] in the subfactor theory.

Finally, it is explained that balanced depth 2 extensions of algebras are the proper analogues of the Galois extensions of fields (i.e., normal and separable field extensions) because they have "finite quantum automorphism groups" with subalgebra of invariants equal to N and which are characterized by a universal property, hence, unique. The role of such a "finite quantum automorphism group" is played by a bialgebroid that is finitely generated projective over its base as a left and a right module (the problem of the existence of the antipode in this bialgebroid is still open). If R is separable, then A is a weak bialgebra; if, moreover, $N \subset M$ is a Frobenius extension, then A is a weak Hopf algebra. In the special case of a separable field extension, the structure of such a universal weak Hopf algebra is written down explicitly.

References

[1] S. Baaj, Représentation régulière du groupe quantique des déplacements de Woronowicz, *Astérisque* 232 (1995), 11–48.

[2] S. Baaj and G. Skandalis, Unitaires multiplicatifs et dualité pour les produits croisés de C^*-algèbres, *Ann. Sci. École Norm. Sup.* (4) 26 (1993), 425–488.

[3] S. Baaj and G. Skandalis, Transformations pentagonales, *C. R. Acad. Sci. Paris Sér.* I *Math.* 327 (1998), 623–628.

[4] S. Baaj, E. Blanchard and G. Skandalis, Unitaires multiplicatifs en dimention finie et leurs sous-objets, *Ann. Inst. Fourier* 49 (1999), 1305–1344.

[5] S. Baaj, G. Skandalis and S. Vaes, Non-semi-regular quantum groups coming from number theory, preprint 2002.

[6] G. Böhm, F. Nill and K. Szlachányi, Weak Hopf algebras I. Integral theory and C^*-structure, *J. Algebra* 221 (1999), 385–438.

[7] G. Böhm and K. Szlachányi, A co-associative C^*-quantum group with nonintegral dimensions, *Lett. Math. Phys.* 35 (1996), 437–456.

[8] G. Böhm and K. Szlachányi, Weak Hopf algebras II. Representation Theory, Dimensions and the Markov Trace, *J. Algebra* 233 (2000), 156–212.

[9] G. Böhm and K. Szlachányi, Weak C^*-Hopf algebras and multiplicative isometries, *J. Operator Theory* 45 (2001), 357–376.

[10] T. Brzeziński and G. Militaru, Bialgebroids, $\times_R$-bialgebras and duality, *J. Algebra* 251 (2002), 279–294.

[11] Yu. A. Chapovsky and L. I. Vainerman, Compact Quantum Hypergroups, *J. Operator Theory* 41 (1999), 1–28.

[12] M. C. David, Paragroupe d'Adrian Ocneanu et algèbre de Kac, *Pacific J. Math.* 172 (1996), 331–363.

[13] V. G. Drinfeld, Quantum groups, *Proceedings ICM Berkeley* (1986), Amer. Math. Soc., Providence, RI, 1987, 798–820.

[14] V. G. Drinfeld, Quasi-Hopf algebras, *Leningrad. Math. J.* 1 (1990), 1419–1457.

[15] M. S. Dijkhuisen and T. H. Koornwinder, CQG algebras: a direct algebraic approach to compact quantum groups, *Lett. Math. Phys.* 32 (1994), 315–330.

[16] E. Effros and Z.-J. Ruan, Discrete quantum groups: The Haar measure, *Internat. J. Math.* 5 (1994), 681–723.

[17] M. Enock, Inclusions irréductibles de facteurs et unitaires multiplicatifs II, *J. Funct. Anal.* 154 (1998), 67–109.

[18] M. Enock, Inclusions of von Neumann algebras and quantum groupoids II, *J. Funct. Anal.* 178 (2000), 156–225.

[19] M. Enock, Quantum groupoids of compact type, *prépublication, Institut de Math. de Jussieu* 328, Mai 2002.

[20] M. Enock and R. Nest, Inclusions of factors, multiplicative unitaires and Kac algebras, *J. Funct. Anal.* 137 (1996), 466–543.

[21] M. Enock and J.-M. Schwartz, Une dualité dans les algèbres de von Neumann, *C. R. Acad. Sci. Paris* 277 (1973), 683–685; *Bull. Soc. Math. France, Suppl., Mém.* 44 (1975), 1–144.

[22] M. Enock and J.-M. Schwartz, *Kac Algebras and Duality of Locally Compact Groups*, Springer-Verlag, Berlin 1992.

[23] M. Enock and L. Vainerman, Deformation of a Kac Algebra by an Abelian Subgroup, *Comm. Math. Phys.* 66 (1993), 619–650.

[24] M. Enock and J. M. Vallin, C^*-algèbres de Kac et algèbres de Kac, *Proc. London. Math. Soc.* 172 (2000), 249–300.

[25] M. Enock and J. M. Vallin, Inclusions of von Neumann algebras and quantum groupoids, *J. Funct. Anal.* 172 (2000), 249–300.

[26] P. Etingof and D. Nikshych, Dynamical quantum groups at roots of 1, *Duke Math. J.* 108 (2001), 135–168.

[27] P. Etingof, D. Nikshych and V. Ostrik, On fusion categories, preprint `math.QA/0203060` (2002).

[28] P. Eymard, L'algèbre de Fourier d'un groupe localement compact, *Bull. Soc. Math. France* 92 (1964), 181–236.

[29] T. Hayashi, A canonical Tannaka duality for finite semisimple tensor categories, preprint `math.QA/9904073` (1999).

[30] E. Hewitt and K. A. Ross, *Abstract Harmonic Analysis* II, Grundlehren Math. Wiss. 152, Springer-Verlag, Berlin–Heidelberg–New York 1970.

[31] M. Jimbo, A q-difference analogue of $U(g)$ and the Yang-Baxter equation, *Lett. Math. Phys.* 10 (1985), 63–69.

[32] V. Jones and V. S. Sunder, *Introduction to Subfactors*, London Math. Soc. Lecture Note Ser. 234, Cambridge University Press, Cambridge 1997.

[33] G. I. Kac, Generalization of the group principle of duality, *Soviet. Math. Dokl.* 2 (1961), 581–584.

[34] G. I. Kac, Ring groups and the principle of duality, I, II, *Trans. Moscow Math. Soc.* (1963), 291–339; ibid. 13 (1965), 94–126.

[35] G. I. Kac, Extensions of groups to ring groups, *Math. USSR-Sb.* 5 (1968), 451–474.

[36] G. I. Kac and V. G. Paljutkin, Finite ring groups, *Trans. Moscow Math. Soc.* 5 (1966), 251–294.

[37] G. I. Kac and V. G. Paljutkin, Example of a ring group generated by Lie groups, *Ukraïn. Mat. Zh.* 16 (1) (1964), 99–105 (in Russian).

[38] G. I. Kac and V. G. Paljutkin, Example of a ring group of order eight, *Uspekhi Mat. Nauk* 20 (1965), 268–269 (in Russian).

[39] G. I. Kac and L. I. Vainerman, Nonunimodular ring groups and Hopf-von Neumann algebras, *Soviet. Math. Dokl.* 14 (1973), 1144–1148; *Math. USSR-Sb.* 23 (1974), 185–214.

[40] L. Kadison and K. Szlachányi, Dual bialgebroids for depth 2 ring extensions, preprint `math.RA/0108067` (2001).

[41] E. Koelink and J. Kustermans, A locally compact quantum group analogue of the normalizer of $SU(1, 1)$ in $SL(2, C)$, preprint `math.QA/0105117` (2001).

[42] M. G. Krein, The principle of duality for a compact group and a square block algebra, *Soviet. Math. Dokl.* 69 (1949), 725–728; Hermitian-positive kernels on homogeneous spaces I, II, *Amer. Math. Soc. Transl. Ser.* 34 (1963), 69–108; ibid. 109–164, translated from the Russian original *Ukraïn. Mat. Zh.* 1 (4) (1949), 64–98; ibid. 2 (1) (1950), 10–59.

[43] J. Kustermans and S. Vaes, Locally compact quantum groups, *Ann. Sci. École Norm. Sup.* 33 (6) (2000), 837–934.

[44] J. Kustermans and S. Vaes, Locally compact quantum groups in the von Neumann algebraic setting, to appear in *Comm. Math. Phys.* (2000).

[45] J. Kustermans and S. Vaes, A simple definition for locally compact quantum groups, *C. R. Acad. Sci. Paris Sér.* I 328 (10) (1999), 871–876.

[46] J. Kustermans and A. Van Daele, C^*-algebraic quantum groups arising from algebraic quantum groups, *Internat. J. Math.* 8 (1997), 1067–1139.

[47] J. Kustermans, S. Vaes, L. Vainerman, A. Van Daele and S. L. Woronowicz, Locally Compact Quantum Groups, in *Lecture Notes for the School on Noncommutative Geometry and Quantum Groups in Warsaw (17–29 September* 2001), Banach Center Publ., Warsaw, to appear.

[48] R. Longo, A duality for Hopf algebras and subfactors I, *Comm. Math. Phys.* 159 (1994), 133–150.

[49] J.-H. Lu, Hopf algebroids and quantum groupoids, *Internat. J. Math.* 7 (1996), 47–70.

[50] S. Majid, Physics for algebraists: Non-commutative and non-cocommutative Hopf algebras by a bicrossproduct construction, *J. Algebra* 130 (1990), 17–64.

[51] S. Majid, Hopf-von Neumann algebra bicrossproducts, Kac algebra bicrossproducts, and the classical Yang-Baxter equations, *J. Funct. Anal.* 95 (1991), 291–319.

[52] S. Majid, More examples of bicrossproduct and double cross product Hopf algebras, *Israel J. Math.* 72 (1990), 133–148.

[53] S. Majid, Foundations of quantum group theory, Cambridge University Press, Cambridge 1995.

[54] T. Masuda and Y. Nakagami, A von Neumann algebraic framework for the duality of the quantum groups, *Publ. Res. Inst. Math. Sci.* 30 (1994), 799–850.

[55] D. Nikshych, K_0-ring and twisting of finite dimensional Hopf algebras, *Comm. Algebra* 26 (1998), 321–342.

[56] D. Nikshych, On the structure of weak Hopf algebras, to appear in *Adv. Math.*, preprint `math.QA/0106010` (2001).

[57] D. Nikshych, V. Turaev and L. Vainerman, Invariants of Knots and 3-manifolds from Quantum Groupoids, to appear in *Topology and its applications*; preprint `math.QA/0006078` (2000).

[58] D. Nikshych and L. Vainerman, Algebraic versions of a finite-dimensional quantum groupoid, in *Hopf algebras and quantum groups* (S. Caenepeel et al., eds.), Lecture Notes in Pure and Appl. Math. 209 , Marcel Dekker, New York 2000, 189–221.

[59] D. Nikshych and L. Vainerman, A characterization of depth 2 subfactors of II_1-factors, *J. Funct. Anal.* 171 (2000), 278–307.

[60] D. Nikshych and L. Vainerman, A Galois correspondence for II_1-factors and quantum groupoids, *J. Funct. Anal.* 178 (2000), 113–142.

[61] D. Nikshych and L. Vainerman, Finite Quantum Groupoids and Their Applications, in

New Directions in Hopf Algebras (S. Montgomery and H.-J. Schneider, eds.), Math. Sci. Res. Inst. Publ. 43, Cambridge University Press, Cambridge 2002, 211–262.

[62] V. Ostrik, Module categories, weak Hopf algebras, and modular invariants, preprint `math.QA/0111139` (2001).

[63] V. G. Paljutkin, Invariant measure of a compact ring group, *Trans. Amer. Math. Soc.* 84 (1969), 89–102.

[64] P. Podleś and S. L. Woronowicz, Quantum deformation of Lorentz group, *Comm. Math. Phys.* 130 (1990), 381–431.

[65] M. Rieffel, Some solvable quantum groups. Operator algebras and topology, Proceedings of the OATE 2 conference, Bucarest, Romania, 1989 (W. B. Arveson et al., eds.), *Pitman Res. Notes Math. Ser.* 270 (1992), 146–159.

[66] P. Schauenburg, Weak Hopf algebras and quantum groupoids, preprint `math.QA/0204180` (2002).

[67] G. Skandalis, Duality for locally compact 'quantum groups' (joint work with S. Baaj), Mathematisches Forschungsinstitut Oberwolfach, Tagungsbericht 46/1991, C*-algebren, 20.10–26.10.1991, 20pp.

[68] L. L. Vaksman and Y. S. Soibel'man, Algebra of functions on the quantum group $SU(2)$, *Func. Anal. Appl.* 22 (1988), 170–181.

[69] W. F. Stinespring, Integration theorem for gauges and duality for unimodular locally compact groups, *Trans. Amer. Math. Soc.* 90 (1959), 15–56.

[70] S. Stratila, *Modular Theory in Operator Algebras*, Editura Academiei/Abacus Press, Bucuresti/Tunbridge Wells, Kent, 1981.

[71] W. Szymanski, Finite index subfactors and Hopf algebra crossed products, *Proc. Amer. Math. Soc.* 120 (1994), 519–528.

[72] M. Takesaki, A characterization of group algebras as a converse of Tannaka-Stinespring-Tatsuuma duality, *Amer. J. Math.* 91 (1969), 529–564.

[73] M. Takesaki, Duality and von Neumann algebras, in *Lectures on operator algebras*, Tulane University Ring and Operator Theory Year 1970–1971, Vol. II, Lecture Notes in Math. 247, Springer-Verlag, Berlin–Heidelberg–New York 1972, 665–785.

[74] M. Takeuchi, Groups of algebras over $A \times \overline{A}$, *J. Math. Soc. Japan* 29 (1977), 459–492.

[75] T. Tannaka, Über den Dualitätssatz der Nichtkommutativen Topologischen Gruppen, *Tôhoku. Math. J.* 45 (1938), 1–12.

[76] N. Tatsuuma, A duality theorem for locally compact groups, I, II, *Proc. Japan Acad.* 41 (1965), 878–882; ibid. 42 (1966), 46–47.

[77] S. Vaes, The unitary implementation of a locally compact quantum group action, *J. Funct. Anal.* 180 (2001), 426–480.

[78] S. Vaes, Examples of locally compact quantum groups through the bicrossed product construction, in *Proceedings of the XIIIth Int. Conf. Math. Phys., London* 2000 (A. Grigoryan, A. Fokas, T. Kibble and B. Zegarlinsky, eds.), International Press of Boston, Sommerville, MA, 2001, 341–348.

[79] S. Vaes, Locally compact quantum groups, Ph. D. Thesis. K.U. Leuven, 2001.

[80] S. Vaes and L. Vainerman, Extensions of locally compact quantum groups and the bicrossed product construction, to appear in *Adv. Math.*, preprint math.QA/0101133 (2001).

[81] L. I. Vainerman, Characterization of Dual Objects to Locally Compact Groups, *Funct. Anal. Appl.* 8 (1974), 75–76.

[82] L. I. Vainerman, Relation between Compact Quantum Groups and Kac Algebras, *Adv. Soviet Math.* 9 (1992), 151–160.

[83] L. Vainerman, 2-Cocycles and Twisting of Kac Algebras, *Comm. Math. Phys.* 191 (1998), 697–721.

[84] J. M. Vallin, Bimodules de Hopf et poids opératoriels de Haar, *J. Operator Theory* 35 (1996), 39–65.

[85] J. M. Vallin, Unitaire pseudo-multiplicatif associé à un groupoïde; applications à la moyennabilité, *J. Operator Theory* 44 (2000), 347–368.

[86] J. M. Vallin, Groupoïdes quantiques finis, *J. Algebra* 26 (2001), 425–488.

[87] A. Van Daele, Discrete quantum groups, *J. Algebra* 180 (1996), 431–444.

[88] A. Van Daele, Multiplier Hopf Algebras, *Trans. Amer. Math. Soc.* 342 (1994), 917–932.

[89] A. Van Daele, An algebraic framework for group duality, *Adv. Math.* 140 (1998), 323–366.

[90] A. Van Daele and S. L. Woronowicz, Duality for the quantum $E(2)$ group, *Pacific J. Math.* 173 (1996), 375–385.

[91] P. Vecsernyes, Larson-Sweedler theorem, grouplike elements, and irreducible modules in weak Hopf algebras, preprint math.QA/0111045 (2001).

[92] S. L. Woronowicz, Compact matrix pseudogroups, *Comm. Math. Phys.* 111 (1987), 613–665.

[93] S. L. Woronowicz, Twisted $SU(2)$ group. An example of a non-commutative differential calculus, *Publ. Res. Inst. Math. Sci.* 23 (1987), 117–181.

[94] S. L. Woronowicz, Tannaka-Krein duality for compact matrix pseudogroups, Twisted $SU(N)$ groups. *Invent. Math.* 93 (1988), 35–76.

[95] S. L. Woronowicz, Quantum $E(2)$ group and its Pontryagin dual, *Lett. Math. Phys.* 23 (1991), 251–263.

[96] S. L. Woronowicz, From multiplicative unitaries to quantum groups, *Internat. J. Math.* 7 (1996), 127–149.

[97] S. L. Woronowicz, Quantum $'az + b'$ group on complex plane, *Internat. J. Math.* 12 (2001), 461–503.

[98] S. L. Woronowicz and S. Zakrzevski, Quantum $'ax + b'$ group, preprint KMMF (1999).

[99] P. Xu, Quantum groupoids, *Comm. Math. Phys.* 216 (2001), 539–581.

Quantum groupoids and pseudo-multiplicative unitaries

Michel Enock

Institut de Mathématiques de Jussieu
Unité Mixte Paris 6 / Paris 7 / CNRS de Recherche 7586, Case 191
Université Pierre et Marie Curie, 75252 Paris Cedex 05, France
email: enock@math.jussieu.fr

Abstract. To any groupoid, equipped with a Haar system, Jean-Michel Vallin had associated several objects (pseudo-multiplicative unitary, Hopf bimodule) in order to generalize, up to the groupoid case, the classical notions of a multiplicative unitary and a Hopf-von Neumann algebra, which were intensively used in the quantum group theory, in the operator algebra setting. In two recent articles (one of them in collaboration with Jean-Michel Vallin), starting with a depth 2 inclusion of von Neumann algebras, we have constructed the same objects, which allowed us to study two "quantum groupoids" dual to each other. Here is a survey of these notions and results, including the announcement of new results about pseudo-multiplicative unitaries.

1 Introduction

The quantum group theory in the operator algebra setting has recently reached a new viewpoint from which the landscape is greater.

First of all, in their theory of "locally compact quantum groups", Kustermans and Vaes [KV] have obtained a beautiful and efficient axiomatisation of quantum groups. Their axioms are simple, easy to verify and cover all known examples. Many results in harmonic analysis seem now to be obtainable in that new setting and this article seems to be the new keystone of the theory.

Secondly, the links between quantum group theory and subfactor theory are now completely clarified ([EN], [E1], [V]): up to some regularity condition, every depth 2 irreducible inclusion of factors is given by an action of a locally compact quantum group on a factor, and vice-versa.

This situation leads several mathematicians to face two new questions:

– How to modify Kustermans and Vaes axioms in order to catch also locally compact groupoids? How does it correspond to what was done by several theoretical physicists ([BSz1], [BSz2], [Sz])?

– What is to be obtained if we deal with non-irreducible depth 2 inclusions of von Neumann algebras?

Of course, the answers to these two questions are closely linked, and many results were found in this direction. It turned out that the tools are completely different in finite- and in infinite-dimensional situation.

In the finite-dimensional situation, after some early work by Yamanouchi [Y2], the most important work is due to Böhm, Nill and Szlachányi [BNS] and Nikshych and Vainerman [NV1], who gave there a general setting to "finite quantum groupoids" and constructed several examples (see also earlier papers [BSz1], [BSz2], [Sz]). In [NV2], the links of this theory with depth 2 non-irreducible inclusions of type II_1 von Neumann factors are given. Another point of view, with multiplicative partial isometries, is due to Vallin ([Val3], [Val 4]) and Böhm and Szlachányi [BSz3].

In the infinite-dimensional situation, Vallin had associated with any locally compact groupoid, equipped with a left Haar system, two objects (Hopf bimodule structure, pseudo-multiplicative unitary), which generalize the usual coproduct and multiplicative unitary associated with a locally compact group ([Val1], [Val 2]). It appeared then clear that, for going from locally compact groups to locally compact groupoids, it was necessary to use the Hilbert space relative tensor product (Connes–Sauvageot tensor product) instead of the usual Hilbert space tensor product, and the "fiber product" of von Neumann algebras instead of the usual von Neumann algebra tensor product.

In [EV], [E2] the structures of the same kind have been obtained starting with non-irreducible depth 2 inclusions of von Neumann algebras.

New results in that theory can be found in [E3] and in [L], the latter will appear soon.

Here we give a survey on "quantum groupoids of infinite dimension", and announce some results, still unpublished. In Section 2 we recall all the preliminaries required, in particular a description of the Connes–Sauvageot tensor product (2.4) and of the fiber product of von Neumann algebras (2.5). In Section 3 we give the definitions of Hopf bimodules (3.1) and of pseudo-multiplicative unitaries (3.2), as well as examples coming from groupoids (3.1, 3.2) and from depth 2 inclusions (3.3). We also discuss the first properties of these objects. In Section 4, inspired by [BS], we develop the theory of quantum groupoids of compact type. Examples are given in Section 5.

For the sake of simplicity, all von Neumann algebras are supposed to be σ-finite.

This article is mostly inspired by the talk I gave at the conference on Quantum groups, Hopf algebras and their applications, which held in Strasbourg on February 21–23, 2002. I would like to thank the organizers of this conference.

2 Preliminaries

2.1 From locally compact groups to quantum groups

Let G be a locally compact group, with a left Haar measure ds. Then, let us recall the two classical objects associated: for any f in $L^\infty(G, ds)$, and s, t in G, let us put:

$$\Gamma_G(f)(s,t) = f(st).$$

This defines a normal injective homomorphism Γ_G from $L^\infty(G, ds)$ into the von Neumann algebra $L^\infty(G \times G, ds \times ds)$, which can (and will) be identified with the von Neumann algebra tensor product $L^\infty(G, ds) \otimes L^\infty(G, ds)$. Then, it satisfies the co-associativity condition:

$$(\Gamma_G \otimes \mathrm{id})\Gamma_G = (\mathrm{id} \otimes \Gamma_G)\Gamma_G.$$

The Haar measure, by integration of any positive f in $L^\infty(G, ds)$, gives a normal semi-finite faithful trace φ_G on $L^\infty(G, ds)$:

$$\varphi_G(f) = \int_G f(s)ds,$$

which satisfies, by left invariance of the Haar measure, the following condition:

$$(\mathrm{id} \otimes \varphi_G)\Gamma(f) = \varphi_G(f)1,$$

for all positive f in $L^\infty(G, ds)$.

On the second hand, for any g in $L^2(G \times G, ds \times ds)$, we define:

$$W_G g(s,t) = g(s, s^{-1}t).$$

The Hilbert space $L^2(G \times G, ds \times ds)$ may (and will) be identified with the Hilbert space tensor product $L^2(G, ds) \otimes L^2(G, ds)$; so, W_G will be considered as a unitary on the Hilbert space $L^2(G, ds) \otimes L^2(G, ds)$, which satisfies the pentagonal equation:

$$W_{G(1,2)} W_{G(1,3)} W_{G(2,3)} = W_{G(2,3)} W_{G(1,2)}.$$

These two objects are linked by the property:

$$\Gamma_G(f) = W_G(1 \otimes f)W_G^*.$$

The quantum group theory in the operator algebra context was developed by studying generalizations of these objects:

On one hand, by considering an operator algebra (for simplicity, let us say a von Neumann algebra) M, with a coassociative coproduct Γ, i.e., an injective normal homomorphism from M to $M \otimes M$, such that:

$$(\Gamma \otimes \mathrm{id})\Gamma = (\mathrm{id} \otimes \Gamma)\Gamma,$$

and a normal semi-finite faithful weight φ on M, left-invariant with respect to Γ, i.e., such that, for any positive x in M:

$$(\mathrm{id} \otimes \varphi)\Gamma(x) = \varphi(x)1.$$

On the other hand a so-called multiplicative unitary, i.e., a Hilbert space $\mathcal{H}$, and a unitary W on $\mathcal{H} \otimes \mathcal{H}$, such that:

$$W_{1,2}W_{1,3}W_{2,3} = W_{2,3}W_{1,2}.$$

Roughly speaking, we can say that the first approach was implemented in [ES], [W1], [MN], [KV], and the second one in [BS], [B], [W2].

These two points of view are complementary: in the first one, the goal is to add extra conditions in order to find a multiplicative unitary. In particular, in the locally compact quantum group theory [KV], it was beautifully done by adding as an axiom the existence of a right-invariant weight on M. In the second point of view, it is rather straightforward to construct two von Neumann algebras with co-mutiplicative coproducts, and the goal is to add extra conditions in order to construct a left-invariant weight. It was done only in the compact case, in [BS], with the assumption of the existence of non-zero fixed vectors for W, which leads to a left (and right)-invariant state on M.

2.2 From locally compact groups to groupoids

A groupoid $\mathcal{G}$ is a set equipped with a partially defined product (i.e., defined on a subset $\mathcal{G}^{(2)}$ of $\mathcal{G}^2$), and an inverse everywhere defined, verifying the following properties:

(i) associativity: if the products xy and yz exist, then the products $(xy)z$ and $x(yz)$ exist and are equal.

(ii) simplification: for all $x \in \mathcal{G}$, if we denote by x^{-1} the inverse of x, then (x, x^{-1}) and (x^{-1}, x) belong to $\mathcal{G}^{(2)}$, and, if (x, y) (resp., (z, x)) belongs to $\mathcal{G}^{(2)}$, then we get $x^{-1}xy = y$ (resp., $zxx^{-1} = z$).

Then we define by the formulae $r(x) = xx^{-1}$ and $s(x) = x^{-1}x$ the range and source applications going from $\mathcal{G}$ to a subset $\mathcal{G}^{(0)}$ of $\mathcal{G}$, called its set of units. Then, we get that:

$$\mathcal{G}^{(2)} = \{(x, y) \in \mathcal{G}^2; s(x) = r(y)\}.$$

A topological groupoid is a groupoid equipped with a topology such that the applications $x \to x^{-1}$ from $\mathcal{G}$ to $\mathcal{G}$, and $(x, y) \to xy$ from $\mathcal{G}^{(2)}$ to $\mathcal{G}$ are continuous. If this topology is locally compact, we shall speak of locally compact groupoids. We shall only consider σ-compact locally compact groupoids.

A left Haar system on such a groupoid is a set of positive Radon measures λ^u, for all $u \in \mathcal{G}^{(0)}$, such that:

(i) the support of λ^u is $\mathcal{G}^u = r^{-1}(\{u\})$.

(ii) for any continuous function f with compact support, the function $u \mapsto \int_C f d\lambda^u$ is continuous on $\mathcal{G}^{(0)}$.

(iii) the system of λ^u is left invariant, which means that, for any continuous function f with compact support, we have:

$$\int_{\mathcal{G}} f(xy)d\lambda^{s(x)}(y) = \int_{\mathcal{G}} f(y)d\lambda^{r(x)}(y).$$

If such a left Haar system exists, then it is possible to construct a right Haar system, by taking the images of the measures λ^u by the application $x \to x^{-1}$. We then get a set of measures denoted by λ_u, whose support is $\mathcal{G}_u = s^{-1}(\{u\})$, which will verify a right invariance property.

Under these hypothesis, the applications r and s are open ([R1], 2.4). Moreover, if μ is a positive Radon measure on $\mathcal{G}^{(0)}$, we shall write ν the measure induced on G, $\nu = \int_{\mathcal{G}} \lambda^u d\mu(u)$. The measure μ will be called quasi-invariant if the measure ν is equivalent to $\nu^{-1} = \int_{\mathcal{G}} \lambda_u d\mu(u)$. If such a measure exists, all the equivalent measures are quasi-invariant; therefore, we may suppose that μ is bounded (see [R1], [R2] and ([C2] II.5) for more details and examples of groupoids). With these hypothesis, it has been shown in [Val 1] that it is possible to define a measure $\nu^2_{s,r}$ on $\mathcal{G}^{(2)}$, and a measure $\nu^2_{r,r}$ on the set $\mathcal{G}^2_r = \{(x, y) \in \mathcal{G}^2, r(x) = r(y)\}$, such that the following holds.

First, for f in $L^\infty(\mathcal{G}, \nu)$, and (s, t) in $\mathcal{G}^{(2)}$, the function $\Gamma_{\mathcal{G}}(f)$, defined on $\mathcal{G}^{(2)}$ by the formula

$$(s, t) \mapsto f(st),$$

belongs to $L^\infty(\mathcal{G}^{(2)}, \nu^2_{s,r})$. Then $\Gamma_{\mathcal{G}}$ is an involutive homomorphism from $L^\infty(\mathcal{G}, \nu)$ into $L^\infty(\mathcal{G}^{(2)}, \nu^2_{s,r})$.

Second, for g in $L^2(\mathcal{G}^{(2)}, \nu^2_{s,r})$ and (s, t) in $\mathcal{G}^2_r$, the function $W_{\mathcal{G}}g$, defined on $\mathcal{G}^2_r$ by the formula

$$W_{\mathcal{G}}g(s, t) = g(s, s^{-1}t),$$

belongs to $L^2(\mathcal{G}^2_r, \nu^2_{r,r})$ and, furthermore, $W_{\mathcal{G}}$ is a unitary from $L^2(\mathcal{G}^{(2)}, \nu^2_{s,r})$ to $L^2(\mathcal{G}^2_r, \nu^2_{r,r})$.

Therefore, in order to generalize these constructions up to groupoids, it is necessary to generalize also the Hilbert space tensor product and the von Neumann algebra tensor product.

2.3 Spatial theory ([C1], [S2])

Let N be a von Neumann algebra, and let ψ be a faithful semi-finite normal weight on N; let $\mathfrak{N}_\psi$, $\mathfrak{M}_\psi$, H_ψ, π_ψ, Λ_ψ, J_ψ, Δ_ψ, ... be the canonical objects of the Tomita–

Takesaki construction associated to the weight ψ. Let α be a non-degenerate normal representation of N on a Hilbert space $\mathcal{H}$. We can as well consider $\mathcal{H}$ as a left N-module, and denote it then ${}_\alpha\mathcal{H}$. Following ([C1], Definition 1), we define the set of ψ-bounded elements of ${}_\alpha\mathcal{H}$ as

$$D({}_\alpha\mathcal{H}, \psi) = \{\xi \in \mathcal{H}; \exists C < \infty, \|\alpha(y)\xi\| \leq C\|\Lambda_\psi(y)\|, \forall y \in \mathfrak{N}_\psi\}.$$

Then, for any ξ in $D({}_\alpha\mathcal{H}, \psi)$, there exists a bounded operator $R^{\alpha,\psi}(\xi)$ from H_ψ to $\mathcal{H}$, defined, for all y in $\mathfrak{N}_\psi$, by

$$R^{\alpha,\psi}(\xi)\Lambda_\psi(y) = \alpha(y)\xi.$$

If there is no ambiguity about the representation α, we shall write $R^\psi(\xi)$ instead of $R^{\alpha,\psi}(\xi)$. This operator belongs to $\mathrm{Hom}_N(H_\psi, \mathcal{H})$; therefore, for any ξ, η in $D({}_\alpha\mathcal{H}, \psi)$, the operator

$$\theta^{\alpha,\psi}(\xi, \eta) = R^{\alpha,\psi}(\xi)R^{\alpha,\psi}(\eta)^*$$

belongs to $\alpha(N)'$; moreover, $D({}_\alpha\mathcal{H}, \psi)$ is dense ([C1], Lemma 2), stable under $\alpha(N)'$, and the linear span generated by the operators $\theta^{\alpha,\psi}(\xi, \eta)$ is a dense ideal in $\alpha(N)'$. We shall denote by $\mathcal{K}_{\alpha,\psi}$ the norm closure of this ideal, which is also a dense ideal in $\alpha(N)'$.

With the same hypothesis, the operator

$$\langle\xi, \eta\rangle_{\alpha,\psi} = R^{\alpha,\psi}(\eta)^*R^{\alpha,\psi}(\xi)$$

belongs to $\pi_\psi(N)'$. Due to the Tomita–Takesaki's theory, this last algebra is equal to $J_\psi\pi_\psi(N)J_\psi$, and therefore is anti-isomorphic to N (or isomorphic to the opposite von Neumann algebra N^o). We shall consider now $\langle\xi, \eta\rangle_{\alpha,\psi}$ as an element of N^o, and the linear span generated by these operators is a dense ideal in N^o.

Let us suppose now that there exists a normal non-degenerate anti-representation β of N on $\mathcal{H}$. We may then as well consider $\mathcal{H}$ as a right N-module, and write it $\mathcal{H}_\beta$, or consider β as a normal non-degenerate representation of the opposite von Neumann algebra N^o, and consider $\mathcal{H}$ as a left N^o-module.

From these remarks, we infer that the set of ψ^o-bounded elements of $\mathcal{H}_\beta$ is:

$$D(\mathcal{H}_\beta, \psi^o) = \{\xi \in \mathcal{H}; \exists C < \infty, \|\beta(y^*)\xi\| \leq C\|\Lambda_\psi(y)\|, \forall y \in \mathfrak{N}_\psi\}$$

and, for any ξ in $D(\mathcal{H}_\beta, \psi^o)$ and y in $\mathfrak{N}_\psi$, the bounded operator $R^{\beta,\psi^o}(\xi)$ is given by the formula:

$$R^{\beta,\psi^o}(\xi)J_\psi\Lambda_\psi(y) = \beta(y^*)\xi.$$

This operator belongs to $\mathrm{Hom}_{N^o}(H_\psi, \mathcal{H})$. Moreover, $D(\mathcal{H}_\beta, \psi^o)$ is dense, stable under $\beta(N)' = P$, and, for all y in P, we have:

$$R^{\beta,\psi^o}(y\xi) = yR^{\beta,\psi^o}(\xi).$$

Then, for any ξ, η in $D(\mathcal{H}_\beta, \psi^o)$, the operator $\theta^{\beta,\psi^o}(\xi, \eta) = R^{\beta,\psi^o}(\xi)R^{\beta,\psi^o}(\eta)^*$ belongs to P, and the linear span generated by these operators is a dense ideal in P;

moreover, the operator-valued product $\langle\xi,\eta\rangle_{\beta,\psi^o} = R^{\beta,\psi^o}(\eta)^* R^{\beta,\psi^o}(\xi)$ belongs to $\pi_\psi(N)$; we shall consider now, for simplicity, that $\langle\xi,\eta\rangle_{\beta,\psi^o}$ belongs to N, and the linear span generated by these operators is a dense ideal in N.

2.4 Relative tensor product ([C1], [S2])

Using the notations of 2.3, let now $\mathcal{K}$ be another Hilbert space on which there exists a non-degenerate representation γ of N. Following Sauvageot ([S2], 2.1), we define the relative tensor product $\mathcal{H}\ {}_\beta\!\underset{\psi}{\otimes}_\gamma\ \mathcal{K}$ as the Hilbert space obtained from the algebraic tensor product $D(\mathcal{H}_\beta, \psi^o) \odot \mathcal{K}$ equipped with the scalar product defined, for ξ_1, ξ_2 in $D(\mathcal{H}_\beta, \psi^o)$, η_1, η_2 in $\mathcal{K}$, by:

$$(\xi_1 \odot \eta_1 | \xi_2 \odot \eta_2) = (\gamma(\langle\xi_1, \xi_2\rangle_{\beta,\psi^o})\eta_1|\eta_2),$$

where we have identified N with $\pi_\psi(N)$ for simplifying the notations.

The image of $\xi \odot \eta$ in $\mathcal{H}\ {}_\beta\!\underset{\psi}{\otimes}_\gamma\ \mathcal{K}$ will be denoted by $\xi\ {}_\beta\!\underset{\psi}{\otimes}_\gamma\ \eta$. We shall use intensively this construction; one should bear in mind that, if we start from another faithful semi-finite normal weight ψ', we get another Hilbert space $\mathcal{H}\ {}_\beta\!\underset{\psi'}{\otimes}_\gamma\ \mathcal{K}$; there exists an isomorphism $U^{\psi,\psi'}_{\beta,\gamma}$ from $\mathcal{H}\ {}_\beta\!\underset{\psi}{\otimes}_\gamma\ \mathcal{K}$ to $\mathcal{H}\ {}_\beta\!\underset{\psi'}{\otimes}_\gamma\ \mathcal{K}$, which is unique up to some functorial property ([S2], 2.6) (but this isomorphism does not send $\xi\ {}_\beta\!\underset{\psi}{\otimes}_\gamma\ \eta$ on $\xi\ {}_\beta\!\underset{\psi'}{\otimes}_\gamma\ \eta$!).

When no confusion is possible about the representation and the anti-representation, we shall write $\mathcal{H} \otimes_\psi \mathcal{K}$ instead of $\mathcal{H}\ {}_\beta\!\underset{\psi}{\otimes}_\gamma\ \mathcal{K}$, and $\xi \otimes_\psi \eta$ instead of $\xi\ {}_\beta\!\underset{\psi}{\otimes}_\gamma\ \eta$.

For any ξ in $D(\mathcal{H}_\beta, \psi^o)$, we define the bounded linear application $\lambda^{\beta,\gamma}_\xi$ from $\mathcal{K}$ to $\mathcal{H}\ {}_\beta\!\underset{\psi}{\otimes}_\gamma\ \mathcal{K}$ by, for all η in $\mathcal{K}$, $\lambda^{\beta,\gamma}_\xi(\eta) = \xi\ {}_\beta\!\underset{\psi}{\otimes}_\gamma\ \eta$. We shall write λ_ξ if no confusion is possible. We get ([EN], 3.10):

$$\lambda^{\beta,\gamma}_\xi = R^{\beta,\psi^o}(\xi) \otimes_\psi 1_{\mathcal{K}},$$

where we use the canonical identification (as left N-modules) of $L^2(N) \otimes_\psi \mathcal{K}$ with $\mathcal{K}$. We have:

$$(\lambda^{\beta,\gamma}_\xi)^* \lambda^{\beta,\gamma}_\xi = \gamma(\langle\xi,\xi\rangle_{\beta,\psi^o}).$$

In ([S2] 2.1), the relative tensor product $\mathcal{H}\ {}_\beta\!\underset{\psi}{\otimes}_\gamma\ \mathcal{K}$ is defined also, if ξ_1, ξ_2 are in $\mathcal{H}$, η_1, η_2 are in $D({}_\gamma\mathcal{K}, \psi)$, by the following formula:

$$(\xi_1 \odot \eta_1 | \xi_2 \odot \eta_2) = (\beta(\langle\eta_1, \eta_2\rangle_{\gamma,\psi})\xi_1|\xi_2),$$

which leads to the definition of a relative flip σ_ψ which is an isomorphism from $\mathcal{H}\ {}_\beta\!\underset{\psi}{\otimes}_\gamma\ \mathcal{K}$ onto $\mathcal{K}\ {}_\gamma\!\underset{\psi^o}{\otimes}_\beta\ \mathcal{H}$, defined, for any ξ in $D(\mathcal{H}_\beta, \psi^o)$, η in $D({}_\gamma\mathcal{K}, \psi)$, by:

$$\sigma_\psi(\xi \otimes_\psi \eta) = \eta \otimes_{\psi^o} \xi.$$

This allows us to define a relative flip ς_ψ from $\mathcal{L}(\mathcal{H}\ {}_\beta\!\underset{\psi}{\otimes}_\gamma\ \mathcal{K})$ to $\mathcal{L}(\mathcal{K}\ {}_\gamma\!\underset{\psi^o}{\otimes}_\beta\ \mathcal{H})$ which sends X in $\mathcal{L}(\mathcal{H}\ {}_\beta\!\underset{\psi}{\otimes}_\gamma\ \mathcal{K})$ onto $\varsigma_\psi(X) = \sigma_\psi X \sigma_\psi^*$. Starting from another faithful semi-finite normal weight ψ', we get a von Neumann algebra $\mathcal{L}(\mathcal{H}\ {}_\beta\!\underset{\psi'}{\otimes}_\gamma\ \mathcal{K})$ which is isomorphic to $\mathcal{L}(\mathcal{H}\ {}_\beta\!\underset{\psi}{\otimes}_\gamma\ \mathcal{K})$, and a von Neumann algebra $\mathcal{L}(\mathcal{K}\ {}_\gamma\!\underset{\psi'^o}{\otimes}_\beta\ \mathcal{H})$ which is isomorphic to $\mathcal{L}(\mathcal{K}\ {}_\gamma\!\underset{\psi^o}{\otimes}_\beta\ \mathcal{H})$; as we get that:

$$\sigma_{\psi'} \circ U^{\psi,\psi'}_{\beta,\gamma} = U^{\psi^o,\psi'^o}_{\gamma,\beta} \circ \sigma_\psi,$$

we see that these isomorphisms exchange ς_ψ and $\varsigma_{\psi'}$. Therefore, the homomorphism ς_ψ can be denoted by ς_N without any reference to a specific weight.

We may define, for any η in $D({}_\gamma\mathcal{K}, \psi)$, an application $\rho^{\beta,\gamma}_\eta$ from $\mathcal{H}$ to $\mathcal{H}\ {}_\beta\!\underset{\psi}{\otimes}_\gamma\ \mathcal{K}$ by $\rho^{\beta,\gamma}_\eta(\xi) = \xi\ {}_\beta\!\underset{\psi}{\otimes}_\gamma\ \eta$. We shall write ρ_η if no confusion is possible. We get that:

$$(\rho^{\beta,\gamma}_\eta)^* \rho^{\beta,\gamma}_\eta = \beta(\langle \eta, \eta \rangle_{\gamma,\psi}).$$

We recall, following ([S2], 2.2b), that, for all ξ in $\mathcal{H}$, η in $D({}_\gamma\mathcal{K}, \psi)$, y in N, analytic with respect to ψ, we have:

$$\beta(y)\xi \otimes_\psi \eta = \xi \otimes_\psi \gamma(\sigma^\psi_{-i/2}(y))\eta.$$

Let x be an element of $\mathcal{L}(\mathcal{H})$, commuting with the right action of N on $\mathcal{H}_\beta$ (i.e., $x \in \beta(N)'$). It is possible to define an operator $x\ {}_\beta\!\underset{\psi}{\otimes}_\gamma\ 1_\mathcal{K}$ on $\mathcal{H}\ {}_\beta\!\underset{\psi}{\otimes}_\gamma\ \mathcal{K}$. In the same way, if y commutes with the left action of N on ${}_\gamma\mathcal{K}$ (i.e. $y \in \gamma(N)'$), it is possible to define $1_\mathcal{H}\ {}_\beta\!\underset{\psi}{\otimes}_\gamma\ y$ on $\mathcal{H}\ {}_\beta\!\underset{\psi}{\otimes}_\gamma\ \mathcal{K}$, and by composition, then it is possible to define $x\ {}_\beta\!\underset{\psi}{\otimes}_\gamma\ y$. If we start from another faithful semi-finite normal weight ψ', the canonical isomorphism $U^{\psi,\psi'}_{\beta,\gamma}$ from $\mathcal{H}\ {}_\beta\!\underset{\psi}{\otimes}_\gamma\ \mathcal{K}$ to $\mathcal{H}\ {}_\beta\!\underset{\psi'}{\otimes}_\gamma\ \mathcal{K}$ sends $x\ {}_\beta\!\underset{\psi}{\otimes}_\gamma\ y$ on $x\ {}_\beta\!\underset{\psi'}{\otimes}_\gamma\ y$ ([S2], 2.3 and 2.6); therefore, this operator can be denoted by $x\ {}_\beta\!\underset{N}{\otimes}_\gamma\ y$ without any reference to a specific weight.

Let us suppose now that $\mathcal{K}$ is a $N - P$ bimodule; that means that there exists a von Neumann algebra P, and a non-degenerate normal anti-representation ϵ of P on $\mathcal{K}$, such that $\epsilon(P) \subset \gamma(N)'$. We shall write then ${}_\gamma\mathcal{K}_\epsilon$. If y belongs to P, we have seen that it is possible to define then the operator $1_{\mathcal{H}} \; {}_{\beta}\!\otimes_{\gamma}\!{}_{\psi} \; \epsilon(y)$ on $\mathcal{H} \; {}_{\beta}\!\otimes_{\gamma}\!{}_{\psi} \; \mathcal{K}$, and we define in this way a non-degenerate normal antirepresentation of P on $\mathcal{H} \; {}_{\beta}\!\otimes_{\gamma}\!{}_{\psi} \; \mathcal{K}$, we shall call again ϵ for simplification. If $\mathcal{H}$ is a $Q - N$ bimodule, then $\mathcal{H} \; {}_{\beta}\!\otimes_{\gamma}\!{}_{\psi} \; \mathcal{K}$ becomes a $Q - P$ bimodule (Connes' fusion of bimodules).

Taking a faithful semi-finite normal weight ν on P, and a left P-module ${}_{\zeta}\mathcal{L}$ (i.e., a Hilbert space $\mathcal{L}$ and a normal non-degenerate representation ζ of P on $\mathcal{L}$), then it is possible to define $(\mathcal{H} \; {}_{\beta}\!\otimes_{\gamma}\!{}_{\psi} \; \mathcal{K}) \; {}_{\epsilon}\!\otimes_{\zeta}\!{}_{\nu} \; \mathcal{L}$. Of course, it is possible also to consider the Hilbert space $\mathcal{H} \; {}_{\beta}\!\otimes_{\gamma}\!{}_{\psi} \; (\mathcal{K} \; {}_{\epsilon}\!\otimes_{\zeta}\!{}_{\nu} \; \mathcal{L})$. It can be shown that these two Hilbert spaces are isomorphic as $\beta(N)' - \zeta(P)'^{o}$-bimodules (the proof, given in ([Val1] 2.1.3) for $N = P$ abelian, can be used, without modification, under these wider conditions). We shall write then $\mathcal{H} \; {}_{\beta}\!\otimes_{\gamma}\!{}_{\psi} \; \mathcal{K} \; {}_{\epsilon}\!\otimes_{\zeta}\!{}_{\nu} \; \mathcal{L}$ without parenthesis, to emphasize this co-associativity property of the relative tensor product.

If the Hilbert spaces $\mathcal{H}$ and $\mathcal{K}$ are finite dimensional, then the relative tensor product $\mathcal{H} \; {}_{\beta}\!\otimes_{\gamma}\!{}_{\psi} \; \mathcal{K}$ can be identified with a subspace of the usual tensor product $\mathcal{H} \otimes \mathcal{K}$.

2.5 Fiber product ([Val1], [EV])

Let us follow the notations of 2.4; let now M_1 be a von Neumann algebra on $\mathcal{H}$, such that $\beta(N) \subset M_1$, and M_2 be a von Neumann algebra on $\mathcal{K}$, such that $\gamma(N) \subset M_2$. The von Neumann algebra generated by all elements $x \; {}_{\beta}\!\otimes_{\gamma}\!{}_{N} \; y$, where x belongs to M_1', and y belongs M_2', will be denoted $M_1' \; {}_{\beta}\!\otimes_{\gamma}\!{}_{N} \; M_2'$ (or $M_1' \otimes_N M_2'$ if no confusion if possible), and will be called the relative tensor product of M_1' and M_2' over N. The commutant of this algebra will be denoted by $M_1 \; {}_{\beta}\!*_{\gamma}\!{}_{N} \; M_2$ (or $M_1 *_N M_2$ if no confusion is possible) and called the fiber product of M_1 and M_2, over N. It is straightforward to verify that, if P_1 and P_2 are two other von Neumann algebras satisfying the same relations with N, we have

$$M_1 *_N M_2 \cap P_1 *_N P_2 = (M_1 \cap P_1) *_N (M_2 \cap P_2).$$

Moreover, we get that $\varsigma_N(M_1 \; {}_{\beta}\!*_{\gamma}\!{}_{N} \; M_2) = M_2 \; {}_{\gamma}\!*_{\beta}\!{}_{N^o} \; M_1$. In particular, we have

$$(M_1 \cap \beta(N)') \; \underset{N}{{}_{\beta}\otimes_{\gamma}} \; (M_2 \cap \gamma(N)') \subset M_1 \; \underset{N}{{}_{\beta}*_{\gamma}} \; M_2$$

and

$$M_1 \; \underset{N}{{}_{\beta}*_{\gamma}} \; \gamma(N) = (M_1 \cap \beta(N)') \; \underset{N}{{}_{\beta}\otimes_{\gamma}} \; 1.$$

More generally, if β is a non-degenerate normal involutive anti-homomorphism from N into a von Neumann algebra M_1, and γ a non-degenerate normal involutive homomorphism from N into a von Neumann algebra M_2, it is possible to define, without any reference to a specific Hilbert space, a von Neumann algebra $M_1 \underset{N}{{}_\beta *_\gamma} M_2$.

Moreover, if now β' is a non-degenerate normal involutive anti-homomorphism from N into another von Neumann algebra P_1, γ' a non-degenerate normal involutive homomorphism from N into another von Neumann algebra P_2, Φ a normal involutive homomorphism from M_1 into P_1 such that $\Phi \circ \beta = \beta'$, and Ψ a normal involutive homomorphism from M_2 into P_2 such that $\Psi \circ \gamma = \gamma'$, it is possible then to define a normal involutive homomorphism (the proof given in ([S1], 1.2.4) in the case when N is abelian, can be extended without modification in the general case):

$$\Phi \underset{N}{{}_\beta *_\gamma} \Psi : M_1 \underset{N}{{}_\beta *_\gamma} M_2 \to P_1 \underset{N}{{}_{\beta'} *_{\gamma'}} P_2.$$

In the case when ${}_\gamma\mathcal{K}_\epsilon$ is a $N - P^o$ bimodule, as explained in 2.4, and ${}_\zeta\mathcal{L}$ a P-module, if $\gamma(N) \subset M_2$ and $\epsilon(P) \subset M_2$, and if $\zeta(P) \subset M_3$, where M_3 is a von Neumann algebra on $\mathcal{L}$, it is possible to consider then $(M_1 \underset{N}{{}_\beta *_\gamma} M_2) \underset{P}{{}_\epsilon *_\zeta} M_3$ and $M_1 \underset{N}{{}_\beta *_\gamma} (M_2 \underset{P}{{}_\epsilon *_\zeta} M_3)$. The co-associativity property for relative tensor products leads then to the isomorphism of these von Neumann algebras, which we shall write now $M_1 \underset{N}{{}_\beta *_\gamma} M_2 \underset{P}{{}_\epsilon *_\zeta} M_3$ without parenthesis.

If M_1 and M_2 are finite-dimensional, the fiber product $M_1 \underset{N}{{}_\beta *_\gamma} M_2$ can be identified with a reduced algebra of $M_1 \otimes M_2$ ([EV] 2.4).

2.6 Slice maps [E2]

Let A be in $M_1 \underset{N}{{}_\beta *_\gamma} M_2$, and let ξ_1, ξ_2 be in $D(\mathcal{H}_\beta, \psi^o)$. Let us define

$$(\omega_{\xi_1,\xi_2} * \mathrm{id})(A) = \lambda^*_{\xi_2} A \lambda_{\xi_1}.$$

We define in this way $(\omega_{\xi_1,\xi_2} * \mathrm{id})(A)$ as a bounded operator on $\mathcal{K}$, which belongs to M_2 and verifies

$$((\omega_{\xi_1,\xi_2} * \mathrm{id})(A)\eta_1|\eta_2) = (A(\xi_1 \underset{\psi}{{}_\beta\otimes_\gamma} \eta_1)|\xi_2 \underset{\psi}{{}_\beta\otimes_\gamma} \eta_2).$$

One should note that $(\omega_{\xi_1,\xi_2} * \mathrm{id})(1) = \gamma(\langle\xi_1, \xi_2\rangle_{\beta,\psi^o})$.

Let us define in the same way, for any η_1, η_2 in $D({}_\gamma\mathcal{K}, \psi)$,

$$(\mathrm{id} * \omega_{\eta_1,\eta_2})(A) = \rho^*_{\eta_2} A \rho_{\eta_1},$$

which belongs to M_1.

Therefore, we have a Fubini formula for these slice maps: for any ξ_1, ξ_2 in $D(\mathcal{H}_\beta, \psi^o)$, η_1, η_2 in $D({}_\gamma\mathcal{K}, \psi)$, we have

$$\langle(\omega_{\xi_1,\xi_2} * \mathrm{id})(A), \omega_{\eta_1,\eta_2}\rangle = \langle(\mathrm{id} * \omega_{\eta_1,\eta_2})(A), \omega_{\xi_1,\xi_2}\rangle.$$

Moreover, if P_2 is a von Neumann algebra, such that

$$\gamma(N) \subset P_2 \subset M_2,$$

and E is a faithful normal conditional expectation from M_2 on P_2, we can define ([E2], 3.4) a faithful conditional expactation $(\mathrm{id} * E)$ from $M_1 \underset{N}{{}_\beta *_\gamma} M_2$ to $M_1 \underset{N}{{}_\beta *_\gamma} P_2$ verifying the following property: for all A in $M_1 \underset{N}{{}_\beta *_\gamma} M_2$ and all ω in M_{1*}^+ such that there exist k_1 in $\mathbb{R}^+$ such that $\omega \circ \beta \leq k_1 \psi^o$, we have

$$(\omega * \mathrm{id})(\mathrm{id} * E)(A) = E((\omega * \mathrm{id})(A)).$$

If ϕ_1 is a normal semi-finite weight on M_1^+, and operator A is a positive element of the fiber product $M_1 \underset{N}{{}_\beta *_\gamma} M_2$, then we may define a positive self-adjoint operator affiliated to M_2, denoted $(\phi_1 * \mathrm{id})(A)$, such that, for all η in $D({}_\gamma L^2(M_2), \psi)$, we have

$$\|(\phi_1 * \mathrm{id})(A)^{1/2}\eta\|^2 = \phi_1(\mathrm{id} * \omega_\eta)(A).$$

Moreover, then, if ϕ_2 is a normal semi-finite weight on M_2^+, we have

$$\phi_2(\phi_1 * \mathrm{id})(A) = \phi_1(\mathrm{id} * \phi_2)(A).$$

Let now P_1 be a von Neumann algebra, such that

$$\beta(N) \subset P_1 \subset M_1,$$

and let Φ_i $(i = 1, 2)$ be a normal faithful semi-finite operator valued weight from M_i to P_i; for any positive operator A in the fiber product $M_1 \underset{N}{{}_\beta *_\gamma} M_2$, there exists a positive self-adjoint operator $(\Phi_1 * \mathrm{id})(A)$ affiliated to $P_1 \underset{N}{{}_\beta *_\gamma} M_2$, such that ([E2], 3.5), for all η in $D({}_\gamma L^2(M_2), \psi)$ and ξ in $D(L^2(P_1)_\beta, \psi^o)$, we have

$$\|(\Phi_1 * \mathrm{id})(A)^{1/2}(\xi \underset{\psi}{{}_\beta \otimes_\gamma} \eta)\|^2 = \|\Phi_1(\mathrm{id} * \omega_\eta)(A)^{1/2}\xi\|^2.$$

If ϕ is a normal semi-finite weight on P, we have

$$(\phi \circ \Phi_1 * \mathrm{id})(A) = (\phi * \mathrm{id})(\Phi_1 * \mathrm{id})(A).$$

We define in the same way a positive self-adjoint operator $(\mathrm{id} * \Phi_2)(A)$ affiliated to $M_1 \underset{N}{{}_\beta *_\gamma} P_2$, and we have

$$(\mathrm{id} * \Phi_2)((\Phi_1 * \mathrm{id})(A)) = (\Phi_1 * \mathrm{id})((\mathrm{id} * \Phi_2)(A)).$$

3 Quantum groupoids

In this section we give the definitions of Hopf bimodules (3.1) and pseudo-multiplicative unitaries (3.2) and the examples coming from groupoids. The results concerning non-irreducible depth 2 inclusions of von Neumann algebra are exposed in 3.3. Moreover, we show how, as in the case of the multiplicative unitaries, it is possible to go from pseudo-multiplicative unitaries to two dual Hopf bimodules structures (3.4). In the converse direction, we announce a result which will appear in the Lesieur's thesis (3.5). In order to follow the Baaj–Skandalis strategy of the study of multiplicative unitaries, we have to define regularity conditions (3.6) for pseudo-multiplicative unitaries.

3.1 Hopf bimodules

Following ([Val1], [EV] 6.5), a quadruplet (N, M, r, s, Γ) will be called a Hopf bimodule, if N, M are von Neumann algebras, r is a faithful non-degenerate representation of N into M, s is a faithful non-degenerate anti-representation of N into M, with commuting ranges, and Γ is an injective involutive homomorphism from M into $M \underset{N}{{}_s*_r} M$ such that, for all X in N

(i) $\Gamma(s(X)) = 1 \underset{N}{{}_s\otimes_r} s(X)$;

(ii) $\Gamma(r(X)) = r(X) \underset{N}{{}_s\otimes_r} 1$;

(iii) Γ satisfies the co-associativity relation:

$$(\Gamma \underset{N}{{}_s*_r} \mathrm{id})\Gamma = (\mathrm{id} \underset{N}{{}_s*_r} \Gamma)\Gamma.$$

This last formula makes sense, due to the two preceding ones and 2.5.

If (N, M, r, s, Γ) is a Hopf bimodule, it is clear that $(N^o, M, s, r, \varsigma_N \circ \Gamma)$ is another Hopf bimodule, we shall call it the symmetrized of the first one (recall that $\varsigma_N \circ \Gamma$ is a homomorphism from M to $M \underset{N^o}{{}_r*_s} M$).

If N is abelian, $r = s$, $\Gamma = \varsigma_N \circ \Gamma$, then the quadruplet (N, M, r, r, Γ) is equal to its symmetrized Hopf bimodule, and we shall say that it is a symmetric Hopf bimodule.

In [Val1] two Hopf bimodules were associated to a locally compact groupoid, having a left Haar system, and a quasi-invariant measure μ.

The first one is $((L^\infty(\mathcal{G}^{(0)}, \mu), L^\infty(\mathcal{G}, \nu), r_{\mathcal{G}}, s_{\mathcal{G}}, \Gamma_{\mathcal{G}})$, where, for g in $L^\infty(\mathcal{G}^{(0)}, \mu)$, we put $r_{\mathcal{G}}(g) = g \circ r$, $s_{\mathcal{G}}(g) = g \circ s$, and $\Gamma_{\mathcal{G}}$ had been defined in 2.2. But now we shall identify $L^\infty(\mathcal{G}^{(2)}, \nu^2_{s,r})$ with $L^\infty(\mathcal{G})_s*_r L^\infty(\mathcal{G})$.

The second one is symmetric; it is constructed on the von Neumann algebra generated by the left regular representation of $\mathcal{G}$. See ([Y1], [Val1]) for more details.

If the algebra M is finite dimensional, we obtain a homomorphism from M into a reduced algebra of $M \otimes M$, or, equivalently, a usual coproduct Γ from M into $M \otimes M$, but with $\Gamma(1) \neq 1$. Therefore, we are then in the situation of a weak $\mathbb{C}^*$-Hopf algebra, described in [BNS].

3.2 Pseudo-multiplicative unitaries [EV]

Let N be a von Neumann algebra and $\mathfrak{H}$ a Hilbert space on which N has a non-degenerate normal representation α and two non-degenerate normal anti-representations β and $\hat{\beta}$. These three applications are supposed to be injective and to commute two by two. Let μ be a normal semi-finite faithful weight on N; we can therefore construct the Hilbert spaces $\mathfrak{H} \underset{\mu}{{}_{\hat{\beta}}\otimes_{\alpha}} \mathfrak{H}$ and $\mathfrak{H} \underset{\mu^o}{{}_{\alpha}\otimes_{\beta}} \mathfrak{H}$. A unitary W from $\mathfrak{H} \underset{\mu}{{}_{\hat{\beta}}\otimes_{\alpha}} \mathfrak{H}$ onto $\mathfrak{H} \underset{\mu^o}{{}_{\alpha}\otimes_{\beta}} \mathfrak{H}$ will be called a pseudo-multiplicative unitary over the base N, with respect to the representation α and the anti-representations β and $\hat{\beta}$, if

(i) W intertwines α, β, $\hat{\beta}$ in the following way:

$$W(\alpha(X) \underset{N}{{}_{\hat{\beta}}\otimes_{\alpha}} 1) = (1 \underset{N^o}{{}_{\alpha}\otimes_{\beta}} \alpha(X))W;$$
$$W(1 \underset{N}{{}_{\hat{\beta}}\otimes_{\alpha}} \hat{\beta}(X)) = (1 \underset{N^o}{{}_{\alpha}\otimes_{\beta}} \hat{\beta}(X))W;$$
$$W(\beta(X) \underset{N}{{}_{\hat{\beta}}\otimes_{\alpha}} 1) = (\beta(X) \underset{N^o}{{}_{\alpha}\otimes_{\beta}} 1)W;$$
$$W(1 \underset{N}{{}_{\hat{\beta}}\otimes_{\alpha}} \beta(X)) = (\hat{\beta}(X) \underset{N^o}{{}_{\alpha}\otimes_{\beta}} 1)W.$$

(ii) The operator W satisfies:

$$(1_{\mathfrak{H}} \underset{N^o}{{}_{\alpha}\otimes_{\beta}} W)(W \underset{N}{{}_{\hat{\beta}}\otimes_{\alpha}} 1_{\mathfrak{H}})$$
$$= (W \underset{N^o}{{}_{\alpha}\otimes_{\beta}} 1_{\mathfrak{H}})(\sigma_{\mu^o} \underset{N^o}{{}_{\alpha}\otimes_{\beta}} 1_{\mathfrak{H}})(1_{\mathfrak{H}} \underset{N^o}{{}_{\alpha}\otimes_{\beta}} W)\sigma_{2\mu}(1_{\mathfrak{H}} \underset{N}{{}_{\hat{\beta}}\otimes_{\alpha}} \sigma_{\mu^o})(1_{\mathfrak{H}} \underset{N}{{}_{\hat{\beta}}\otimes_{\alpha}} W).$$

In that formula, the first σ_{μ^o} is the relative flip defined in 2.4 from $\mathfrak{H} \underset{\mu^o}{{}_{\alpha}\otimes_{\hat{\beta}}} \mathfrak{H}$ to $\mathfrak{H} \underset{\mu}{{}_{\hat{\beta}}\otimes_{\alpha}} \mathfrak{H}$, and the second is the relative flip from $\mathfrak{H} \underset{\mu^o}{{}_{\alpha}\otimes_{\beta}} \mathfrak{H}$ to $\mathfrak{H} \underset{\mu}{{}_{\beta}\otimes_{\alpha}} \mathfrak{H}$; while $\sigma_{2\mu}$ is the relative flip from $\mathfrak{H} \underset{\mu}{{}_{\hat{\beta}}\otimes_{\alpha}} \mathfrak{H} \underset{\mu}{{}_{\beta}\otimes_{\alpha}} \mathfrak{H}$ to $\mathfrak{H} \underset{\mu^o}{{}_{\alpha}\otimes_{\beta}} (\mathfrak{H} \underset{\mu}{{}_{\hat{\beta}}\otimes_{\alpha}} \mathfrak{H})$. The index 2 is written to recall that the flip "turns" around the second relative tensor product and the parenthesis are written to recall that, in such a situation, associativity is not valid because the anti-representation β is here acting on the second leg of $\mathfrak{H} \underset{\mu}{{}_{\hat{\beta}}\otimes_{\alpha}} \mathfrak{H}$.

All the properties supposed in (i) allow us to write such a formula, which will be called the "pentagonal relation".

If we start from another normal semi-finite faithful weight μ' on N, we may define, using 2.4, another unitary $W^{\mu'} = U^{\mu^o,\mu'^o}_{\alpha,\beta} W U^{\mu',\mu}_{\hat\beta,\alpha}$ from $\mathfrak{H} {}_{\hat\beta}\otimes_{\alpha}{}_{\mu'} \mathfrak{H}$ onto $\mathfrak{H} {}_{\alpha}\otimes_{\beta}{}_{\mu'^o} \mathfrak{H}$. The formulae which link these isomorphisms between relative tensor product Hilbert spaces and the relative flips allow us to check that this operator $W^{\mu'}$ is also pseudo-multiplicative; this can be resumed in saying that a pseudo-multiplicative unitary does not depend on the choice of the weight on N.

Let us check that we get the same duality construction as for multiplicative unitaries: more precisely, if W is a pseudo-multiplicative unitary over the base N, with respect to the representation α, and the anti-representations β and $\hat\beta$, then, the unitary $\widehat{W} = \sigma_{\mu^o} W^* \sigma_\mu$ is a pseudo-multiplicative unitary over N, with respect to the representation α, and the anti-representations $\hat\beta$ and β.

Now let us show the fundamental example coming from groupoids: the Hilbert space $L^2(\mathcal{G}^{(2)}, \nu^2_{s,r})$ introduced in 2.2 can be identified with the relative tensor product $L^2(\mathcal{G}, \nu) {}_{s_\mathcal{G}}\otimes_{r_\mathcal{G}}{}_{\mu} L^2(\mathcal{G}, \nu)$, and the Hilbert space $L^2(\mathcal{G}^2_r, \nu^2_{r,r})$ – with the relative tensor product $L^2(\mathcal{G}, \nu) {}_{r_\mathcal{G}}\otimes_{r_\mathcal{G}}{}_{\mu} L^2(\mathcal{G}, \nu)$. Then, the unitary $W_\mathcal{G}$ defined in 2.2 can be interpreted [Val2] as a pseudo-multiplicative unitary over the base $L^\infty(\mathcal{G}^{(0)}, \mu)$, with respect to the representation $r_\mathcal{G}$, and anti-representation $s_\mathcal{G}$ and $r_\mathcal{G}$ (as here the base is abelian, the notions of a representation and an anti-representation coincide, and the commutation property is fulfilled), where $r_\mathcal{G}$ and $s_\mathcal{G}$ has been defined in 3.1.

If the Hilbert space H is finite dimensional, W is a unitary from some subspace of $H \otimes H$ to another subspace of the same finite dimensional Hilbert space. We can therefore consider W as a partial isometry, and W is then a multiplicative partial isometry, as studied in [Val3] and [Val4].

3.3 Depth 2 inclusions of von Neumann algebras

The same objects can be constructed starting with depth 2 inclusions of von Neumann algebras ([EV], [E2]).

Let $M_0 \subset M_1$ be an inclusion of von Neumann algebras; let ψ_i be a faithful semi-finite normal weight on M_i $(i = 0, 1)$; we shall write H_1 instead of H_{ψ_1}, J_1 instead of J_{ψ_1}, etc. By restriction of the standard representations of M_1 and M_1^o, H_1 is canonically a $M_0 - M_0^o$ bimodule. More precisely, we shall call r the inclusion of M_0 into M_1, and s the anti-homomorphism defined, for any x in M_0, by $s(x) = J_1 x^* J_1$, which gives this bimodule structure. More generally, we shall note that j_1 the mirroring application on H_1 given by J_1, defined, for any x in $\mathcal{L}(H_1)$, by $j_1(x) = J_1 x^* J_1$. Following ([J], 3.1.5(i)), the von Neumann algebra $M_2 = j_1(M_0') = \mathrm{End}_{M_0^o}(H_1)$ will be called the basic construction for the inclusion $M_0 \subset M_1$ (or we shall say the inclusion $M_0 \subset M_1 \subset M_2$ is standard).

Then, we may repeat the procedure and construct on H_2 the basic construction $M_3 = j_2(M_1')$. Let ψ_3 be a faithful semi-finite normal weight on M_3; we shall write H_3 instead of H_{ψ_3}, J_3 instead of J_{ψ_3}, etc. Following [PP], we define the Jones' tower associated to the inclusion $M_0 \subset M_1$ as the sequence

$$M_0 \subset M_1 \subset M_2 \subset M_3 \subset M_4 \subset \cdots$$

defined by recurrence by successive basic constructions.

Following ([GHJ] 4.6.4), we shall say that the inclusion $M_0 \subset M_1$ is of depth 2 if the inclusion $M_0' \cap M_1 \subset M_0' \cap M_2 \subset M_0' \cap M_3$ is standard. This means that there exists a faithful normal representation π_1 of $M_0' \cap M_3$ on the Hilbert space $L^2(M_0' \cap M_2)$, such that $\pi_1(M_0' \cap M_3) = \tilde{j}_2(M_0' \cap M_1)$, where $\tilde{j}_2$ is the mirroring defined on the Hilbert space $L^2(M_0' \cap M_2)$.

Let us go back to an inclusion of von Neumann algebras $M_0 \subset M_1$, and let us suppose that there exists a faithful semi-finite normal operator-valued weight T_1 from M_1 to M_0, ψ_0 a faithful semi-finite normal weight on M_0, and $\psi_1 = \psi_0 \circ T_1$.

Then ([EN] 10.6), for all x in $\mathfrak{N}_{T_1}$, a in $\mathfrak{N}_{\psi_0}$, xa belongs to $\mathfrak{N}_{T_1} \cap \mathfrak{N}_{\psi_1}$; moreover, the application $\Lambda_{\psi_0}(a) \to \Lambda_{\psi_1}(xa)$ can be extended to an element $\Lambda_{T_1}(x)$ of $\mathrm{Hom}_{M_0^o}(H_0, H_1)$; then, Λ_{T_1} is an injective $M_1 - M_0$-module morphism from $\mathfrak{N}_{T_1}$ to $\mathrm{Hom}_{M_0^o}(H_0, H_1)$ such that, for all x, y in $\mathfrak{N}_{T_1}$,

$$\Lambda_{T_1}(y)^* \Lambda_{T_1}(x) = \pi_{\psi_0}(T_1(y^*x)).$$

We get ([EN] 10.7) that, for all x, y in $\mathfrak{N}_{T_1}$, $\Lambda_{T_1}(x)\Lambda_{T_1}(y)^*$ belongs to the basic construction M_2; more precisely, the von Neumann algebra M_2 is the weak closed subspace generated by these operators. If z belongs to $\mathfrak{N}_{T_1} \cap \mathfrak{N}_{\psi_1}$, we have ([EN], 10.6):

$$\Lambda_{T_1}(x)\Lambda_{T_1}(y)^* \Lambda_{\psi_1}(z) = \Lambda_{\psi_1}(xT_1(y^*z)).$$

Using the Haagerup's construction ([St], 12.11), it is possible ([EN], 10.1) to define a canonical operator-valued weight from M_2 to M_1, such that ([EN], 10.7):

$$T_2(\Lambda_{T_1}(x)\Lambda_{T_1}(y)^*) = xy^*.$$

The operator-valued weight T_2 will be called the basic construction made from T_1. This construction can be repeated in constructing T_3 from M_3 to M_2, etc.

We shall now consider the set of double intertwiners $\mathrm{Hom}_{M_0, M_1^o}(H_1, H_2)$, and, more precisely, if χ is a normal semi-finite faithful weight on $M_0' \cap M_1$, the subset Hom_χ, defined as

$$\mathrm{Hom}_\chi = \{x \in \mathrm{Hom}_{M_0, M_1^o}(H_1, H_2) / \chi(x^*x) < \infty\}.$$

This set is clearly a pre-Hilbert space, and we shall denote by $\mathfrak{H}$ its completion, and Λ_χ the canonical injection of Hom_χ into $\mathfrak{H}$.

For all $a \in \mathfrak{N}_{T_2} \cap M_0'$, $\Lambda_{T_2}(a)$ belongs to $\mathrm{Hom}_{M_0, M_1^o}(H_1, H_2)$, and, for any e in $\mathfrak{N}_\chi$, $\Lambda_{T_2}(a)e$ belongs to Hom_χ. So, if $\mathfrak{N}_{T_2} \cap M_0'$ is not reduced to $\{0\}$, Hom_χ (and $\mathfrak{H}$) are not reduced to $\{0\}$.

If the restriction $\tilde{T}_2$ of T_2 to $M_0' \cap M_2$ is semi-finite, and if we write $\chi_2 = \chi \circ \tilde{T}_2$, which is a normal semi-finite faithful weight on $M_0' \cap M_2$, we obtain in this way an injection I from $L^2(M_0' \cap M_2)$ into $\mathfrak{H}$, defined, for all a in $\mathfrak{N}_{\chi_2}$, by

$$I\Lambda_{\chi_2}(a) = \Lambda_\chi(\Lambda_{T_2}(a)).$$

If, in addition to, the restriction of T_3 to $M_1' \cap M_3$ is semi-finite, then we can prove ([EV], 3.8) that this isometry I is surjective, and we shall identify $\mathfrak{H}$ with $L^2(M_0' \cap M_2)$. We shall say that T_1 is regular if the restriction of T_2 to $M_0' \cap M_2$ is semi-finite, and the restriction of T_3 to $M_1' \cap M_3$ is semi-finite

Let us suppose now that the inclusion $M_0 \subset M_1$ is of depth 2. So, by definition, there exists a normal faithful representation π of $M_0' \cap M_3$ on the Hilbert space $L^2(M_0' \cap M_2)$, such that $\pi(M_0' \cap M_3) = J_{\chi_2}\pi_{\chi_2}(M_0' \cap M_1)' J_{\chi_2}$. The restriction of π to $M_0' \cap M_2$ is π_{χ_2}; moreover, π can be easily described using the identification of $L^2(M_0' \cap M_2)$ with $\mathfrak{H}$ ([EV], 3.2(ii), 3.9, 3.10); we have, for all X in $M_0' \cap M_3$ and x in Hom_χ:

$$\pi(X)\Lambda_\chi(x) = \Lambda_\chi(Xx).$$

With these hypothesis, we had constructed in [EV] a pseudo-multiplicative unitary on the Hilbert space $\mathfrak{H} = L^2(M_0' \cap M_2)$ over the base $(M_0' \cap M_1)^o$, with respect to a representation s, and two anti-representations r and $\hat{r}$ of $(M_0' \cap M_1)^o$.

Here, r is the restriction of π_{χ_2} to $M_0' \cap M_1$, $\hat{r}$ is the isomorphism of $M_0' \cap M_1$ onto $M_2' \cap M_3$ given by $j_2 \circ j_1$, composed with the restriction of π to $M_2' \cap M_3$, and s is the anti-representation of $M_0' \cap M_1$ given, for all x in $M_0' \cap M_1$, by $s(x) = J_{\chi_2} x^* J_{\chi_2}$, which sends $M_0' \cap M_1$ onto $J_{\chi_2}\pi_{\chi_2}(M_0' \cap M_1) J_{\chi_2}$, which, due to the depth 2 condition, is equal to $\pi_1(M_0' \cap M_3)'$.

Moreover, if the operator-valued weight T_1 is adapted to χ (i.e., if $\sigma_t^{T_1} = \sigma_t^\chi$ for all t in $\mathbb{R}$), we have discovered in [E2] analytic properties for this pseudo-multiplicative unitary, similar to the Woronowicz' manageability. We shall say that T_1 is adapted if there exists some normal semi-finite faithful weight χ on $M_0' \cap M_1$, such that T_1 is adapted to χ. We shall then consider the pseudo-multiplicative unitary constructed with the help of this auxiliary weight χ.

3.4 From a pseudo-multiplicative unitary to two Hopf bimodules ([EV], [E2])

For ξ_1 in $D(\mathfrak{H}_{\hat{\beta}}, \mu^o)$, η_1 in $D({}_\alpha\mathfrak{H}, \mu)$, the operator $(\lambda_{\eta_1}^{\alpha,\beta})^* W \lambda_{\xi_1}^{\hat{\beta},\alpha}$ will be written as $(\omega_{\xi_1,\eta_1} * \mathrm{id})(W)$; we have therefore, for all ξ_2, η_2 in $\mathfrak{H}$,

$$((\omega_{\xi_1,\eta_1} * \mathrm{id})(W)\xi_2|\eta_2) = (W(\xi_1 \;{}_{\hat{\beta}}\!\otimes_{\alpha}\!{}_{\mu}\; \xi_2)|\eta_1 \;{}_{\alpha}\!\otimes_{\beta}\!{}_{\mu^o}\; \eta_2),$$

and, using the intertwining property of W with respect to $\hat{\beta}$, we get that $(\omega_{\xi_1,\eta_1} * \mathrm{id})(W)$ belongs to $\hat{\beta}(N)'$.

We shall write $A_n(W)$ (resp., $A_w(W)$) for the norm closure (resp., weak closure) of the linear span of these operators. We have $A_n(W) \subset A_w(W) \subset \hat{\beta}(N)'$.

For ξ_2 in $D({}_\alpha\mathfrak{H}, \mu)$, η_2 in $D(\mathfrak{H}_\beta, \mu^o)$, the operator $(\rho_{\eta_2}^{\alpha,\beta})^* W \rho_{\xi_2}^{\hat{\beta},\alpha}$ will be written $(\mathrm{id} * \omega_{\xi_2,\eta_2})(W)$; we have therefore, for all ξ_1, η_1 in $\mathfrak{H}$

$$((\mathrm{id} * \omega_{\xi_2,\eta_2})(W)\xi_1|\eta_1) = (W(\xi_1 \underset{\mu}{{}_{\hat{\beta}}\otimes_\alpha} \xi_2)|\eta_1 \underset{\mu^o}{{}_\alpha\otimes_\beta} \eta_2),$$

and, using the intertwining property of W with respect to β, we get that $(\mathrm{id} * \omega_{\xi_2,\eta_2})(W)$ belongs to $\beta(N)'$.

We shall write $\widehat{A}_n(W)$ (resp., $\widehat{A}_w(W)$) for the norm (resp., weak) closure of the linear span of these operators. We have $\widehat{A}_n(W) \subset \widehat{A}_w(W) \subset \beta(N)'$.

With the notations of 3.2, we have $\widehat{A}_n(W) = A_n(\widehat{W})^*$ and $\widehat{A}_w(W) = A_w(\widehat{W})^*$.

Then we have:

Proposition ([E3]). *The norm closed subspaces $A_n(W)$ and $A_n(W)^*$ (resp., the weakly closed subspaces $A_w(W)$ and $A_w(W)^*$) are non-degenerate algebras.*

Following ([EV], 6.1 and 6.5), we shall denote by $\mathcal{A}$ the von Neumann algebra generated by $A_w(W)$ and by $\widehat{\mathcal{A}}$ the von Neumann algebra generated by $\widehat{A}_w(W)$.

In ([EV], 6.3 and 6.5), using the pentagonal equation, we got that $(N^o, \mathcal{A}, \alpha, \beta, \Gamma)$ and $(N, \widehat{\mathcal{A}}, \hat{\beta}, \alpha, \widehat{\Gamma})$ are Hopf bimodules, where Γ and $\widehat{\Gamma}$ are defined, for any x in $\mathcal{A}$ and y in $\widehat{\mathcal{A}}$, by

$$\Gamma(x) = W(x \underset{N}{{}_{\hat{\beta}}\otimes_\alpha} 1)W^*,$$

$$\widehat{\Gamma}(y) = W^*(1 \underset{N^o}{{}_\alpha\otimes_\beta} y)W.$$

Moreover, we get that $\alpha(N) \subset \mathcal{A} \cap \widehat{\mathcal{A}}$, $\beta(N) \subset \mathcal{A}$, $\hat{\beta}(N) \subset \widehat{\mathcal{A}}$ and, for all x in N:

$$\Gamma(\alpha(x)) = 1 \underset{N^o}{{}_\alpha\otimes_\beta} \alpha(x),$$

$$\Gamma(\beta(x)) = \beta(x) \underset{N^o}{{}_\alpha\otimes_\beta} 1,$$

$$\widehat{\Gamma}(\alpha(x)) = \alpha(x) \underset{N}{{}_{\hat{\beta}}\otimes_\alpha} 1,$$

$$\widehat{\Gamma}(\hat{\beta}(x)) = 1 \underset{N}{{}_{\hat{\beta}}\otimes_\alpha} \hat{\beta}(x).$$

Let us take the notations of 2.2; the von Neumann algebra $\mathcal{A}(W_{\mathcal{G}})$ is equal to the von Neumann algebra $\mathcal{L}(\mathcal{G})$ ([Val2], 3.2.6 and 3.2.7); using ([Val2], 3.1.1), we get that the Hopf bimodule homomorphism Γ defined on $\mathcal{L}(\mathcal{G})$ by $W_{\mathcal{G}}$ is the usual Hopf bimodule homomorphism $\widehat{\Gamma_{\mathcal{G}}}$ studied in [Y1] and [Val1], and recalled in 2.2. The von Neumann algebra $\widehat{\mathcal{A}(W_{\mathcal{G}})}$ is equal to the von Neumann algebra $L^\infty(\mathcal{G}, \nu)$ ([Val2], 3.2.6 and 3.2.7); using ([Val2], 3.1.1), we get that the Hopf bimodule homomorphism $\widehat{\Gamma}$ defined on $L^\infty(\mathcal{G}, \nu)$ by $W_{\mathcal{G}}$ is equal to the usual Hopf bimodule homomorphism $\Gamma_{\mathcal{G}}$ studied in [Val1], and recalled in 2.2.

Now let us take a depth 2 inclusion equipped with a regular operator-valued weight, as described in 3.3 and let W be then the pseudo-multiplicative unitary associated. In [EV] it is shown that the von Neumann algebra $\mathcal{A}(W)$ is then equal to $\pi_1(M_0' \cap M_2)'$, and that the bimodule homomorphism Γ sends this algebra to

$$\pi(M_0' \cap M_2)' \underset{M_0' \cap M_1}{{}_s*_{\hat{r}}} \pi(M_0' \cap M_2)'.$$

The von Neumann algebra $\widehat{\mathcal{A}(W)}$ is then equal to $\pi_1(M_1' \cap M_3)'$, and the coproduct $\widehat{\Gamma}$ sends it to $\pi(M_1' \cap M_3)' \underset{(M_0' \cap M_1)^o}{{}_r*_s} \pi(M_1' \cap M_3)'$.

When the operator-valued weight is adapted to the weight χ defined on $M_0' \cap M_1$, then we have discovered in [E2] analytic properties of W which allows to define "antipodes" on $\mathcal{A}(W)$ and $\widehat{\mathcal{A}(W)}$, and, by polar decomposition, "unitary antipodes", which are involutive anti-isomorphisms. So, just by composition with the mirroring given by the Tomita–Takesaki construction, it is then possible to put these structures on $M_0' \cap M_2$ and $M_1' \cap M_3$, respectively.

More precisely ([E2], 8.2), there exists a normal faithful homomorphism Γ from $M_0' \cap M_2$ to the fiber product $(M_0' \cap M_2) \underset{M_0' \cap M_1}{{}_{j_1}*_{\mathrm{id}}} (M_0' \cap M_2)$, where j_1 is the mirroring $x \mapsto J_1 x^* J_1$ defined on $\mathcal{L}(H_1)$ (cf. 3.3), and id is for the imbedding of $M_0' \cap M_1$ into $M_0' \cap M_2$, such that, for all a in $M_0' \cap M_1$:

$$\Gamma(a) = a \underset{M_0' \cap M_1}{{}_{j_1}\otimes_{\mathrm{id}}} 1,$$

$$\Gamma(j_1(a)) = 1 \underset{M_0' \cap M_1}{{}_{j_1}\otimes_{\mathrm{id}}} j_1(a),$$

$$(\Gamma * \mathrm{id})\Gamma = (\mathrm{id} * \Gamma)\Gamma,$$

which gives to $(M_0' \cap M_1, M_0' \cap M_2, \mathrm{id}, j_1)$ the structure of a Hopf bimodule, as defined in 3.1. The structure on $M_1' \cap M_3$ can be just obtained by considering the inclusion $M_1 \subset M_2$.

Moreover, in this situation, the restriction $\tilde{T}_2$ of T_2 to $M_0' \cap M_2$ is a normal semi-finite faithful operator-valued weight from $M_0' \cap M_2$ to $M_0' \cap M_1$, which satisfies, for all positive X in $M_0' \cap M_2$,

$$(\mathrm{id}_{j_1} *_{\mathrm{id}} \tilde{T}_2)\Gamma(X) = \tilde{T}_2(X),$$

and $S = j_1 \tilde{T}_2 j_1$ is a normal semi-finite faithful operator-valued weight from $M_0' \cap M_2$ to $j_1(M_0' \cap M_1)$, which satisfies, for all positive X in $M_1' \cap M_2$,

$$(S_{j_1} *_{\mathrm{id}} \mathrm{id})\Gamma(X) = S(X).$$

3.5 From a Hopf bimodule to a pseudo-multiplicative unitary [L]

Following the ideas of Kusterman and Vaes in their axiomatization of locally compact quantum groups [KV], F. Lesieur has obtained the following:

Let (N, M, r, s, Γ) be a Hopf bimodule (3.1), i.e., N and M are von Neumann algebras, r (resp., s) is a faithful non-degenerate representation (resp., anti-representation) in M, and Γ is an injective involutive homomorphism from M into $M \ {}_s{*}_r\, M$ (with N under $*$) such that, for all n in N,

$$\Gamma(s(n)) = 1 \ \underset{N}{{}_s\otimes_r} \ s(n),$$

$$\Gamma(r(n)) = r(n) \ \underset{N}{{}_s\otimes_r} \ 1,$$

which satisfies the co-associativity condition:

$$(\Gamma \ \underset{N}{{}_s{*}_r} \ \mathrm{id})\Gamma = (\mathrm{id} \ \underset{N}{{}_s{*}_r} \ \Gamma)\Gamma.$$

Now let us suppose now that there exists a normal semi-finite faithful operator-valued weight S from M to $s(N)$, such that, for all positive x in M,

$$(S \ \underset{N}{{}_s{*}_r} \ \mathrm{id})\Gamma(x) = \Gamma \circ S(x) = 1 \ \underset{N}{{}_s\otimes_r} \ S(x),$$

and a normal semi-finite faithful operator-valued weight T from M to $r(N)$, such that, for all positive x in M,

$$(\mathrm{id} \ \underset{N}{{}_s{*}_r} \ T)\Gamma(x) = \Gamma \circ T(x) = T(x) \ \underset{N}{{}_s\otimes_r} \ 1.$$

Then F. Lesieur ([L]) has constructed, for any normal semi-finite weight μ on N, a pseudo-multiplicative unitary:

$$W : L^2(M) \ \underset{\mu^o}{{}_{\hat{r}}\otimes_s} \ L^2(M) \to L^2(M) \ \underset{\mu}{{}_s\otimes_r} \ L^2(M),$$

where, for all n in N, we have $\hat{r}(n) = J_{\mu \circ S} s(n^*) J_{\mu \circ S}$.

3.6 Regularity conditions for a pseudo-multiplicative unitary [E3]

For any η_1, ξ_2 in $D({}_\alpha\mathfrak{H}, \mu)$, let us write $(\mathrm{id} * \omega_{\xi_2, \eta_1})(\sigma_{\mu^o} W)$ for $(\lambda_{\eta_1}^{\alpha,\beta})^* W \rho_{\xi_2}^{\hat{\beta},\alpha}$. We have therefore, for all ξ_1 and η_2 in $\mathfrak{H}$,

$$((\mathrm{id} * \omega_{\xi_2,\eta_1})(\sigma_{\mu^o} W)\xi_1 | \eta_2) = (W(\xi_1 \ \underset{\mu}{{}_\beta\otimes_\alpha} \ \xi_2) | \eta_1 \ \underset{\mu^o}{{}_\alpha\otimes_\beta} \ \eta_2).$$

We can verify easily that this operator can be written also as $(\omega_{\xi_2,\eta_1} * \mathrm{id})(W\sigma_\mu)$.

Using the intertwining property of W with respect to α, we get that

$$(\mathrm{id} * \omega_{\xi_2,\eta_1})(\sigma_{\mu^o} W)$$

belongs to $\alpha(N)'$. If ξ belongs to $D({}_\alpha\mathfrak{H}, \mu)$, we shall write $(\mathrm{id} * \omega_\xi)(\sigma_{\mu^o} W)$ instead of $(\mathrm{id} * \omega_{\xi,\xi})(\sigma_{\mu^o} W)$.

We shall denote by $C_n(W)$ (resp., $C_w(W)$) the norm (resp., weak) closure of the linear span of these operators; we have $C_n(W) \subset C_w(W) \subset \alpha(N)'$.

Using 3.2, we get that $C_n(\widehat{W}) = C_n(W)^*$ and $C_w(\widehat{W}) = C_w(W)^*$.

Then we get:

Proposition ([E3]). (i) *The norm closed subspace $C_n(W)$ and the weakly closed subspace $C_n(W)$ are algebras.*

(ii) *If $C_n(W)$ is a *-algebra, so are $A_n(W)$ and $\widehat{A}_n(W)$. If $C_w(W)$ is a *-algebra, so are $A_w(W)$ and $\widehat{A}_w(W)$, and then $A_w(W)$ is the von Neumann algebra $\mathcal{A}$ and $\widehat{A}_w(W)$ is the von Neumann algebra $\widehat{\mathcal{A}}$.*

This leads, following the strategy described in [BS], to the following:

Definitions. With the definitions of 3.2, we shall say that the pseudo-multiplicative unitary W is weakly regular, if $C_w(W) = \alpha(N)'$. Then we get that the weakly closed algebras $A_w(W)$ and $\widehat{A}_w(W)$ are the von Neumann algebras, respectively $\mathcal{A}$ and $\widehat{\mathcal{A}}$, generated by the "right leg" (resp., the "left leg") of W, introduced in 3.4.

We shall say that the pseudo-multiplicative unitary W is norm regular, if $C_n(W) = \mathcal{K}_{\alpha,\mu}$, where $\mathcal{K}_{\alpha,\mu}$ has been defined in 2.3 as the norm closure of the linear set generated by all operators $\theta^{\alpha,\mu}(\xi_1, \xi_2)$, where ξ_1, ξ_2 belong to $D({}_\alpha\mathfrak{H}, \mu)$. Then we get that the weakly closed algebra $A_n(W)$ (resp., $\widehat{A}_n(W)$) is the $\mathbb{C}^*$-algebra generated by the "right leg" (resp., the "left leg") of W. It is clear that a norm regular pseudo-multiplicative unitary W is weakly regular.

Let us come back to the basic examples of pseudo-multiplicative unitaries; in the groupoid case, we get:

Proposition. *Let us take the notations of 2.2. The pseudo-multiplicative unitary $W_{\mathcal{G}}$ is norm regular.*

In the depth 2 case, we get:

Proposition. *Let $M_0 \subset M_1$ be a depth 2 inclusion of von Neumann algebras, and let T_1 be a regular faithful normal semi-finite operator-valued weight T_1 from M_1 to M_0, adapted to a normal faithful semi-finite weight χ on $M_0' \cap M_1$. Then, the pseudo-multiplicative unitary W constructed in* [EV], [E2] *(see* 3.3*) is weakly regular.*

4 Quantum groupoids of compact type [E3]

In 4.1, following [BS], we define fixed vectors for a pseudo-multiplicative unitary W. We introduce a suitable notion of normalisation and, as for multiplicative unitaries, we get then a formula for the orthogonal projection on the closure of the linear space of fixed vectors. In 4.2, we obtain then, if W is weakly regular, a left (resp., a right) invariant faithful conditional expectation on the Hopf bimodule structure constructed by the "left leg" of W as well as Heisenberg type relations between the von Neumann algebras $\mathcal{A}$ and $\widehat{\mathcal{A}}$ constructed by the left and right legs of W. In 4.3, we obtain a relative tensor product decomposition of the Hilbert space H and show that the unitary W "lives" on $L^2(\mathcal{A})$. In 4.4, we get a co-unit for the Hopf bimodule structure constructed by the "right leg" of W. Finally, in 4.5, we obtain two generalizations of the Baaj's theorem concerning regularity of compact multiplicative unitaries. All proofs can be found in [E3].

4.1 Fixed vectors

Let W be a pseudo-multiplicative unitary over the base N, with respect to the representation α and the anti-representations β and $\hat{\beta}$. Let μ be a normal semi-finite faithful weight on N. A vector ξ in $\mathfrak{H}$ will be called fixed by W, with respect to μ, if:

(i) the vector ξ belongs to $D(\mathfrak{H}_{\hat{\beta}}, \mu^o) \cap D({}_\alpha\mathfrak{H}, \mu)$,

(ii) we have, for all η in $\mathfrak{H}$,

$$W(\xi \;{}_{\hat{\beta}}\!\underset{\mu}{\otimes}_{\alpha}\; \eta) = \xi \;{}_{\alpha}\!\underset{\mu^o}{\otimes}_{\beta}\; \eta.$$

Thanks to the intertwining property of W with respect to β, it is clear that, if ξ is fixed and x belongs to N, then $\beta(x)\xi$ is fixed. More generally, if b belongs to $\widehat{\mathcal{A}}'$, we obtain that $b\xi$ is fixed.

Proposition. *Let ξ and ξ' be two vectors fixed by W, with respect to a normal semi-finite faithful weight μ on N. Then we have:*

(i) $\alpha(\langle\xi, \xi'\rangle_{\hat{\beta}.\mu^o}) = \beta(\langle\xi, \xi'\rangle_{\alpha.\mu}) = (\omega_{\xi,\xi'} * \mathrm{id})(W) = (\omega_{\xi',\xi} * \mathrm{id})(W)^*$,

(ii) $\langle\xi, \xi'\rangle_{\hat{\beta}.\mu^o} = \langle\xi, \xi'\rangle_{\alpha.\mu} \in Z(N)$.

Let W be a pseudo-multiplicative unitary over the base N, with respect to the representation α and the anti-representations β and $\hat{\beta}$. Let ξ be a vector fixed by W, with respect to a normal semi-finite faithful weight μ on N; we shall say that ξ is normalized if

$$\langle\xi, \xi\rangle_{\hat{\beta}.\mu^o} = \langle\xi, \xi\rangle_{\alpha.\mu} = 1.$$

We have then $(\omega_{\xi,\xi} * \mathrm{id})(W) = 1$, and $1 \in A_n(W)$.

Proposition. *Let ξ be a normalized fixed vector by W, with respect to a normal semi-finite faithful weight μ on N. Then:*

(i) *For all x in N, we have*

$$x^*x = \langle \alpha(x)\xi, \alpha(x)\xi \rangle_{\hat{\beta},\mu^o},$$

$$xx^* = \langle \hat{\beta}(x)\xi, \hat{\beta}(x)\xi \rangle_{\alpha,\mu}.$$

(ii) *We have $\mu = \omega_\xi \circ \alpha = \omega_\xi \circ \hat{\beta}$. The positive forms $\omega_\xi \circ \alpha$ and $\omega_\xi \circ \hat{\beta}$ are therefore faithful and equal, and they do not depend on the choice of the fixed and normalized vector ξ. We shall call μ the canonical faithful positive form on N.*

Let W be a pseudo-multiplicative unitary over the base N, with respect to the representation α and the anti-representations β and $\hat{\beta}$; let ξ be a vector fixed and normalized by W, with respect to a normal semi-finite faithful weight μ on N; then we shall say that ξ is binormalized if, moreover, ξ belongs to $D(\mathfrak{H}_\beta, \mu^o)$, and $\langle \xi, \xi \rangle_{\beta,\mu^o} = 1$.

Theorem. *Let us take the notations of 3.2. Let ξ be a vector fixed and binormalized by W, with respect to the canonical normal faithful positive form μ on N; then $(\mathrm{id} * \omega_\xi)(W)$ is the projection on the closed subspace $\mathfrak{F}$ generated by the fixed vectors. More precisely, if ζ belongs to $D({}_\alpha\mathfrak{H}, \mu)$ (resp., $D(\mathfrak{H}_{\hat{\beta}}, \mu^o)$), then $p\zeta$ belongs to $D({}_\alpha\mathfrak{H}, \mu) \cap D(\mathfrak{H}_{\hat{\beta}}, \mu^o)$, and is fixed by W.*

Proposition. *Let us take the notations of 3.2. Let ξ be a vector fixed and binormalized by W, with respect to the canonical normal faithful positive form μ on N; then we have $\omega_\xi \circ \beta = \omega_\xi \circ \hat{\beta} = \omega_\xi \circ \alpha = \mu$. Therefore, $\omega_\xi \circ \beta$ is faithful, and does not depend on the choice of the fixed binormalized vector ξ.*

Let W be a pseudo-multiplicative unitary over the base N, with respect to the representation α and the anti-representations β and $\hat{\beta}$; we shall say that W is of "compact type" if there exists a vector ξ fixed and binormalized by W, with respect to the normal faithful positive form $\mu = \omega_\xi \circ \alpha = \omega_\xi \circ \hat{\beta} = \omega_\xi \circ \beta$ on N. We shall call μ the canonical normal positive form on N.

We shall say that W is of "discrete type" if $\widehat{W} = \sigma_{\mu^o} W^* \sigma_\mu$ is of compact type.

4.2 Weakly regular pseudo-multiplicative unitaries of compact type

In the following subsections 4.2, 4.3 and 4.4 we consider a weakly regular pseudo-multiplicative unitary W over the base N, with respect to the representation α and the anti-representations β and $\hat{\beta}$, of compact type (4.1). Let $\mathcal{A}$ (resp., $\widehat{\mathcal{A}}$) be the von Neumann algebra generated by the "right leg" (resp., the "left leg") of W, and let ξ

be a fixed and binormalized vector and μ be the canonical normal positive form on N (4.1).

Theorem. *Using the above mentioned data, we have:*

(i) *The state ω_ξ on $\mathcal{A}$ is faithful and does not depend on the choice of the vector ξ. We shall denote it ω and call it the Haar positive form of the Hopf bimodule $(\mathcal{A}, \Gamma)$.*

(ii) *The application F_ξ defined by $F_\xi(X) = (\mathrm{id}\ {}_\alpha{*}_\beta\ \omega_\xi)\Gamma(X)$, for all X in $\mathcal{A}$, is a* $_{N^o}$ *faithful conditional expectation from $\mathcal{A}$ onto $\beta(N)$, which does not depend on the choice of the fixed and binormalized vector ξ. Moreover, if F is a conditional expectation from $\mathcal{A}$ onto $\beta(N)$ such that*

$$(\mathrm{id} * F)\Gamma = \Gamma \circ F,$$
$$\omega \circ F = \omega,$$

then $F = F_\xi$, and F is therefore faithful. This unique conditional expectation will be called the right Haar conditional expectation of the Hopf bimodule $(\mathcal{A}, \Gamma)$.

(iii) *The application E_ξ defined by $E_\xi(X) = (\omega_\xi * \mathrm{id})\Gamma(X)$, for all X in $\mathcal{A}$, is a faithful conditional expectation from $\mathcal{A}$ onto $\alpha(N)$, which does not depend on the choice of the fixed and binormalized vector ξ. Moreover, if E is a conditional expectation from $\mathcal{A}$ onto $\alpha(N)$ such that*

$$(E * \mathrm{id})\Gamma = \Gamma \circ E,$$
$$\omega \circ E = \omega,$$

then $E = E_\xi$, and E is therefore faithful. This unique conditional expectation will be called the left Haar conditional expectation of the Hopf bimodule $(\mathcal{A}, \Gamma)$.

Theorem. *With the same hypothesis, we have:*

$$\mathcal{A} \cap \widehat{\mathcal{A}} = \alpha(N),$$
$$\mathcal{A} \cap \widehat{\mathcal{A}}' = \beta(N),$$
$$\mathcal{A}' \cap \widehat{\mathcal{A}} = \hat{\beta}(N).$$

4.3 Standard form of a weakly regular pseudo-multiplicative unitary of compact type

Theorem. *With the data mentioned at the beginning of* 4.2, *and with the notations of* 4.2, *let us consider the GNS construction $(H_\omega, \Lambda_\omega, \pi_\omega)$ for the faithful Haar positive form ω introduced in* 4.2 *and identify it with $L^2(\mathcal{A})$. Let γ (resp., δ) be the non* *degenerate representation (resp., anti-representation) of N on* $L^2(\mathcal{A})$ *defined, for all n in N, by*

$$\gamma(n) = J_\omega \beta(n^*) J_\omega,$$
$$\delta(n) = J_\omega \alpha(n^*) J_\omega.$$

Let $\mathfrak{F}$ *be the closed subspace generated by the fixed vectors; the anti-representation* β *leaves* $\mathfrak{F}$ *invariant, and its restriction to* $\mathfrak{F}$ *will be denoted by* $\beta_|$.

(i) *Then the application U, defined for x in* $\mathcal{A}$ *and* η *fixed by*

$$U(\Lambda_\omega(x) \;{}_\gamma\!\underset{\mu^o}{\otimes}{}_{\beta_|}\; \eta) = x\eta,$$

is a unitary from $L^2(\mathcal{A}) \;{}_\gamma\!\underset{\mu^o}{\otimes}{}_{\beta_|}\; \mathfrak{F}$ *onto* $\mathfrak{H}$, *such that, for all n in N:*

$$\alpha(n)U = U(\pi_\omega \circ \alpha(n) \;{}_\gamma\!\underset{N^o}{\otimes}{}_{\beta_|}\; 1),$$
$$\beta(n)U = U(\pi_\omega \circ \beta(n) \;{}_\gamma\!\underset{N^o}{\otimes}{}_{\beta_|}\; 1),$$
$$\hat{\beta}(n)U = U(\delta(n) {}_\gamma\!\underset{N^o}{\otimes}{}_{\beta_|} 1).$$

(ii) *We have, for all x in* $\mathcal{A}$,

$$U^*xU = \pi_\omega(x) \;{}_\gamma\!\underset{N^o}{\otimes}{}_{\beta_|}\; 1.$$

Moreover, there exists a faithful normal representation π *of* $\widehat{\mathcal{A}}$ *on* $L^2(\mathcal{A})$, *such that* $\pi(\widehat{\mathcal{A}}) \subset \gamma(N)'$; *for all y in* $\widehat{\mathcal{A}}$, *we have* $U^*yU = \pi(y) \;{}_\gamma\!\underset{N^o}{\otimes}{}_{\beta_|}\; 1$. *Moreover,* π *is such that* $\pi \circ \alpha = \pi_\omega \circ \alpha$ *and* $\pi \circ \hat{\beta} = \pi_\omega \circ \hat{\beta}$.

Proposition ([L]). *Let us take the notations of* 4.3; *then, for any x in* $\mathcal{A}$ *and* η *in* $L^2(\mathcal{A})$, *the formula*

$$W_s(\Lambda_\omega(x) \;{}_\delta\!\underset{\mu}{\otimes}{}_{\pi_\omega\circ\alpha}\; \eta) = \Gamma(x)(\Lambda_\omega(1) \;{}_{\pi_\omega\circ\alpha}\!\underset{\mu^o}{\otimes}{}_{\pi_\omega\circ\beta}\; \eta)$$

defines a weakly regular pseudo-multiplicative unitary of compact type, over the base N, with respect to the representation $\pi_\omega \circ \alpha$, *and the anti-representations* $\pi_\omega \circ \beta$ *and* δ.

Moreover, $W_s \;{}_{\gamma\otimes\gamma}\!\underset{N^o\otimes N^o}{\otimes}{}_{\beta_|\otimes\beta_|}\; 1$, *is canonically isomorphic to W.*

4.4 Weakly regular pseudo-multiplicative unitaries of discrete type

Theorem. *With the hypothesis and notations of* 4.2 *and* 4.3, *we have:*

(i) *The state* ω_ξ *on* $\widehat{\mathcal{A}}$ *does not depend on the choice of the vector* ξ. *Let us denote* $\hat{\epsilon} = \omega_{\xi|\widehat{\mathcal{A}}}$; *then we have*

$$(\hat{\epsilon} * \mathrm{id})\widehat{\Gamma} = (\mathrm{id} * \hat{\epsilon})\widehat{\Gamma} = \mathrm{id}\,.$$

We shall call $\hat{\epsilon}$ the counit positive form of $(\widehat{\mathcal{A}}, \widehat{\Gamma})$.

(ii) *We have $\hat{\epsilon} = \omega_{\Lambda_\omega(1)} \circ \pi$, where π has been introduced in* 4.3.

Theorem. *With the hypothesis and notations of* 4.2 *and* 4.3, *we have:*

(i) *The G.N.S. representation $\pi_{\hat{\epsilon}}$ of $\widehat{\mathcal{A}}$ associated to $\hat{\epsilon}$, is a bimodule representation of $\widehat{\mathcal{A}}$ on $L^2(N)$.*

(ii) *For any x in $\widehat{\mathcal{A}}$, we have:*

$$(\pi_{\hat{\epsilon}} \; {}_{\hat{\beta}}\underset{N}{*}{}_{\alpha} \; \mathrm{id})\widehat{\Gamma}(x) = (\mathrm{id} \; {}_{\hat{\beta}}\underset{N}{*}{}_{\alpha} \; \pi_{\hat{\epsilon}})\widehat{\Gamma}(x) = x,$$

where we made the canonical identifications of $L^2(N) \underset{\mu}{\otimes}{}_{\alpha} \mathfrak{H}$ and of $\mathfrak{H} \; {}_{\hat{\beta}}\underset{\mu}{\otimes} L^2(N)$ with $\mathfrak{H}$. We shall call $\pi_{\hat{\epsilon}}$ the counit representation of $\widehat{\mathcal{A}}$.

4.5 Baaj's theorem

In this subsection, we consider a weakly regular pseudo-multiplicative unitary W over the base N, with respect to the representation α and the anti-representations β and $\hat{\beta}$, of compact type (4.1), and let μ be the canonical normal positive form on N.

Theorem. *With the above mentioned hypothesis, let us suppose that the base N is a sum of type I factors. Then the following statements are equivalent:*

(i) *W is weakly regular, i.e., $C_w(W) = \alpha(N)'$.*

(ii) *$\mathcal{K}_{\alpha,\mu} \subset C_n(W)$.*

(iii) *W is norm regular, i.e., $C_n(W) = \mathcal{K}_{\alpha,\mu}$.*

Theorem. *With the above mentioned hypothesis, let us suppose that the base N is abelian. Then the following statements are equivalent:*

(i) *W is weakly regular, i.e., $C_w(W) = \alpha(N)'$.*

(ii) *$\mathcal{K}_{\alpha,\mu} \subset C_n(W)$.*

(iii) *W is norm regular, i.e., $C_n(W) = \mathcal{K}_{\alpha,\mu}$.*

5 Examples

In this section, we give examples of pseudo-multiplicative unitaries of compact type. First we show (5.1), in the groupoid case, that we recover then the étale (resp., proper) groupoids. In the depth 2 case (5.2), we recover the inclusions which are equipped, on some level of the Jones' tower, with a conditional expectation. We finish (5.3) with other examples (quantum pair, transformation quantum group).

5.1 The groupoid case

Let $\mathcal{G}$ be a locally compact groupoid, equipped with a Haar system, having a quasi-invariant measure μ on its space of units $\mathcal{G}^{(0)}$, and r and s the range and source maps from $\mathcal{G}$ to $\mathcal{G}^{(0)}$ (2.2); let $W_{\mathcal{G}}$ be the pseudo-multiplicative unitary over the base $L^\infty(\mathcal{G}^{(0)}, \mu)$, with respect to the representation $r_{\mathcal{G}}$, and anti-representation $s_{\mathcal{G}}$ and $r_{\mathcal{G}}$ (3.2).

Proposition. *With the above mentioned notations, the following statements are equivalent:*

(i) *$W_{\mathcal{G}}$ is of compact type, in the sense of* 4.1 *(or, equivalently, $\sigma W_{\mathcal{G}}^*\sigma$ is discrete);*

(ii) *G is "étale", in the sense of* [C2] *(or r-discrete in the sense of* [R1]*), i.e., if $\mathcal{G}^{(0)}$ is an open subset of $\mathcal{G}$.*

Proposition. *With the above mentioned notations, the following statements are equivalent:*

(i) *$W_{\mathcal{G}}$ is of discrete type, in the sense of* 4.1 *(or, equivalently, $\sigma W_{\mathcal{G}}\sigma$ is compact);*

(ii) *G is "proper", in the sense of* [AR], *i.e., if the map $(r, s) : \mathcal{G} \to \mathcal{G}^{(0)} \times \mathcal{G}^{(0)}$ is proper in the usual sense, i.e., if the inverse image of every compact in $\mathcal{G}^{(0)} \times \mathcal{G}^{(0)}$ is compact.*

5.2 Depth 2 inclusions

In this subsection, we study the case of the pseudo-multiplicative unitary constructed from a depth 2 inclusion $M_0 \subset M_1$, with a regular operator-valued weight T_1 from M_1 to M_0, as mentioned in 3.3.

If T_2 is a conditional expectation, then its restriction to $M_1' \cap M_2$ is a conditional expectation from $M_1' \cap M_2$ onto $Z(M_1)$, and using the mirroring j_1, we obtain that there exists a conditional expectation from $M_1' \cap M_2$ onto $Z(M_1)$. Then, lifting any normal faithful state on $Z(M_1)$, we obtain a normal faithful state χ on $M_0' \cap M_1$, such that T_1 is adapted to χ. We shall denote $\chi_2 = \chi \circ \tilde{T}_2$, where $\tilde{T}_2$ is the restriction of T_2 to $M_0' \cap M_2$, which is a faithful state on $M_0' \cap M_2$.

We shall denote by W the pseudo-multiplicative unitary constructed in 3.3 starting with this data.

Theorem. *The vector $\Lambda_{\chi_2}(1)$ is fixed and binormalized for W.*

Conversely, we obtain:

Theorem. *Let $M_0 \subset M_1$ be a depth 2 inclusion of von Neumann algebras, equipped with a regular operator-valued weight T_1 from M_1 to M_0. If the pseudo-multiplicative unitary W, constructed in 3.3, is of compact type, then:*

(i) *There exists a normal faithful conditional expectation $\mathcal{F}$ from M_2 onto M_1.*

(ii) *The operator-valued weight T_1 is adapted, in the sense of 3.3.*

(iii) *The conditional expectation $\mathcal{F}$ is regular and adapted, and there exists an invertible positive operator a affiliated to $Z(M_0) \cap Z(M_1)$, such that, for all positive X in M_2, we have $T_2(X) = \mathcal{F}(aX)$.*

Examples. *Actions of a compact quantum group.* We shall follow the notations of [V], in which all the necessary results concerning actions of locally compact quantum groups (in their von Neumann version) are given.

Let (M, Δ) be a (von Neumann) locally compact quantum group, and let α be its left action on a von Neumann algebra N, i.e., a normal injective homomorphism $\alpha : N \to M \otimes N$, such that $(\mathrm{id} \otimes \alpha)\alpha = (\Delta \otimes \mathrm{id})\alpha$.

Let $(\widehat{M}, \widehat{\Delta})$ be the dual of (M, Δ), and let $\widehat{\varphi}$ be a left Haar weight on (M, Δ); then the crossed-product $M_\alpha \ltimes N$ is the von Neumann algebra generated by $\alpha(N)$ and $\widehat{M} \otimes \mathbb{C}$. It is proved in [V] that the inclusion $\alpha(N) \subset M_\alpha \ltimes N$ is of depth 2, and that there exists a regular operator-valued weight T from $M_\alpha \ltimes N$ onto $\alpha(N)$. Moreover, the von Neumann algebra constructed by the basic construction from the inclusion $\alpha(N) \subset M_\alpha \ltimes N$, is $\mathcal{L}(L^2(M)) \otimes N$.

So, we get, that the pseudo-multiplicative unitary obtained from this depth 2 inclusion is of compact type if and only if (M, Δ) is a compact quantum group.

So, any action of a compact quantum group on a von Neumann algebra gives a pseudo-multiplicative unitary of compact type; this unitary is defined on the relative tensor product of two copies of the Hilbert space $\mathfrak{H} = L^2(\alpha(N)' \cap \mathcal{L}(L^2(M)) \otimes N)$, relatively to the base $\alpha(N)' \cap M_\alpha \ltimes N$.

5.3 Other examples

The quantum pair. This is a simple example of a quantum groupoid of compact type, constructed starting with a factor N having a faithful normal state ω.

Let us consider the Hilbert space $H = L^2(N) \otimes L^2(N)$, the representation α of N on H and the anti-representations β and $\hat{\beta}$ of N on H given, for $x \in N$, by:

$$\alpha(x) = 1 \otimes x,$$
$$\beta(x) = J_\omega x^* J_\omega \otimes 1,$$
$$\hat{\beta}(x) = 1 \otimes J_\omega x^* J_\omega,$$

and the operator:

$$W : H \underset{\omega}{{}_{\hat{\beta}}\otimes_\alpha} H \to H \underset{\omega^o}{{}_\alpha\otimes_\beta} H$$

given, for all ξ, η_1 and η_2 in $L^2(N)$ and x in N, by:

$$W((\xi \otimes \Lambda_\omega(x)) \underset{\omega}{{}_{\hat{\beta}}\otimes_\alpha} (\eta_1 \otimes \eta_2)) = (\xi \otimes \Lambda_\omega(1)) \underset{\omega^o}{{}_\alpha\otimes_\beta} (\eta_1 \otimes x\eta_2).$$

It is straightforward to show that both Hilbert spaces $H \underset{\omega}{{}_{\hat{\beta}}\otimes_\alpha} H$ and $H \underset{\omega^o}{{}_\alpha\otimes_\beta} H$ are isomorphic to $L^2(N) \otimes L^2(N) \otimes L^2(N)$, and, in fact, W appears just as the composition of these isomorphisms, which proves that W is a well defined pseudo-multiplicative unitary.

It can be proved that W is norm regular: in fact, $C_n(W)$ is the spatial $\mathbb{C}^*$-tensor product of the compact operators on $L^2(N)$ with N, considered as a $\mathbb{C}^*$-algebra. And $C_w(W)$ is the von Neumann algebra $\mathcal{L}(L^2(N)) \otimes N$. It is straightforward to show that any vector of the form $\xi \otimes \Lambda_\omega(1)$ is fixed, for all ξ in $L^2(N)$. Moreover, one gets that $\Lambda_\omega(1) \otimes \Lambda_\omega(1)$ is binormalized.

The von Neumann algebra $A_w(W)$ is then $N \otimes N^o$, and the Haar conditional expectations are, for X in $N \otimes N^o$, as follows:

$$E(X) = (\omega \otimes \mathrm{id})(X),$$
$$F(X) = (\mathrm{id} \otimes \omega^o)(X).$$

The coproduct Γ on $N \otimes N^o$ sends $N \otimes N^o$ into $(N \otimes N^o)_\alpha *_\beta (N \otimes N^o)$, which is isomorphic to $N \otimes N^o$. In fact, the coproduct is nothing other but the identity.

The von Neumann algebra $\widehat{A(W)}$ is equal to $\mathbb{C} \otimes \mathcal{L}(L^2(N))$; the coproduct $\widehat{\Gamma}$ on $\mathcal{L}(L^2(N))$ sends $\mathcal{L}(L^2(N))$ into $\mathcal{L}(L^2(N))$ and is the identity map. The counit sends also $\mathcal{L}(L^2(N))$ onto itself, and is again the identity.

This example can be generalized to any von Neumann algebra M, taking $H = L^2(M) \otimes_{Z(M)} L^2(M)$; we will obtain then a compact type groupoid structure on $M \otimes_{Z(M)} M^o$, and a discrete type groupoid structure on $(M \cup M')''$, with M as a base. This last example has been constructed in ([EV]8.4).

The transformation quantum group. This is a quantum analog of the classical groupoid construction starting with an action of a locally compact group G on a locally compact space X.

Let A be a commutative von Neumann algebra, equipped with a positive faithful state μ, (M, Δ) a (von Neumann) locally compact quantum group, W the multiplicative unitary associated with (M, Δ) and α an action of (M, Δ) on A (see 5.2 for the definitions). Let us define now the following representations of A on the Hilbert space $\mathcal{H} = L^2(A) \otimes L^2(M)$:

$$\beta(x) = x \otimes 1,$$
$$\hat{\beta}(x) = (J_A \otimes J_M)\alpha(x^*)(J_A \otimes J_M).$$

Then, the relative tensor product $\mathcal{H} {}_{\hat{\beta}}\underset{\mu}{\otimes}{}_\alpha \mathcal{H}$ is isomorphic ([S2], 3.1) to $L^2(\alpha(A)')$. Using ([V], 3.7), we get that $\alpha(A)'$ is isomorphic to $\mathcal{L}(L^2(M)) \otimes A'$. Therefore, the relative tensor product $\mathcal{H} {}_{\hat{\beta}}\underset{\mu}{\otimes}{}_\alpha \mathcal{H}$ is isomorphic to $L^2(A) \otimes L^2(M) \otimes L^2(M)$.

On the other hand, it is clear ([S2], 2.5) that the relative tensor product $\mathcal{H} {}_\alpha\underset{\mu}{\otimes}{}_\beta \mathcal{H}$ is also isomorphic to $L^2(A) \otimes L^2(M) \otimes L^2(M)$. It is then possible to verify that $1_{L^2(A)} \otimes W$ is a pseudo-multiplicative unitary over the base A, with respect to α, β, and $\hat{\beta}$.

If (M, Δ) is a compact quantum group and φ is the left (and right) Haar state on (M, Δ), then it is easy to check that $\Lambda_\mu(1) \otimes \Lambda_\varphi(1)$ is a fixed binormalized vector for this pseudo-multiplicative unitary.

References

[AR] C. Anantharaman-Delaroche and J. Renault, Amenable Groupoids, *Monograph. Enseign. Math.* 36, L'Enseignement Mathématique, Genève 2000.

[B] S. Baaj, Représentation régulière du groupe quantique des déplacements de Woronowicz, *Astérisque* 232 (1995), 11–48.

[BS] S. Baaj and G. Skandalis, Unitaires multiplicatifs et dualité pour les produits croisés de $\mathbb{C}^*$-algèbres, *Ann. Sci. École Norm. Sup.* (4) 26 (1993), 425–488.

[BNSz] G. Böhm, F. Nill and K. Szlachányi, Weak Hopf algebras I. Integral theory and $\mathbb{C}^*$-structure, *J. Algebra* 221 (1999), 385–438.

[BSz1] G. Böhm and K. Szlachányi, A co-associative $\mathbb{C}^*$-Quantum group with non-integral dimensions, *Lett. Math. Phys.* 38 (1996), 437–456.

[BSz2] G. Böhm and K. Szlachányi, Weak $\mathbb{C}^*$-Hopf algebras: the co-associative symmetry of non-integral dimensions, in *Quantum Groups and Quantum spaces*, Banach Center Publ. 40 (1997), 9–19.

[BSz3] G. Böhm and K. Szlachányi, Weak $\mathbb{C}^*$-Hopf algebras and multiplicative isometries, *J. Operator Theory* 45 (2001), 357–376.

[C1] A. Connes, On the spatial theory of von Neumann algebras, *J. Funct. Anal.* 35 (1980), 153–164.

[C2] A. Connes, *Noncommutative Geometry*, Academic Press, San Diego, CA, 1994.

[E1] M. Enock, Inclusions irréductibles de facteurs et unitaires multiplicatifs II, *J. Funct. Anal.* 154 (1998), 67–109.

[E2] M. Enock, Inclusions of von Neumann algebras and quantum groupoids II, *J. Funct. Anal.* 178 (2000), 156–225.

[E3] M. Enock, Quantum groupoids of compact type, *Prépublication Inst. Math. de Jussieu* 328, Mai 2000, 104pp.

[EN] M. Enock and R. Nest, Inclusions of factors, multiplicative unitaries and Kac algebras, *J. Funct. Anal.* 137 (1996), 466–543.

[ES] M. Enock and J.-M. Schwartz, *Kac Algebras and Duality of Locally Compact Groups*, Springer-Verlag, Berlin 1992.

[EV] M. Enock and J.-M. Vallin, Inclusions of von Neumann algebras and quantum groupoids, *J. Funct. Anal.* 172 (2000), 249–300.

[GHJ] F. M. Goodman, P. de la Harpe and V. F. R. Jones, *Coxeter Graphs and Towers of Algebras*, Publ. Res. Inst. Math. Sci. 14, Springer-Verlag, New York 1989.

[KV] J. Kustermans and S. Vaes, Locally compact quantum groups, *Ann. Sci. École Norm. Sup.* 33 (2000), 837–934.

[L] F. Lesieur, work in progress.

[MN] T. Masuda and Y. Nakagami, A von Neumann algebra framework for the duality of the quantum groups, *Publ. Res. Inst. Math. Sci.* 30 (1994), 799–850.

[NV1] D. Nikshych and L. Vaĭnerman, Algebraic versions of a finite-dimensional quantum groupoid, in *Hopf algebras and quantum groups* (S. Caenepeel et al., eds.), Lecture Notes Pure Appl. Math. 209 (2000), 189–221.

[NV2] D. Nikshych and L. Vaĭnerman, A characterization of depth 2 subfactors of II_1 factors, *J. Funct. Anal.* 171 (2000), 278–307.

[PP] M. Pimsner and S. Popa, Iterating the basic construction, *Trans. Amer. Math. Soc.* 310 (1988), 127–133.

[R1] J. Renault, *A Groupoid Approach to* $\mathbb{C}^*$*-Algebras*, Lecture Notes in Math. 793, Springer-Verlag, Berlin–Heidelberg–New York 1980.

[R2] J. Renault, The Fourier algebra of a measured groupoid and its multipliers, *J. Funct. Anal.* 145 (1997), 455–490.

[S1] J.-L. Sauvageot, Produit tensoriel de Z-modules et applications, in *Operator Algebras and their Connections with Topology and Ergodic Theory* (H. Araki et al., eds.), Lecture Notes in Math. 1132, Springer-Verlag, Berlin 1985, 468–485.

[S2] J.-L. Sauvageot, Sur le produit tensoriel relatif d'espaces de Hilbert, *J. Operator Theory* 9 (1983), 237–352.

[St] Ş. Strătilă, Modular theory in operator algebras, Editura Academiei/Abacus Press, Bucuresti/Tunbridge Wells, Kent, 1981.

[Sz] K. Szlachányi, Weak Hopf algebras, in *Operator Algebras and Quantum Field Theory* (S. Doplicher, R. Longo, J. E. Roberts, L. Zsido, eds.), International Press, Cambridge, MA, 1996, 621–632.

[V] S. Vaes, The unitary implementation of a locally compact quantum group action, *J. Funct. Anal.* 180 (2001), 426–480.

[Val1] J.-M. Vallin, Bimodules de Hopf et Poids Opératoriels de Haar, *J. Operator Theory* 35 (1996), 39–65

[Val2] J.-M. Vallin, Unitaire pseudo-multiplicatif associé à un groupoïde; applications à la moyennabilité, *J. Operator Theory* 44 (2000), 347–368.

[Val3] J.-M. Vallin, Groupoïdes quantiques finis, *J. Algebra* 239 (2001), 215–261.

[Val4] J.-M. Vallin, Multiplicative partial isometries and finite quantum groupoids, in *Locally compact quantum groups and groupoids* (L. Vainerman, ed.), IRMA Lectures in Math. Theoret. Phys. 2, Walter de Gruyter, Berlin–New York 2003, 189–227.

[W1] S. L. Woronowicz, Compact quantum groups, in *Symétries quantiques* (A. Connes et al., eds.), North-Holland, Amsterdam 1998, 845–884

[W2] S. L. Woronowicz, From multiplicative unitaries to quantum groups, *Internat. J. Math.* 7 (1996), 127–149.

[Y1] T. Yamanouchi, Duality for actions and coactions of measured groupoids on von Neumann algebras, *Mem. Amer. Math. Soc.* 101 (1993), 1–109.

[Y2] T. Yamanouchi, Duality for generalized Kac algebras and a characterization of finite groupoid algebras, *J. Algebra* 163 (1994), 9–50.

Quantum $\widetilde{\mathrm{SU}}(1, 1)$ and its Pontryagin dual

Erik Koelink and Johan Kustermans *

Faculteit ITS
Technische Universiteit Delft
Afdeling Toegepaste Wiskundige Analyse
Mekelweg 4, 2628CD Delft, The Netherlands
email: `h.t.koelink@its.tudelft.nl`

Departement Wiskunde
Katholieke Universiteit Leuven
Celestijnenlaan 200B, 3001 Heverlee, Belgium
email: `johan.kustermans@wis.kuleuven.ac.be`

Abstract. The study of quantum SU(1, 1) started in 1990 when the quantized universal enveloping algebra of $\mathfrak{su}(1,1)$ was investigated. Around the same time it was shown that quantum SU(1, 1) does not exist as a locally compact quantum group. In 1994 it started to emerge that quantum SU(1, 1) is not the group that should be deformed into a locally quantum group but rather the normalizer of quantum SU(1, 1) inside SL(2, $\mathbb{C}$). This new quantum group $\widetilde{\mathrm{SU}}_q(1,1)$ was finally constructed in 2001. In the first half of this paper we give an overview of its definition. In the second half of this paper we look into the Pontryagin dual of $\widetilde{\mathrm{SU}}_q(1,1)$ and discuss how this dual relates to the quantized universal enveloping algebra of $\mathfrak{su}(1,1)$ and its *-representations.

Introduction

One of the most important and simplest non-compact Lie groups is the SU(1, 1) group, which is isomorphic to SL(2, $\mathbb{R}$). In 1990, one of the first attempts to construct a quantum version of SU(1, 1) was made in [16] and [17] by T. Masuda, K. Mimachi, Y. Nakagami, M. Noumi, Y. Saburi and K. Ueno and independently, in [13] by L. Vaksman and L. Korogodskii. Let us give a quick overview [16] and [17]. Their starting point is a real form $\mathcal{U}_q(\mathfrak{su}(1,1))$ of the quantum universal enveloping algebra $\mathcal{U}_q(\mathfrak{sl}(2,\mathbb{C}))$ (defined in [16, Eq. (1.9)]). Intuitively, one should view upon $\mathcal{U}_q(\mathfrak{su}(1,1))$ as a quantum universal enveloping algebra of the 'quantum Lie algebra' of the still to be constructed locally compact quantum group $\mathrm{SU}_q(1,1)$. The dual $\mathcal{A}$ of $\mathcal{U}_q(\mathfrak{su}(1,1))$ is turned into some sort of topological Hopf *-algebra ([16, Sec. 2] and [17, Eq. (2.6)]). In a next step the coordinate Hopf *-algebra $A_q(\mathrm{SU}(1,1))$ is given as

*Post-doctoral researcher of the Fund for Scientific Research – Flanders (Belgium) (F.W.O.)

a *-subalgebra of $\mathcal{A}$ that also inherits the comultiplication, co-unit and antipode from $\mathcal{A}$ ([16, Eq. (2.7)] and [17, Eq. (0.9)]).

In this philosophy, they first introduce infinite dimensional infinitesimal representations of quantum $\mathcal{U}_q(\mathfrak{su}(1,1))$ (see [17, Eq. (1.1)]) which they then exponentiate to infinite dimensional unitary corepresentations of $\mathcal{A}$ (see [17, Eq. (1.2)]) providing hereby the quantum analogues of the discrete, principal unitary and complementary series of $\mathrm{SU}(1,1)$ but also a new strange series of corepresentations.

In [17, Sec. 3], the authors try, as a first attempt, to construct quantum $\mathrm{SU}(1,1)$ as a locally compact quantum group but their claims are undermined by the result, proven by S.L. Woronowicz in 1991, that showed that quantum $\mathrm{SU}(1,1)$ does not exist as a locally compact quantum group (see [21, Thm. 4.1 and Sec. 4.C]). This non-existence of quantum $\mathrm{SU}(1,1)$ as a locally compact group was considered a setback to the theory of quantum groups in the operator algebra approach.

In 1994, Korogodskii (see [12]) showed how the problems surrounding quantum $\mathrm{SU}(1,1)$ could be resolved. Recall that $\mathrm{SU}(1,1)$ is the linear Lie group

$$\{X \in \mathrm{SL}(2,\mathbb{C}) \mid X^*UX = U\},$$

where $U = \begin{pmatrix} 1 & 0 \\ 0 & -1 \end{pmatrix}$. Instead of looking at q-deformations of $\mathrm{SU}(1,1)$, Korogodskii studied q-deformations of the linear Lie group $\widetilde{\mathrm{SU}}(1,1)$. Here, $\widetilde{\mathrm{SU}}(1,1)$ denotes the Lie group $\{X \in \mathrm{SL}(2,\mathbb{C}) \mid X^*UX = U \text{ or } X^*UX = -U\}$ which is in fact the normalizer of $\mathrm{SU}(1,1)$ in $\mathrm{SL}(2,\mathbb{C})$.

Woronowicz picked up on the ideas of Korogodskii and provided the next important step by constructing quantum $\widetilde{\mathrm{SU}}(1,1)$ on the Hilbert space level (see [22]) except for the fact that the property that corresponds to the coassociativity of the comultiplication had not been established.

In 2001, the authors constructed quantum $\widetilde{\mathrm{SU}}(1,1)$ as a full blown locally compact quantum group, thereby heavily relying on the theory of q-hypergeometric functions. For a detailed account of this construction we refer to [6]. In the first half of this paper we will give a description of this new example.

In more recent work [7] a further investigation of the dual of quantum $\widetilde{\mathrm{SU}}(1,1)$ and the direct integral decomposition of the multiplicative unitary, as a corepresentation of quantum $\widetilde{\mathrm{SU}}(1,1)$, has been undertaken. The starting point of this analysis is the realization of $\mathcal{U}_q(\mathfrak{su}(1,1))$ on the GNS-space $\mathcal{K}$ of the Haar weight of quantum $\widetilde{\mathrm{SU}}(1,1)$.

The Casimir element of $\mathcal{U}_q(\mathfrak{su}(1,1))$ is an operator in $\mathcal{K}$ which has several self-adjoint extensions, only one of which is relevant to the dual of quantum $\widetilde{\mathrm{SU}}(1,1)$ as a locally compact group. The spectral decomposition of this self-adjoint extension provides a description of $\mathcal{K}$, in terms of q-special functions, that is more natural to the dual quantum group.

This new description of $\mathcal{K}$ gives rise to 2 important applications. Firstly, the construction of two extra operators that, together with the generators E and K of $\mathcal{U}_q(\mathfrak{su}(1,1))$, generate the von Neumann algebra of the dual quantum group and thereby providing a clearer picture of this von Neumann algebra. Secondly, the direct integral decomposition of the multiplicative unitary into irreducible corepresentations.

These irreducible corepresentations of quantum $\widetilde{SU}(1,1)$ divide into two families of corepresentations. The *-representations of $\mathcal{U}_q(\mathfrak{su}(1,1))$ that are mentioned in this paragraph, were introduced in [17]. One kind of corepresentation of quantum $\widetilde{SU}(1,1)$ is obtained by exponentiating a positive discrete series, a negative discrete series and strange series *-representation of $\mathcal{U}_q(\mathfrak{su}(1,1))$ and combining them together. The second kind of corepresentations arise as a combination of two exponentiations of principal unitary series *-representations of $\mathcal{U}_q(\mathfrak{su}(1,1))$.

All these results concerning the dual of quantum $\widetilde{SU}(1,1)$ are discussed in the second half of this paper.

Let us recall the definition of a locally compact quantum group in the von Neumann algebraic setting as defined in [15]. There is also an equivalent notion in the C*-algebra framework (see [14]) but the von Neumann algebra approach better suits our needs.

Definition 1. Consider a von Neumann algebra M together with a unital normal *-homomorphism $\Delta : M \to M \otimes M$ such that $(\Delta \otimes \iota)\Delta = (\iota \otimes \Delta)\Delta$. Assume moreover the existence of

1. a normal semi-finite faithful weight φ on M that is left invariant: $\varphi((\omega \otimes \iota)\Delta(x)) = \varphi(x)\omega(1)$ for all $\omega \in M_*^+$ and $x \in \mathcal{M}_\varphi^+$.

2. a normal semi-finite faithful weight ψ on M that is right invariant: $\psi((\iota \otimes \omega)\Delta(x)) = \psi(x)\omega(1)$ for all $\omega \in M_*^+$ and $x \in \mathcal{M}_\psi^+$.

Then we call the pair (M, Δ) a von Neumann algebraic quantum group.

For a discussion about the consequences of this definition and related notations we refer to [15, Sec. 1].

The construction of quantum $\widetilde{SU}(1,1)$ and the study of its dual hinges on the theory of q-hypergeometric functions. In the next part of the introduction we fix the necessary notation and terminology involved. The set of all natural numbers, not including 0, is denoted by $\mathbb{N}$. Also, $\mathbb{N}_0 = \mathbb{N} \cup \{0\}$.

Fix a number $0 < q < 1$. Let $a \in \mathbb{C}$. If $k \in \mathbb{N}_0 \cup \{\infty\}$, the q-shifted factorial $(a;q)_k \in \mathbb{C}$ is defined as $(a;q)_k = \prod_{i=0}^{k-1}(1 - q^i a)$ (so $(a;q)_0 = 1$). We also use the notation $(a_1, \ldots, a_m; q)_k = (a_1;q)_k \ldots (a_m;q)_k$ if $a_1, \ldots, a_m \in \mathbb{C}$ and $k \in \mathbb{N}_0 \cup \{\infty\}$.

Let $k, l \in \mathbb{N}_0$ so that $k \le l + 1$. Let $a_1, \ldots, a_k \in \mathbb{C}$ and $b_1, \ldots, b_l \in \mathbb{C} \setminus q^{-\mathbb{N}_0}$. For $z \in \mathbb{C}$ satisfying $|z| < 1$, one defines

$$
{}_k\varphi_l\left(\begin{matrix} a_1, \ldots, a_k \\ b_1, \ldots, b_l \end{matrix} ; q, z \right) = \sum_{n=0}^{\infty} \frac{(a_1, \ldots, a_k; q)_n}{(b_1, \ldots, b_l, q)_n} \left[(-1)^n q^{\frac{n(n-1)}{2}} \right]^{l-k+1} \frac{z^n}{(q;q)_n}.
$$

This series has convergence radius 1 if $k = l + 1$ and is not terminating, and has convergence radius ∞ if $k \le l$. The above notation is extended to all $z \in \mathbb{C}$ if $k \le l$ or if one of the numbers $a_1, \ldots, a_k$ belongs to $q^{-\mathbb{N}_0}$ (in this last case, the series terminates and becomes a finite sum).

There are also other important extensions of the above notation, for instance in the case $k = 2$ and $l = 1$. The function

$$\{ z \in \mathbb{C} \mid |z| < 1 \} \to \mathbb{C} : z \mapsto {}_2\varphi_1 \left(\begin{matrix} a_1, a_2 \\ b_1 \end{matrix} ; q, z \right)$$

has a unique analytic extension to $\mathbb{C} \setminus [1, \infty[$. For $z \in \mathbb{C} \setminus [1, \infty[$ the value of this analytic extension in z is also denoted by ${}_2\varphi_1 \left(\begin{matrix} a_1, a_2 \\ b_1 \end{matrix} ; q, z \right)$.

We will also use a slight modification of this notation. If $a, b, z \in \mathbb{C}$, we define

$$\Psi \left(\begin{matrix} a \\ b \end{matrix} ; q, z \right) = \sum_{n=0}^{\infty} \frac{(a; q)_n \, (b q^n; q)_\infty}{(q ; q)_n} (-1)^n q^{\frac{1}{2}n(n-1)} z^n. \tag{1}$$

If $b \notin q^{-\mathbb{N}_0}$, then $\Psi \left(\begin{matrix} a \\ b \end{matrix} ; q, z \right) = (b; q)_\infty \, {}_1\varphi_1 \left(\begin{matrix} a \\ b \end{matrix} ; q, z \right)$. See [4] for an extensive treatment on q-hypergeometric functions.

The analysis of the dual of quantum $\widetilde{\mathrm{SU}}(1,1)$ depends heavily on Al-Salam & Chihara polynomials and little q-Jacobi functions. In this paper we will not write down the precise formulas involved because this would only clutter the exposition. On the other hand do we not want to leave the reader completely in the dark as to the role of these q-special functions. Therefore we include the precise definition of these q-special functions. In this paper we will use the notation $\mu(y) = \frac{1}{2}(y + y^{-1})$ for all $y \in \mathbb{C} \setminus \{0\}$. Of course, $\mu(e^{i\theta}) = \cos\theta$ for all $\theta \in \mathbb{R}$.

The Al-Salam & Chihara polynomials are Askey–Wilson polynomials with two of the four parameters equal to zero (see [4, Eq. (7.5.2)] with $c = d = 0$).

Definition 2. Consider $a, b \in \mathbb{R}\setminus\{0\}$ so that $ab < 1$. For $n \in \mathbb{N}_0$, the orthonormal Al-Salam & Chihara polynomial $p_n(\,.\,; a, b \,|\, q) : \mathbb{C} \to \mathbb{C}$ is defined so that

$$p_n(\mu(y); a, b \,|\, q) = a^{-n} \sqrt{\frac{(ab ; q)_n}{(q; q)_n}} \; {}_3\varphi_2 \left(\begin{matrix} q^{-n}, & ay, & a/y \\ & ab, & 0 \end{matrix} ; q, q \right)$$

for all $y \in \mathbb{C}\setminus\{0\}$.

Note that this is a terminating series. Also note that if $k \in \mathbb{N}$, then

$$(ay ; q)_k \, (a/y ; q)_k = \prod_{i=0}^{k-1} \left(1 + q^{2i} a^2 - 2q^i a \, \mu(y)\right)$$

so that $p_n(\mu(y))$ is indeed a polynomial in $\mu(y)$ of degree n.

Definition 3. Consider $a, b \in \mathbb{R}\setminus\{0\}$ so that $0 < ab < 1$. For every $n \in \mathbb{Z}$ we define the renormalized and reparametrized little q-Jacobi function $j_n(\,.\,; a, b \,|\, q) : \mathbb{C} \to \mathbb{C}$

so that

$$j_n(\mu(y); a, b \,|\, q) = (ab\,; q)_\infty |b|^{-n} \sqrt{\frac{(-q^{n+1}\, a/b; q)_\infty}{(-q^{n+1}; q)_\infty}}\; {}_2\varphi_1\left(\begin{matrix} by,\ b/y \\ ab \end{matrix} \;; q, -q^{-n} \right)$$

for all $y \in \mathbb{C}\setminus\{0\}$.

The definition of the little q-Jacobi functions in the literature (see e.g. [10]) involves an extra parameter but here and in [7], this parameter is the same throughout so we left it out of the definition.

Let us end this introduction with establishing some notations and conventions. If f is a function, the domain of f will be denoted by $D(f)$. If X is a set, the identity mapping on X will be denoted by ι_X and most of the time even by ι. The set of all complex valued functions on X is denoted by $\mathcal{F}(X)$, the set of all elements in $\mathcal{F}(X)$ having finite support, is denoted by $\mathcal{K}(X)$. Let V be a vector space and S a subset of V. Then $\langle S \rangle$ denotes the linear span of S in V.

Let S, T be two linear operators acting in a Hilbert space H. We say that $S \subseteq T$ if $D(S) \subseteq D(T)$ and $S(v) = T(v)$ for all $v \in D(S)$.

The symbol $\odot$ will be used to denote the algebraic tensor product of vector spaces and linear mappings. The symbol $\otimes$ on the other hand will denote the tensor product of Hilbert spaces, von Neumann algebras and sufficiently continuous linear mappings.

Let $K \subseteq -q^{\mathbb{Z}} \cup q^{\mathbb{Z}}$ and consider a function $f : \mathbb{T} \times K \to \mathbb{C}$. Then we set $f(\lambda, x) = 0$ for all $\lambda \in \mathbb{T}$ and $x \in \left(-q^{\mathbb{Z}} \cup q^{\mathbb{Z}}\right) \setminus K$. (2)

1 The Hopf * algebra underlying quantum $\widetilde{\mathrm{SU}}(1,1)$

In order to resolve the problems surrounding quantum SU(1, 1), Korogodskii proposed in [12] to construct the quantum version of $\widetilde{\mathrm{SU}}(1,1)$. He implicitly suggested the use of the following Hopf *-algebra, the Hopf *-algebra itself was explicitly introduced by Woronowicz in [22].

Throughout this paper, we fix a number $0 < q < 1$. Define $\mathcal{A}$ to be the unital *-algebra generated by elements α_0, γ_0 and e_0 and relations

$$\begin{aligned}
&\alpha_0^\dagger \alpha_0 - \gamma_0^\dagger \gamma_0 = e_0 && \alpha_0 \alpha_0^\dagger - q^2 \gamma_0^\dagger \gamma_0 = e_0 \\
&\gamma_0^\dagger \gamma_0 = \gamma_0 \gamma_0^\dagger && \\
&\alpha_0 \gamma_0 = q\, \gamma_0 \alpha_0 && e_0^\dagger = e_0 \\
&\alpha_0 \gamma_0^\dagger = q\, \gamma_0^\dagger \alpha_0 && e_0^2 = 1 \\
&\alpha_0 e_0 = e_0 \alpha_0 && \\
&\gamma_0 e_0 = e_0 \gamma_0, &&
\end{aligned}$$

where † denotes the $*$-operation on $\mathcal{A}$. There exists a unique unital $*$-homomorphism $\Delta_0 : \mathcal{A} \to \mathcal{A} \odot \mathcal{A}$ such that

$$\begin{aligned}\Delta_0(\alpha_0) &= \alpha_0 \otimes \alpha_0 + q\,(e_0\,\gamma_0^\dagger) \otimes \gamma_0 \\ \Delta_0(\gamma_0) &= \gamma_0 \otimes \alpha_0 + (e_0\alpha_0^\dagger) \otimes \gamma_0 \\ \Delta_0(e_0) &= e_0 \otimes e_0.\end{aligned} \tag{1.1}$$

The pair $(\mathcal{A}, \Delta_0)$ turns out to be a Hopf $*$-algebra with co-unit ε_0 and antipode S_0 determined by

$$\begin{aligned} S_0(\alpha_0) &= e_0\,\alpha_0^\dagger & \qquad \varepsilon_0(\alpha_0) &= 1 \\ S_0(\alpha_0^\dagger) &= e_0\,\alpha_0 & \varepsilon_0(\gamma_0) &= 0 \\ S_0(\gamma_0) &= -q\,\gamma_0 & \varepsilon_0(e_0) &= 1 \\ S_0(\gamma_0^\dagger) &= -\frac{1}{q}\,\gamma_0^\dagger & & \\ S_0(e_0) &= e_0. & & \end{aligned}$$

If one takes $q = 1$ in the above description, one gets the Hopf $*$-algebra of polynomial functions on $\widetilde{\mathrm{SU}}(1,1)$. A simple calculation reveals that

$$\widetilde{\mathrm{SU}}(1,1) = \left\{ \begin{pmatrix} a & c \\ \epsilon\,\overline{c} & \epsilon\,\overline{a} \end{pmatrix} \mid a, b \in \mathbb{C}, \epsilon \in \{-1, 1\} \text{ s.t. } |a|^2 - |c|^2 = \epsilon \right\}$$

The elements α_0, γ_0 and e_0 can then be realized as the complex valued functions on $\widetilde{\mathrm{SU}}(1,1)$ given by

$$\alpha_0 \begin{pmatrix} a & c \\ \epsilon\,\overline{c} & \epsilon\,\overline{a} \end{pmatrix} = a, \quad \gamma_0 \begin{pmatrix} a & c \\ \epsilon\,\overline{c} & \epsilon\overline{a} \end{pmatrix} = c, \quad e_0 \begin{pmatrix} a & c \\ \epsilon\,\overline{c} & \epsilon\,\overline{a} \end{pmatrix} = \epsilon$$

and $\mathcal{A}$ is the unital $*$-algebra of complex valued functions on $\widetilde{\mathrm{SU}}(1,1)$ generated by α_0, γ_0 and e_0.

Let us now go back to the case $0 < q < 1$. As always we want to represent this Hopf $*$-algebra $\mathcal{A}$ by possibly unbounded operators in some Hilbert space in order to produce a locally compact quantum group in the sense of Definition 1 of the introduction. Korogodskii classified the well-behaved irreducible representations of $\mathcal{A}$ in [12, Prop. 2.4]. Roughly speaking, the representation of $\mathcal{A}$ is obtained by gluing together these irreducible representations. The representation we use here is a slight variation of the one introduced by Woronowicz in [22]. For this purpose we define

$$I_q = \{\,-q^k \mid k \in \mathbb{N}\,\} \cup \{\,q^k \mid k \in \mathbb{Z}\,\}.$$

Let $\mathbb{T}$ denote the group of complex numbers of modulus 1. We will consider the uniform measure on I_q and the normalized Haar measure on $\mathbb{T}$. Our $*$-representation of $\mathcal{A}$ will act in the Hilbert space $\mathcal{H}$ defined by

$$\mathcal{H} = L^2(\mathbb{T}) \otimes L^2(I_q).$$

If $p \in -q^{\mathbb{Z}} \cup q^{\mathbb{Z}}$, we define $\delta_p \in \mathcal{F}(I_q)$ such that $\delta_p(x) = \delta_{x,p}$ for all $x \in I_q$ (note that $\delta_p = 0$ if $p \notin I_q$). The family $(\delta_p \mid p \in I_q)$ is the natural orthonormal basis of $L^2(I_q)$. We let ζ denote the identity function on $\mathbb{T}$. Recall the natural orthonormal basis $(\zeta^m \mid m \in \mathbb{Z})$ for $L^2(\mathbb{T})$.

Instead of looking at the algebra $\mathcal{A}$ as the abstract algebra generated by generators and relations we will use an explicit realization of this algebra as linear operators on the dense subspace E of $\mathcal{H}$ defined by $E = \langle \zeta^m \otimes \delta_x \mid m \in \mathbb{Z}, x \in I_q \rangle \subseteq \mathcal{H}$. Of course, E inherits the inner product from $\mathcal{H}$. Let $\mathcal{L}^+(E)$ denote the $*$-algebra of adjointable operators on E (see [19, Prop. 2.1.8]), i.e.

$$\mathcal{L}^+(E) = \{\, T \in \mathrm{End}(E) \mid \exists T^\dagger \in \mathrm{End}(E), \forall v, w \in E : \langle Tv, w\rangle = \langle v, T^\dagger w\rangle \,\},$$

so $\dagger$ denotes the $*$-operation in $\mathcal{L}^+(E)$. If $T \in \mathcal{L}^+(E)$, $T^\dagger \subseteq T^*$ where T^* is the usual adjoint of T as an operator in the Hilbert space $\mathcal{H}$. It also follows that T is a closable operator in $\mathcal{H}$.

Define linear operators α_0, γ_0, e_0 in $\mathcal{L}^+(E)$ such that

$$\begin{aligned}
\alpha_0(\zeta^m \otimes \delta_p) &= \sqrt{\mathrm{sgn}(p) + p^{-2}}\; \zeta^m \otimes \delta_{qp}\\
\gamma_0(\zeta^m \otimes \delta_p) &= p^{-1}\; \zeta^{m+1} \otimes \delta_p\\
e_0(\zeta^m \otimes \delta_p) &= \mathrm{sgn}(p)\; \zeta^m \otimes \delta_p
\end{aligned}$$

for all $p \in I_q$, $m \in \mathbb{Z}$.

Then $\mathcal{A}$ is the $*$-subalgebra of $\mathcal{L}^+(E)$ generated by α_0, γ_0 and e_0. Since $\mathcal{L}^+(E) \odot \mathcal{L}^+(E)$ is canonically embedded in $\mathcal{L}^+(E \odot E)$, we obtain $\mathcal{A} \odot \mathcal{A}$ as a $*$-subalgebra of $\mathcal{L}^+(E \odot E)$. As such, $\Delta_0(\alpha_0)$, $\Delta_0(\gamma_0)$ and $\Delta_0(e_0)$ defined in Eqs. (1.1) belong to $\mathcal{L}^+(E \odot E)$.

2 The von Neumann algebra underlying quantum $\widetilde{\mathrm{SU}}(1,1)$

In this section we introduce the von Neumann algebra acting on $\mathcal{H}$ that underlies the von Neumann algebraic version of the quantum group $\widetilde{\mathrm{SU}}_q(1,1)$ (see Definition 1 of the introduction).

In order to get into the framework of operator algebras, we need to introduce the topological versions of the algebraic objects α_0, γ_0 and e_0 as possibly unbounded operators in the Hilbert space $\mathcal{H}$. So let α denote the closure of α_0, γ the closure of γ_0 and e the closure of e_0, all as linear operators in $\mathcal{H}$. So e is a bounded linear operator on $\mathcal{H}$, whereas α and γ are unbounded, closed, densely defined linear operators in $\mathcal{H}$. Note also that α^* is the closure of $\alpha_0^\dagger$ and that γ^* is the closure of $\gamma_0^\dagger$. Note that γ is normal.

Consider $p \in -q^{\mathbb{Z}} \cup q^{\mathbb{Z}}$. We define a translation operator T_p on $\mathcal{F}(\mathbb{T} \times I_q)$ such that for $f \in \mathcal{F}(\mathbb{T} \times I_q)$, $\lambda \in \mathbb{T}$ and $x \in I_q$, we have that $(T_p f)(\lambda, x) = f(\lambda, px)$. By discussion (2) in the introduction, we get that $(T_p f)(\lambda, x) = 0$ if $px \notin I_q$. If $p, t \in I_q$ and $g \in \mathcal{F}(\mathbb{T})$, then $T_p(g \otimes \delta_t) = g \otimes \delta_{p^{-1}t}$, thus, $T_p(g \otimes \delta_t) = 0$ if $p^{-1}t \notin I_q$.

Notation 2.1. For $p \in -q^{\mathbb{Z}} \cup q^{\mathbb{Z}}$ we define the partial isometry $\rho_p \in B(\mathcal{H})$ as the one that is induced by T_p. Let us single out the following special case.

Define $u = \rho_{-1}$, which is a self-adjoint partial isometry on $B(\mathcal{H})$.

Let us recall the following natural terminology. If $T_1, \ldots, T_n$ are closed, densely defined linear operators in $\mathcal{H}$, the von Neumann algebra N on $\mathcal{H}$ generated by $T_1, \ldots, T_n$ is the one such that

$$N' = \{ x \in B(\mathcal{H}) \mid xT_i \subseteq T_i x \text{ and } xT_i^* \subseteq T_i^* x \text{ for } i = 1, \ldots, n \}.$$

Almost by definition, N is the smallest von Neumann algebra acting on $\mathcal{H}$ so that $T_1, \ldots, T_n$ are affiliated with M in the von Neumann algebraic sense.

It is now very tempting to define the von Neumann algebra underlying quantum $\widetilde{\mathrm{SU}}_q(1,1)$ as the von Neumann algebra generated by α, γ and e. However, for reasons that will become clear later (see the discussion in the beginning of the next section and the remark after Proposition 3.5), the underlying von Neumann algebra will be the one generated by α, γ, e and u (the necessity of the element u was first observed by Woronowicz in [22]).

Proposition 2.2. *We define M to be the von Neumann algebra on $\mathcal{H}$ generated by α, γ, e and u. Then $M = L^\infty(\mathbb{T}) \otimes B(L^2(I_q))$.*

The following picture of M turns out to be the most useful one. For every $p, t \in I_q$ and $m \in \mathbb{Z}$ we define $\Phi(m, p, t) \in B(\mathcal{H})$ so that for $x \in I_q$ and $r \in \mathbb{Z}$,

$$\Phi(m, p, t)\,(\zeta^r \otimes \delta_x) = \delta_{x,t}\; \zeta^{m+r} \otimes \delta_p.$$

Define $M^\circ = \langle\, \Phi(m, p, t) \mid m \in \mathbb{Z},\, p, t \in I_q \,\rangle$. Using the above equation, it is obvious that $(\, \Phi(m, p, t) \mid m \in \mathbb{Z},\, p, t \in I_q \,)$ is a linear basis of M°.

The multiplication and *-operation are easily expressed in terms of these basis elements:

$$\Phi(m_1, p_1, t_1)\,\Phi(m_2, p_2, t_2) = \delta_{p_2, t_1}\; \Phi(m_1 + m_2, p_1, t_2)$$
$$\Phi(m, p, t)^* = \Phi(-m, t, p)$$

for all $m, m_1, m_2 \in \mathbb{Z}$, $p, p_1, p_2, t, t_1, t_2 \in I_q$. So we see that M° is a σ-weakly dense sub*-algebra of M.

3 The comultiplication on quantum $\widetilde{\mathrm{SU}}(1, 1)$

In this section we introduce the comultiplication of $\widetilde{\mathrm{SU}}_q(1,1)$. In the first part we start with a motivation for the formulas appearing in Definition 3.1. Although the discussion is not really needed in the build up of $\widetilde{\mathrm{SU}}_q(1,1)$, it is important and clarifying to know how we arrived at the formulas in Definition 3.1.

Our purpose is to define a comultiplication $\Delta : M \to M \otimes M$. Assume for the moment that this has already been done. It is natural to require Δ to be closely related to the comultiplication Δ_0 on $\mathcal{A}$ as defined in Eqs. (1.1). The least that we expect is $\Delta_0(T_0) \subseteq \Delta(T)$ and $\Delta_0(T_0^\dagger) \subseteq \Delta(T)^*$ for $T = \alpha, \gamma, e$. In the rest of this discussion we will focus on the inclusion $\Delta_0(\gamma_0^\dagger \gamma_0) \subseteq \Delta(\gamma^*\gamma)$, where $\Delta_0(\gamma_0^\dagger \gamma_0) \in \mathcal{L}^+(E \odot E)$.

Because $\gamma^*\gamma$ is self-adjoint, the element $\Delta(\gamma^*\gamma)$ would also be self-adjoint. So the hunt is on for self-adjoint extensions of the explicit operator $\Delta_0(\gamma_0^\dagger \gamma_0)$. Unlike in the case of quantum $E(2)$ (see [21]), the operator $\Delta_0(\gamma_0^\dagger \gamma_0)$ is not essentially self-adjoint. But it was already known in [12] that $\Delta_0(\gamma_0^\dagger \gamma_0)$ has self-adjoint extensions (this follows easily because the operator in (3.1) commutes with complex conjugation, implying that the deficiency spaces are isomorphic).

Although $\Delta_0(\gamma_0^\dagger \gamma_0)$ has a self-adjoint extension, it is not unique. We have to make a choice for this self-adjoint extension, but we cannot extract the information necessary to make this choice from α, γ and e alone. This is why we do not work with the von Neumann algebra $M^\diamond$ that is generated by α, γ and e alone but with M which has the above extra extension information contained in the element u. These kind of considerations were already present in [23] and were also introduced in [22] for quantum $\widetilde{\mathrm{SU}}(1,1)$. In [6], this principle is only lurking in the background but it is treated in a fundamental and rigorous way in [22]. In order to deal with this, Woronowicz develops a nice theory of balanced extensions of operators that is comparable to the theory of self-adjoint extensions of symmetric operators.

Now we get into slightly more detail in our discussion about the extension of $\Delta_0(\gamma_0^\dagger \gamma_0)$. But first we introduce the following auxiliary function

$$\kappa : \mathbb{R} \to \mathbb{R} : x \mapsto \kappa(x) = \operatorname{sgn}(x)\, x^2.$$

Define a linear map $L : \mathcal{F}(\mathbb{T} \times I_q \times \mathbb{T} \times I_q) \to \mathcal{F}(\mathbb{T} \times I_q \times \mathbb{T} \times I_q)$ such that

$$\begin{aligned}
&(Lf)(\lambda, x, \mu, y) = \\
&[\,x^{-2}(\operatorname{sgn}(y) + y^{-2}) + (\operatorname{sgn}(x) + q^2 x^{-2})\, y^{-2}\,]\, f(\lambda, x, \mu, y) \\
&+ \operatorname{sgn}(x)\, q^{-1}\, \bar{\lambda}\mu\, x^{-1} y^{-1} \sqrt{(\operatorname{sgn}(x) + x^{-2})(\operatorname{sgn}(y) + y^{-2})}\; f(\lambda, qx, \mu, qy) \\
&+ \operatorname{sgn}(x)\, q\, \lambda\bar{\mu}\, x^{-1} y^{-1} \sqrt{(\operatorname{sgn}(x) + q^2 x^{-2})(\operatorname{sgn}(y) + q^2 y^{-2})}\; f(\lambda, q^{-1}x, \mu, q^{-1}y)
\end{aligned}$$

for all $\lambda, \mu \in \mathbb{T}$ and $x, y \in I_q$. A straightforward calculation shows that $\Delta_0(\gamma_0^\dagger \gamma_0)\, f = L(f)$ for all $f \in E \odot E$. From this, it is a standard exercise to check that $f \in D(\Delta_0(\gamma_0^\dagger \gamma_0)^*)$ and $\Delta_0(\gamma_0^\dagger \gamma_0)^* f = L(f)$ if $f \in \mathcal{L}^2(\mathbb{T} \times I_q \times \mathbb{T} \times I_q)$ and $L(f) \in \mathcal{L}^2(\mathbb{T} \times I_q \times \mathbb{T} \times I_q)$ (without any difficulty, one can even show that $D(\Delta_0(\gamma_0^\dagger \gamma_0)^*)$ consists precisely of such elements f).

If $\theta \in -q^{\mathbb{Z}} \cup q^{\mathbb{Z}}$, we define $\ell'_\theta = \{ (\lambda, x, \mu, y) \in \mathbb{T} \times I_q \times \mathbb{T} \times I_q \mid y = \theta x \}$ and consider $L^2(\ell'_\theta)$ naturally embedded in $L^2(\mathbb{T} \times I_q \times \mathbb{T} \times I_q)$. It follows easily from the above discussion that $\Delta_0(\gamma_0^\dagger \gamma_0)^*$ leaves $L^2(\ell'_\theta)$ invariant. Thus, if T is a self-adjoint extension of $\Delta_0(\gamma_0^\dagger \gamma_0)$, the obvious inclusion $T \subseteq \Delta_0(\gamma_0^\dagger \gamma_0)^*$ implies that T also leaves $L^2(\ell'_\theta)$ invariant.

Therefore every self-adjoint extension T of $\Delta_0(\gamma_0^\dagger \gamma_0)$ is obtained by choosing a self-adjoint extension T_θ of the restriction of $\Delta_0(\gamma_0^\dagger \gamma_0)$ to $L^2(\ell'_\theta)$ for every $\theta \in -q^{\mathbb{Z}} \cup q^{\mathbb{Z}}$ and setting $T = \oplus_{\theta \in -q^{\mathbb{Z}} \cup q^{\mathbb{Z}}} T_\theta$.

Therefore fix $\theta \in -q^{\mathbb{Z}} \cup q^{\mathbb{Z}}$. Define $J_\theta = \{ z \in I_{q^2} \mid \kappa(\theta)\, z \in I_{q^2} \}$ which is a q^2-interval around 0 (bounded or unbounded towards ∞). On J_θ we define a measure ν_θ such that $\nu_\theta(\{x\}) = |x|$ for all $x \in J_\theta$.

Define the linear operator $L_\theta : \mathcal{F}(J_\theta) \to \mathcal{F}(J_\theta)$ such that

$$\begin{aligned} \theta^2 x^2 \, (L_\theta f)(x) = & -\sqrt{(1+x)(1+\kappa(\theta)\,x)}\, f(q^2 x) \\ & - q^2 \sqrt{(1+q^{-2}x)(1+q^{-2}\kappa(\theta)\,x)}\, f(q^{-2}x) \\ & + [(1+\kappa(\theta)\,x) + q^2(1+q^{-2}x)]\, f(x) \end{aligned} \tag{3.1}$$

for all $f \in \mathcal{F}(J_\theta)$ and $x \in J_\theta$.

Then, an easy verification reveals that $\Delta_0(\gamma_0^\dagger \gamma_0)\restriction_{\mathcal{K}(\ell'_\theta)}$ is unitarily equivalent to $1 \odot L_\theta \restriction_{\mathcal{K}(J_\theta)}$. So our problem is reduced to finding self-adjoint extensions of $L_\theta \restriction_{\mathcal{K}(J_\theta)}$. This operator $L_\theta \restriction_{\mathcal{K}(J_\theta)}$ is a second order q-difference operator for which eigenfunctions in terms of q-hypergeometric functions are known.

We can use a reasoning similar to the one in [8, Sec. 2] to get hold of the self-adjoint extensions of $L_\theta \restriction_{\mathcal{K}(J_\theta)}$: Let $\beta \in \mathbb{T}$. Then we define a linear operator $L_\theta^\beta : D(L_\theta^\beta) \subseteq L^2(J_\theta, \nu_\theta) \to L^2(J_\theta, \nu_\theta)$ such that $D(L_\theta^\beta)$ consists of all $f \in L^2(J_\theta, \nu_\theta)$ for which

$$L_\theta(f) \in L^2(J_\theta, \nu_\theta), \quad f(0+) = \beta\, f(0-) \text{ and } (D_q f)(0+) = \beta\, (D_q f)(0-)$$

and L_θ^β is the restriction of L_θ to $D(L_\theta^\beta)$. Here, D_q denotes the Jackson derivative, that is, $(D_q f)(x) = (f(qx) - f(x))/(q-1)x$ for $x \in J_\theta$. Also, $f(0+) = \beta\, f(0-)$ is an abbreviated form of saying that the limits $\lim_{x \uparrow 0} f(x)$ and $\lim_{x \downarrow 0} f(x)$ exist and $\lim_{x \downarrow 0} f(x) = \beta \lim_{x \uparrow 0} f(x)$.

Then L_θ^β is a self-adjoint extension of $L_\theta \restriction_{\mathcal{K}(J_\theta)}$.

It is tempting to use the extension L_θ^1 to construct our final self-adjoint extension for $\Delta_0(\gamma_0^\dagger \gamma_0)$ (although there is no apparent reason for this choice). However, in order to obtain a coassociative comultiplication, it turns out that we have to use the extension $L_\theta^{\operatorname{sgn}(\theta)}$ to construct our final self-adjoint extension. This is reflected in the fact that the expression $s(x, y)$ appears in the formula for a_p in Definition 3.1.

This all would be only a minor achievement if we could not go any further. But the results and techniques used in the theory of q-hypergeometric functions will even allow us to find an explicit orthonormal basis consisting of eigenvectors of $L_\theta^{\operatorname{sgn}(\theta)}$.

These eigenvectors are, up to a unitary transformation, obtained by restricting the functions a_p in Definition 3.1 to ℓ_θ, which is introduced after this definition. The special case $\theta = 1$ was already known to Korogodskii (see [12, Prop. A.1]).

In order to compress the formulas even further, we introduce three other auxiliary functions in order to compress the notation.

(1) $\chi : -q^{\mathbb{Z}} \cup q^{\mathbb{Z}} \to \mathbb{Z}$ such that $\chi(x) = \log_q(|x|)$ for all $x \in -q^{\mathbb{Z}} \cup q^{\mathbb{Z}}$,

(2) $\nu : -q^{\mathbb{Z}} \cup q^{\mathbb{Z}} \to \mathbb{R}^+$ such that $\nu(t) = q^{\frac{1}{2}(\chi(t)-1)(\chi(t)-2)}$ for all $t \in -q^{\mathbb{Z}} \cup q^{\mathbb{Z}}$.

(3) Another auxiliary function $s : \mathbb{R}_0 \times \mathbb{R}_0 \to \{-1, 1\}$ is defined such that

$$s(x,y) = \begin{cases} -1 & \text{if } x > 0 \text{ and } y < 0 \\ 1 & \text{if } x < 0 \text{ or } y > 0 \end{cases}$$

We will also use the normalization constant $c_q = (\sqrt{2}\,q\,(q^2, -q^2; q^2)_\infty)^{-1}$. Recall the special functions introduced in Eq. (1) of the introduction.

Definition 3.1. If $p \in I_q$, we define a function $a_p : I_q \times I_q \to \mathbb{R}$ such that for all $x, y \in I_q$, the value $a_p(x, y)$ is given by

$$c_q\, s(x,y)\, (-1)^{\chi(p)}\, (-\mathrm{sgn}(y))^{\chi(x)}\, |y|\, \nu(py/x)$$
$$\times \sqrt{\frac{(-\kappa(p), -\kappa(y); q^2)_\infty}{(-\kappa(x); q^2)_\infty}}\ \Psi\left(\begin{matrix} -q^2/\kappa(y) \\ q^2\kappa(x/y) \end{matrix} ; q^2, q^2\kappa(x/p)\right)$$

if $\mathrm{sgn}(xy) = \mathrm{sgn}(p)$ and $a_p(x,y) = 0$ if $\mathrm{sgn}(xy) \neq \mathrm{sgn}(p)$.

The extra vital information that we need is contained in the following proposition. For $\theta \in -q^{\mathbb{Z}} \cup q^{\mathbb{Z}}$ we define $\ell_\theta = \{ (x,y) \in I_q \times I_q \mid y = \theta x \}$. See [1] and [2].

Proposition 3.2. *Consider $\theta \in -q^{\mathbb{Z}} \cup q^{\mathbb{Z}}$. Then the family $(a_p\restriction_{\ell_\theta} \mid p \in I_q$ such that $\mathrm{sgn}(p) = \mathrm{sgn}(\theta))$ is an orthonormal basis for $\ell^2(\ell_\theta)$.*

This proposition is used to define the comultiplication on M. It is also essential to the proof of the left invariance of the Haar weight.

Let us also mention the nice symmetry in $a_p(x,y)$ with respect to interchanging x, y and p:

Proposition 3.3. *If $x, y, p \in I_q$, then*

$$a_p(x,y) = (-1)^{\chi(yp)}\, \mathrm{sgn}(x)^{\chi(x)}\, |y/p|\, a_y(x,p)$$
$$a_p(x,y) = \mathrm{sgn}(p)^{\chi(p)}\, \mathrm{sgn}(x)^{\chi(x)}\, \mathrm{sgn}(y)^{\chi(y)}\, a_p(y,x)$$
$$a_p(x,y) = (-1)^{\chi(xp)}\, \mathrm{sgn}(y)^{\chi(y)}\, |x/p|\, a_x(p,y).$$

Now we produce the eigenvectors of our self-adjoint extension of $\Delta_0(\gamma_0^\dagger \gamma_0)$ (see the remarks after the proof of Proposition 3.6). We will use these eigenvectors to define a unitary operator that will induce the comultiplication. The dependence of $F_{r,s,m,p}$ on r,s and p is chosen in such a way that Proposition 3.6 is true.

Definition 3.4. Consider $r, s \in \mathbb{Z}$, $m \in \mathbb{Z}$ and $p \in I_q$. We define the element $F_{r,s,m,p} \in \mathcal{H} \otimes \mathcal{H}$ such that

$$F_{r,s,m,p}(\lambda, x, \mu, y) = \begin{cases} a_p(x, y)\, \lambda^{r+\chi(y/p)}\, \mu^{s-\chi(x/p)} & \text{if } y = \operatorname{sgn}(p)\, q^m\, x \\ 0 & \text{otherwise} \end{cases}$$

for all $x, y \in I_q$ and $\lambda, \mu \in \mathbb{T}$.

Now we are ready to introduce the comultiplication of quantum $\widetilde{\mathrm{SU}}_q(1,1)$.

Proposition 3.5. *Define the unitary transformation* $V : \mathcal{H} \otimes \mathcal{H} \to L^2(\mathbb{T}) \otimes L^2(\mathbb{T}) \otimes \mathcal{H}$ *such that* $V(F_{r,s,m,p}) = \zeta^r \otimes \zeta^s \otimes \zeta^m \otimes \delta_p$ *for all* $r, s \in \mathbb{Z}$, $m \in \mathbb{Z}$ *and* $p \in I_q$. *Then there exists a unique injective normal *-homomorphism* $\Delta : M \to M \otimes M$ *such that* $\Delta(a) = V^*(1_{L^2(\mathbb{T})} \otimes 1_{L^2(\mathbb{T})} \otimes a)V$ *for all* $a \in M$.

The requirement that $\Delta(M) \subseteq M \otimes M$ is the primary reason for introducing the extra generator u. We cannot work with the von Neumann algebra $M^\diamond$ that is generated by α, γ and e alone, because $\Delta(M^\diamond) \nsubseteq M^\diamond \otimes M^\diamond$.

This definition of Δ and the operators α and γ imply easily that the space $\langle\, F_{r,s,m,p} \mid r, s \in \mathbb{Z}, m \in \mathbb{Z}, p \in I_q \,\rangle$ is a core for $\Delta(\alpha)$, $\Delta(\gamma)$ and

$$\begin{aligned} \Delta(\alpha)\, F_{r,s,m,p} &= \sqrt{\operatorname{sgn}(p) + p^{-2}}\; F_{r,s,m,pq} \\ \Delta(\gamma)\, F_{r,s,m,p} &= p^{-1}\; F_{r,s,m+1,p}. \end{aligned} \tag{3.2}$$

for $r, s \in \mathbb{Z}$, $m \in \mathbb{Z}$ and $p \in I_q$.

Recall the linear operators $\Delta_0(\alpha_0)$, $\Delta_0(\gamma_0)$ acting on $E \odot E$ (Eqs. (1.1)). Also recall the distinction between $*$ and $\dagger$. The next proposition shows that Δ and Δ_0 are related in a natural way.

Proposition 3.6. *The following inclusions hold:* $\Delta_0(\alpha_0) \subseteq \Delta(\alpha)$, $\Delta_0(\alpha_0)^\dagger \subseteq \Delta(\alpha)^*$, $\Delta_0(\gamma_0) \subseteq \Delta(\gamma)$ *and* $\Delta_0(\gamma_0)^\dagger \subseteq \Delta(\gamma)^*$. *Moreover* $\Delta(e) = e \otimes e$.

This proposition implies also that $\Delta(\gamma^*\gamma)$ is an extension of $\Delta_0(\gamma_0^\dagger\gamma_0)$. We also know that $\langle\, F_{r,s,m,p} \mid r, s \in \mathbb{Z}, m \in \mathbb{Z}, p \in I_q \,\rangle$ is a core for $\Delta(\gamma^*\gamma)$ and $\Delta(\gamma^*\gamma)\, F_{r,s,m,p} = p^{-2}\, F_{r,s,m,p}$ for $r, s, m \in \mathbb{Z}$, $p \in I_q$. Using this information one can indeed show that $\Delta(\gamma^*\gamma)\restriction_{L^2(\ell'_\theta)}$ is unitarily equivalent to $1 \otimes L_\theta^{\operatorname{sgn}(\theta)}$ for all $\theta \in -q^{\mathbb{Z}} \cup q^{\mathbb{Z}}$, but we will not make any use of this fact in this paper.

4 Quantum $\widetilde{\mathrm{SU}}(1,1)$ as a locally compact quantum group.

Now we can state the main result of [6]. Verifying the coassociativity of Δ as in Definition 1 of the introduction turns out to be the most difficult property to check. Producing the Haar weight is not that difficult (and goes back to [9]) but proving its invariance requires some work.

Theorem 4.1. *The pair (M, Δ) is a unimodular locally compact quantum group.*

We define $\widetilde{\mathrm{SU}}_q(1,1) = (M, \Delta)$ and also refer to $\widetilde{\mathrm{SU}}_q(1,1)$ as quantum $\widetilde{\mathrm{SU}}(1,1)$.

Let us give an explicit formula for the Haar weight. Since $M = L^\infty(\mathbb{T}) \otimes B(L^2(I_q))$ we can consider the trace Tr on M given by $\mathrm{Tr} = \mathrm{Tr}_{L^\infty(\mathbb{T})} \otimes \mathrm{Tr}_{B(L^2(I_q))}$, where $\mathrm{Tr}_{L^\infty(\mathbb{T})}$ and $\mathrm{Tr}_{B(L^2(I_q))}$ are the canonical traces on $L^\infty(\mathbb{T})$ and $B(L^2(I_q))$ which we choose to be normalized in such a way that $\mathrm{Tr}_{L^\infty(\mathbb{T})}(1) = 1$ and $\mathrm{Tr}_{B(L^2(I_q))}(P) = 1$ for every rank one projection P in $B(L^2(I_q))$.

Given a weight η on M, we use the following standard concepts from weight theory:

$$\mathcal{M}_\eta^+ = \{\, x \in M^+ \mid \eta(x) < \infty \,\}, \quad \mathcal{M}_\eta = \text{linear span of } \mathcal{M}_\eta^+$$

and

$$\mathcal{N}_\eta = \{\, x \in M \mid \eta(x^*x) < \infty \,\}.$$

Next we introduce a GNS-construction for the trace Tr. Define

$$\mathcal{K} = H \otimes L^2(I_q) = L^2(\mathbb{T}) \otimes L^2(I_q) \otimes L^2(I_q).$$

If $m \in \mathbb{Z}$ and $p, t \in -q^{\mathbb{Z}} \cup q^{\mathbb{Z}}$, we set $f_{m,p,t} = \zeta^m \otimes \delta_p \otimes \delta_t \in \mathcal{K}$ if $p, t \in I_q$ and $f_{m,p,t} = 0$ otherwise. Now define

(1) a linear map $\Lambda_{\mathrm{Tr}} : \mathcal{N}_{\mathrm{Tr}} \to \mathcal{K}$ such that $\Lambda_{\mathrm{Tr}}(a) = \sum_{p \in I_q} (a \otimes 1_{L^2(I_q)}) f_{0,p,p}$ for $a \in \mathcal{N}_{\mathrm{Tr}}$.

(2) a unital *-homomorphism $\pi : M \to B(\mathcal{K})$ such that $\pi(a) = a \otimes 1_{L^2(I_q)}$ for all $a \in M$.

Then $(\mathcal{K}, \pi, \Lambda_{\mathrm{Tr}})$ is a GNS-construction for Tr.

Now we are ready to define the weight that will turn out to be left- and right invariant with respect to Δ. Use the remarks before [14, Prop. 1.15] to define a linear map $\Lambda = (\Lambda_{\mathrm{Tr}})_{\gamma^*\gamma} : D(\Lambda) \subseteq M \to \mathcal{K}$.

Definition 4.2. We define the faithful normal semi-finite weight φ on M as $\varphi = \mathrm{Tr}_{\gamma^*\gamma}$. By definition, $(\mathcal{K}, \pi, \Lambda)$ is a GNS-construction for φ.

So, on a formal level, $\varphi(x) = \mathrm{Tr}(x\,\gamma^*\gamma)$ and $\Lambda(x) = \Lambda_{\mathrm{Tr}}(x\,|\gamma|)$. See [20] for more details about the exact definition. This definition of φ is of course compatible with the usual construction of absolutely continuous weights (see [18]). So we already know that the modular automorphism group σ^φ of φ is such that $\sigma_s^\varphi(x) = |\gamma|^{2is}\, x\, |\gamma|^{-2is}$ for all $x \in M$ and $s \in \mathbb{R}$.

As for any locally compact quantum group we can consider the polar decomposition $S = R\,\tau_{-\frac{i}{2}}$ of the antipode S of (M, Δ). Here, R is an anti-*-automorphism of M and τ is a σ-weakly continuous one parameter group on M so that R and τ commute. In this example, the following formulas hold:

$$S(\Phi(m,p,t)) = \operatorname{sgn}(p)^{\chi(p)} \operatorname{sgn}(t)^{\chi(t)} (-1)^m q^m\; \Phi(m,t,p)$$
$$R(\Phi(m,p,t)) = \operatorname{sgn}(p)^{\chi(p)} \operatorname{sgn}(t)^{\chi(t)} (-1)^m\; \Phi(m,t,p)$$
$$\tau_s(\Phi(m,p,t)) = q^{2mis}\; \Phi(m,p,t)$$
$$\sigma_s^\varphi(\Phi(m,p,t)) = |p^{-1}t|^{2is}\; \Phi(m,p,t)$$

for all $m \in \mathbb{Z}$, $p, t \in I_q$ and $s \in \mathbb{R}$.

To any locally compact quantum group one can associate a multiplicative unitary through the left invariance of the left Haar weight. In this example (and this happens also in other examples) we go the other way around. First we use the orthogonality relations involving the functions a_p (see Proposition 3.2) to produce a partial isometry.

Proposition 4.3. *There exists a unique surjective partial isometry W on $\mathcal{K} \otimes \mathcal{K}$ such that*

$$W^*(f_{m_1,p_1,t_1} \otimes f_{m_2,p_2,t_2}) = \sum_{\substack{y,z \in I_q \\ \operatorname{sgn}(p_2t_2)(yz/p_1)q^{m_2} \in I_q}} |t_2/y|\, a_{t_2}(p_1, y)\, a_{p_2}(z, \operatorname{sgn}(p_2t_2)(yz/p_1)q^{m_2})$$
$$\times\; f_{m_1+m_2-\chi(p_1p_2/t_2z), z, t_1} \otimes f_{\chi(p_1p_2/t_2z), \operatorname{sgn}(p_2t_2)(yz/p_1)q^{m_2}, y}$$

for all $m_1, m_2 \in \mathbb{Z}$ and $p_1, p_2, t_1, t_2 \in I_q$.

In a next step one connects this partial isometry with the weight φ by showing that $(\omega\pi \otimes \iota)\Delta(a) \in \mathcal{N}_\varphi$ and $\Lambda((\omega\pi \otimes \iota)\Delta(a)) = (\omega \otimes \iota)(W^*)\,\Lambda(a)$ for all $\omega \in B(K)_*$ and $a \in \mathcal{N}_\varphi$. In turn, this is used to prove the left invariance of φ so that (M, Δ) satisfies Definition 1 of the introduction and W is the multiplicative unitary naturally associated to (M, Δ):

$$W^*(\Lambda(x) \otimes \Lambda(y)) = (\Lambda \otimes \Lambda)(\Delta(y)(x \otimes 1))$$

for all $x, y \in \mathcal{N}_\varphi$. In fact, this formula was used in [6] to obtain the defining formula for W in Proposition 4.3.

From the general theory of locally compact quantum groups we know that all of the information concerning (M, Δ) is contained in W in the following way:

(1) $\pi(M)$ is the σ-weak closure, in $B(\mathcal{K})$, of $\{\, (\iota \otimes \omega)(W^*) \mid \omega \in B(\mathcal{K})_* \,\}$,

(2) $(\pi \otimes \pi)\Delta(x) = W^*(1 \otimes \pi(x))W$ for all $x \in M$.

As a matter of fact, if $m \in \mathbb{Z}$ and $p, t \in I_q$, a concrete element $\omega \in B(\mathcal{K})_*$ can be produced so that $\Phi(m,p,t) = (\iota \otimes \omega)(W^*)$.

In general one associates to a von Neumann algebraic quantum group a C*-algebraic quantum group (A, Δ) by requiring that $\pi(A)$ is the norm closure of the algebra $\{ (\iota \otimes \omega)(W^*) \mid \omega \in B(\mathcal{K})_* \}$ and simply restricting the comultiplication Δ from M to A.

In order to describe the C*-algebra A in this specific case, we will use the following notation. For $f \in C(\mathbb{T} \times I_q)$ and $x \in I_q$ we define $f_x \in C(\mathbb{T})$ so that $f_x(\lambda) = f(\lambda, x)$ for all $\lambda \in \mathbb{T}$.

Proposition 4.4. *Denote by $\mathcal{C}$ the C*-algebra of all functions $f \in C(\mathbb{T} \times I_q)$ such that* (1) *f_x converges uniformly to 0 as $x \to 0$ and* (2) *f_x converges uniformly to a constant function as $x \to \infty$. Then, A is the norm closed linear span, in $B(\mathcal{H})$, of the set $\{ \rho_p M_f \mid f \in \mathcal{C}, p \in -q^{\mathbb{Z}} \cup q^{\mathbb{Z}} \}$.*

Thus, each operator $\Phi(m, p, t)$ belongs to A but the C*-algebra A is not generated by these operators.

5 The Pontryagin dual of quantum $\widetilde{\mathrm{SU}}(1,1)$

The Hopf * algebra $\mathcal{U}_q(\mathfrak{su}(1,1))$ and its pairing with the coordinate algebra $A(\mathrm{SU}_q(1,1))$ have been well studied (see e.g. [13], [16], [17]). Due to the existence of this pairing, one expects that the generators of $\mathcal{U}_q(\mathfrak{su}(1,1))$ give rise to closed operators in $\mathcal{K}$ that are affiliated to the dual von Neumann algebra $\hat{M}$. As explained in this section, this turns out to be the case.

Define the dense subspace D of $\mathcal{K}$ as $D = \langle f_{m,p,t} \mid m \in \mathbb{Z}, p, t \in I_q \rangle$. This space D inherits the inner product of $\mathcal{K}$ so that we can look at the space of adjointable operators $\mathcal{L}^+(D)$ for D. Formal calculations based on Formulae (1.11), (1.12) of [17] and some educated guesses learn us that the Hopf * algebra $\mathcal{U}_q(\mathfrak{su}(1,1))$ in our framework should be realized by the following operators in $\mathcal{L}^+(D)$.

Definition 5.1. We define operators E_0, K_0 in $\mathcal{L}^+(D)$ so that $(q - q^{-1})\, E_0\, f_{m,p,t}$ equals

$$\begin{aligned} &\operatorname{sgn}(t)\, q^{-\frac{m-1}{2}}\, |p/t|^{\frac{1}{2}} \sqrt{1+\kappa(q^{-1}t)}\; f_{m-1,p,q^{-1}t} \\ &\quad - \operatorname{sgn}(p)\, q^{\frac{m-1}{2}}\, |t/p|^{\frac{1}{2}} \sqrt{1+\kappa(p)}\; f_{m-1,qp,t} \end{aligned}$$

and $K_0\, f_{m,p,t} = q^{-\frac{m}{2}}\, |p/t|^{\frac{1}{2}}\, f_{m,p,t}$ for all $m \in \mathbb{Z}$, $p, t \in I_q$.

One easily checks that $K_0^\dagger = K_0$ and that $(q - q^{-1})\, E_0^-\, f_{m,p,t}$ equals

$$\begin{aligned} &\operatorname{sgn}(t)\, q^{-\frac{m+1}{2}}\, |p/t|^{\frac{1}{2}} \sqrt{1+\kappa(t)}\; f_{m+1,p,qt} \\ &\quad - \operatorname{sgn}(p)\, q^{\frac{m+1}{2}}\, |t/p|^{\frac{1}{2}} \sqrt{1+\kappa(q^{-1}p)}\; f_{m+1,q^{-1}p,t} \end{aligned}$$

for all $m \in \mathbb{Z}$, $p, t \in I_q$. Also note that K_0 is invertible in $\mathcal{L}^+(D)$.

Rather straightforward calculations reveal the following result.

Proposition 5.2. *Let $\mathcal{U}$ denote the unital *-subalgebra of $\mathcal{L}^+(D)$ generated by E_0, K_0 and K_0^{-1}. Then, $\mathcal{U}$ is the universal unital *-algebra generated by elements E_0 and K_0 and relations*

$$K_0^\dagger = K_0, \quad K_0 \text{ is invertible in } \mathcal{U}, \quad K_0\, E_0 = q\, E_0\, K_0$$

and

$$E_0^\dagger\, E_0 - E_0\, E_0^\dagger = \frac{K_0^2 - K_0^{-2}}{q - q^{-1}}.$$

As a consequence, we assume for the rest of this paper that $\mathcal{U}_q\big(\mathfrak{su}(1,1)\big) = \mathcal{U}$. An essential role in the representation theory of $\mathcal{U}$ is played by the Casimir operator, see Eq. (1.9) of [17]. We will use a slightly renormalized version (and terminology) of the operator used in [17].

Definition 5.3. We define the Casimir element $\Omega_0 \in \mathcal{U}$ as

$$\Omega_0 = \frac{1}{2}\,\big(\,(q - q^{-1})^2\, E_0^\dagger\, E_0 - q\, K_0^2 - q^{-1}\, K_0^{-2}\big).$$

If C_0 denotes the element introduced in Eq. (1.9) of [17], one has $\Omega_0 = -\frac{1}{2}\,(q - q^{-1})^2\, C_0 - 1$. The Casimir element Ω_0 is self-adjoint and generates together with 1 the center of $\mathcal{U}$. The renormalization is chosen in such a way that the continuous spectrum of the relevant self-adjoint extension of Ω_0 is given by $[-1, 1]$.

The dual locally compact quantum group $(\hat{M}, \hat{\Delta})$ of (M, Δ) is defined through the multiplicative unitary W as follows (see [15]) :

(1) $\hat{M}$ is the σ-weak closure, in $B(\mathcal{K})$, of $\{\, (\omega \otimes \iota)(W^*) \mid \omega \in B(\mathcal{K})_* \,\}$.

(2) $\hat{\Delta}(x) = \Sigma W (x \otimes 1) W^* \Sigma$ for all $x \in \hat{M}$,

where Σ denotes the flip operator on $\mathcal{K} \otimes \mathcal{K}$. The presence of this flip operator is not essential but assures that the dual weight construction applied to a left Haar weight of (M, Δ) produces a left (as opposed to a right) Haar weight of $(\hat{M}, \hat{\Delta})$.

Because of the importance of the Casimir element to the representation theory of $\mathcal{U}$, it will not come as a big surprise that it will also play a vital role in analyzing the dual von Neumann algebra $\hat{M}$.

The operators K_0, E_0 are not affiliated to $\hat{M}$, we need to extend them to closed operators to get hold of the operators that are relevant to $\hat{M}$. Clearly, K_0 is essentially self-adjoint so it is clear what extension of K_0 to use. At this stage, it is not clear what kind of extension of E_0 we need to use, but Proposition 5.5 and the comments thereafter show that the simplest possible extension turns out to be the right one.

Definition 5.4. We define the densely defined, closed, linear operators E and K in $\mathcal{K}$ as the closures of E_0 and K_0 respectively. Thus, K is an injective self-adjoint operator in $\mathcal{K}$.

With a little bit more trouble (involving E_0) than one would expect one can show that

Proposition 5.5. *The operators E and K are affiliated to $\hat{M}$ in the von Neumann algebraic sense.*

One can moreover prove that E^* is the closure of $E_0^\dagger$. As a consequence, E is the unique closed linear operator in $\mathcal{K}$ such that $E_0 \subseteq E$ and $E_0^\dagger \subseteq E^*$, which is the analogue of the fact that K is the unique self-adjoint operator in $\mathcal{K}$ that extends K_0.

We also need to consider the relevant closed extension of the Casimir element Ω_0, but the situation is more subtle than that for the extensions of E_0 and K_0. The operators E^*E and K strongly commute (in the sense that their spectral projections commute), which justifies the following definition:

Definition 5.6. We define the Casimir operator Ω as the closure (as an operator in $\mathcal{K}$) of the linear operator

$$\frac{1}{2}\big((q-q^{-1})^2\, E^*E \;-\; q\,K^2 - q^{-1}\,K^{-2}\,\big).$$

Thus, Ω is a self-adjoint operator in $\mathcal{K}$ that is affiliated to $\hat{M}$.

The Casimir operator Ω is not the closure of the Casimir element Ω_0 but one can prove that Ω is the unique self-adjoint extension of Ω_0 that is affiliated to $\hat{M}$.

As is to be expected, the Casimir operator Ω strongly commutes with K and E but this is not true for every element in $\hat{M}$, a slight refinement is needed. For this purpose, we introduce the closed subspaces $\mathcal{K}_+$ and $\mathcal{K}_-$ of $\mathcal{K}$ as

$$\mathcal{K}_+ = \langle\, f_{m,p,t} \mid \operatorname{sgn}(p) = \operatorname{sgn}(t)\,\rangle \quad \text{and} \quad \mathcal{K}_- = \langle\, f_{m,p,t} \mid \operatorname{sgn}(p) \neq \operatorname{sgn}(t)\,\rangle$$

and the self adjoint subspaces $\hat{M}_+$ and $\hat{M}_-$ of $\hat{M}$ as

$$\hat{M}_+ = \{\, x \in \hat{M} \mid x\,\mathcal{K}_\pm \subseteq \mathcal{K}_\pm \,\} \quad \text{and} \quad \hat{M}_- = \{\, x \in \hat{M} \mid x\,\mathcal{K}_\pm \subseteq \mathcal{K}_\mp \,\}.$$

One can easily characterize $\hat{M}_+$ and $\hat{M}_-$ by slices of W:

Lemma 5.7. *We have that $\hat{M}_\pm$ is the σ-weak closure of*

$$\langle\, (\omega_{f_{m_1,p_1,t_1},f_{m_2,p_2,t_2}} \otimes \iota)(W^*) \mid m_1, m_2 \in \mathbb{Z},\, p_1, p_2, t_1, t_2 \in I_q,\, \operatorname{sgn}(p_1 p_2) = \pm\,\rangle.$$

It follows that $\hat{M} = \hat{M}_+ + \hat{M}_-$ and this provides $\hat{M}$ with a $\mathbb{Z}_2$-grading. The commutation relations involving the Casimir operator now become

Proposition 5.8. *If $x \in \hat{M}_+$ and $y \in \hat{M}_-$, then*

$$x\,\Omega \subseteq \Omega\,x \quad \textit{and} \quad y\,\Omega \subseteq -\,\Omega\,y.$$

Notice that this implies that Ω^2 is affiliated to the center of $\hat{M}$. In the next section we describe the spectral decomposition of Ω in terms of special functions. This results in a description of $\mathcal{K}$ that is better suited for the needs of $\hat{M}$ than our original definition of $\mathcal{K}$ was.

The graded commutation relations above also show that $\hat{M}$ can not be generated by E and K alone (which is to be expected). The extra generators for $\hat{M}$ will naturally emerge from this new description of $\mathcal{K}$.

In two subsequent sections we will indicate how this description is also used to get hold of the corepresentations of (M, Δ) connected to the multiplicative unitary W.

6 The spectral decomposition of the Casimir operator

Let $p \in q^{\mathbb{Z}}$, $m \in \mathbb{Z}$, $\varepsilon, \eta \in \{-, +\}$ and define

$$J(p, m, \varepsilon, \eta) = \{\, z \in I_q \mid \varepsilon\, \eta\, q^m\, p\, z \in I_q \text{ and } \operatorname{sgn}(z) = \varepsilon \,\}$$

and the closed subspace $\mathcal{K}(p, m, \varepsilon, \eta)$ of $\mathcal{K}$ as

$$\mathcal{K}(p, m, \varepsilon, \eta) = [\, f_{-m, \varepsilon\eta q^m p z, z} \mid z \in J(p, m, \varepsilon, \eta)\,].$$

The space $D \cap \mathcal{K}(p, m, \varepsilon, \eta)$ is invariant under Ω_0 and the restriction of Ω_0 to $D \cap \mathcal{K}(p, m, \varepsilon, \eta)$ is given by

$$\begin{aligned} &2\, \Omega_0\, f_{-m, \varepsilon\eta q^m\, pz, z} \\ &\quad = -\, \varepsilon\, \eta\, \sqrt{1 + \kappa(z)}\, \sqrt{1 + \varepsilon\, \eta\, q^{2m}\, p^2\, \kappa(z)}\; f_{-m, \varepsilon\eta q^m p\,(qz), qz} \\ &\qquad -\, \varepsilon\, \eta\, \sqrt{1 + \kappa(q^{-1}z)}\, \sqrt{1 + \varepsilon\, \eta\, q^{2m}\, p^2\, \kappa(q^{-1}z)}\; f_{-m, \varepsilon\eta q^m p\,(q^{-1}z), q^{-1}z} \\ &\qquad + q^{-1} p\, (\varepsilon\, \eta + q^{2m})\, \kappa(z)\; f_{-m, \varepsilon\eta q^m p\, z, z}, \end{aligned}$$

which is clearly unitarily equivalent to a Jacobi (or tridiagonal) operator. It turns out that the self-adjoint extensions of these kind of Jacobi operators and their spectral decompositions are known within the theory of special functions:

(1) $\varepsilon = -$ or $\eta = -$: Then, this restriction is bounded and has a unique bounded self-adjoint extension. The spectral decomposition of this extension is described in terms of Al-Salam & Chihara polynomials (see Definition 2 of the introduction).

(2) $\varepsilon = \eta = +$: If $m \neq 0$, this restriction is essentially self-adjoint. If $m = 0$, this restriction has different self-adjoint extensions, and one selects the (to us) relevant self-adjoint extension by putting on asymptotical behavior conditions on the functions in its domain (see [5], for the exceptional case $m = 0$ extra necessary details have been added in the appendix of [7]). In both cases the spectral decomposition of the self-adjoint extension is described in terms of little q-Jacobi functions (see Definition 3 of the introduction).

Let us characterize these spectral decompositions precisely. For every $p \in q^{\mathbb{Z}}$, $m \in \mathbb{Z}$ and $\varepsilon, \eta \in \{-, +\}$, we define the subset $D(p, m, \varepsilon, \eta) \subseteq -q^{-\mathbb{N}} \cup q^{-\mathbb{N}}$ as follows

$$D(p,m,++) = \{ -q^{1+2r} p \mid r \in \mathbb{Z},\ r \geq \max\{0,m\},\ q^{1+2r} p > 1 \} \\ \cup \{ -q^{1+2r} p^{-1} \mid r \in \mathbb{Z},\ r \geq \max\{0,-m\},\ q^{1+2r} p^{-1} > 1 \} \\ \cup \{ q^{1+2k} p \mid k \in \mathbb{Z},\ q^{1+2k} p > 1 \}$$

Note that at most one of the two first finite sets is non-empty, while the third set is infinite. Furthermore, if $\varepsilon \neq \eta$,

$$D(p,m,\varepsilon,\eta) = \{ q^{1+2r} p^{-\varepsilon} \mid r \in \mathbb{N}_0,\ q^{1+2r} p^{-\varepsilon} > 1 \} \\ \cup \{ -q^{1+2r} p^{-\varepsilon} \mid r \in \mathbb{Z},\ r \geq -\varepsilon m,\ q^{1+2r} p^{-\varepsilon} > 1 \},$$

Moreover,

$$D(p,m,--) = \{ -q^{1+2r} p\, p_0 \mid r \in \mathbb{N}_0,\ q^{1+2r} p\, p_0 > 1 \} \\ \cup \{ -q^{1+2(r+m)} p\, p_0 \mid r \in \mathbb{N}_0,\ q^{1+2(r+m)} p\, p_0 > 1 \},$$

where $p_0 = \min\{1, q^{-2m} p^{-2}\}$. Also in this last case, at most one of the two finite sets is non-empty.

We define the sets $\sigma_d(p,m,\varepsilon,\eta) \subseteq I(p,m,\varepsilon,\eta) \subseteq \mathbb{R}$ as

$$\sigma_d(p,m,\varepsilon,\eta) = \mu(D(p,m,\varepsilon,\eta)) \text{ and } I(p,m,\varepsilon,\eta) = [-1,1] \cup \sigma_d(p,m,\varepsilon,\eta).$$

On $I(p,m,\varepsilon,\eta)$ we consider the natural measure that agrees with the Lebesgue measure on $[-1,1]$ and with the counting measure on $\sigma_d(p,m,\varepsilon,\eta)$.

There exists moreover a unique family of continuous real-valued functions

$$(g_z(\,.\,; p,m,\varepsilon,\eta))_{z\in J(p,m,\varepsilon,\eta)} \in \mathcal{L}^2(I(p,m,\varepsilon,\eta) \setminus \{-1,1\})$$

so that for each $x \in I(p,m,\varepsilon,\eta) \setminus \{-1,1\}$ the following holds.

(1) If $z \in J(p,m,\varepsilon,\eta)$,

$$\begin{aligned} &2x\, g_z(x; p,m,\varepsilon,\eta) \\ &= -\varepsilon\eta \sqrt{1+\kappa(z)} \sqrt{1+\varepsilon\eta q^{2m} p^2 \kappa(z)}\ g_{qz}(x; p,m,\varepsilon,\eta) \\ &\quad - \varepsilon\eta \sqrt{1+\kappa(q^{-1}z)} \sqrt{1+\varepsilon\eta q^{2m} p^2 \kappa(q^{-1}z)}\ g_{q^{-1}z}(x; p,m,\varepsilon,\eta) \\ &\quad + q^{-1} p\,(\varepsilon\eta + q^{2m})\,\kappa(z)\, g_z(x; p,m,\varepsilon,\eta), \end{aligned} \tag{6.1}$$

where we use the convention that $g_{q^{-1}z}(x; p,m,\varepsilon,\eta) = 0$ if $q^{-1}z \notin J(p,m,\varepsilon,\eta)$.

(2) If $\varepsilon = \eta = +$, the family $\big(g_z(x; p,m,\varepsilon,\eta)\big)_{z\in J(p,m,\varepsilon,\eta)}$ is ℓ^2 as $z \to \infty$.

(3) If $\theta \in (0,\pi)$, there exists a function $c : J(p,m,\varepsilon,\eta) \to \mathbb{C}$ so that

- $g_z(\cos\theta; p,m,\varepsilon,\eta) = \Re\big(c(z)\,(-\varepsilon\eta e^{i\theta})^{-\chi(z)}\big)$ for all $z \in J(m,p,\varepsilon,\eta)$,
- $c(z) \to \sqrt{\frac{2}{\pi \sin\theta}}$ as $z \to 0$.

(4) If $x = \mu(y) \in \sigma_d(p, m, \varepsilon, \eta)$ for some $y \in -q^{-\mathbb{N}} \cup q^{-\mathbb{N}}$, the family

$$\big(g_z(x;\, p, m, \varepsilon, \eta) \big)_{z \in J(p,m,\varepsilon,\eta)}$$

belongs to $\ell^2(J(p, m, \varepsilon, \eta))$, has ℓ^2-norm equal to 1 and $g_z(x;\, p, m, \varepsilon, \eta)\ (-\varepsilon\, \eta\, y)^{\chi(z)}$ converges to a positive number as $z \to 0$.

We see that the family $\big(g_z(x;\, p, m, \varepsilon, \eta) \big)_{z \in J(p,m,\varepsilon,\eta)}$ satisfies the 3 terms recurrence relation (6.1). If $\varepsilon = \eta = +$, then $J(p, m, \varepsilon, \eta) = I_q^+$ and the space of families $(a_z)_{z \in J(p,m,\varepsilon,\eta)}$ satisfying this recurrence relation is two dimensional. Thus, in order to determine a unique solution, we need to impose two independent extra conditions. If $\varepsilon = -$ or $\eta = -$, then $J(p, m, \varepsilon, \eta)$ is of the form $\{\, z \in I_q \mid |z| \le t \text{ and } \operatorname{sgn}(z) = \epsilon \,\}$ for some $t \in q^{\mathbb{Z}}$ and $\epsilon \in \{-1, 1\}$. In this case, the space of families $(a_z)_{z \in J(p,m,\varepsilon,\eta)}$ satisfying this recurrence relation is one dimensional so we need only one extra condition to determine a unique solution.

In [7] the functions $g_z(\,.\,;\, p, m, \varepsilon, \eta)$ are explicitly defined in terms of Al-Salam & Chihara polynomials and little q-Jacobi functions. Apart for needing these q-special functions to define $g_z(\,.\,;\, p, m, \varepsilon, \eta)$, we also use their concrete expressions to further analyze their properties. However, in order to discuss the results concerning quantum $\widetilde{\mathrm{SU}}(1,1)$ in this paper we will not need and not use these concrete expressions.

Let us now formalize the spectral decomposition of the Casimir operator.

Definition 6.1. We define the unitary operator

$$\Upsilon : \mathcal{K} \to \bigoplus_{p \in q^{\mathbb{Z}}, m \in \mathbb{Z}, \varepsilon, \eta \in \{-,+\}} L^2\big(I(p, m, \varepsilon, \eta)\big)$$

so that for $p \in q^{\mathbb{Z}}$, $m \in \mathbb{Z}$ and $\varepsilon, \eta \in \{-, +\}$, we have that

$$\Upsilon\big(\mathcal{K}(p, m, \varepsilon, \eta)\big) = L^2\big(I(p, m, \varepsilon, \eta)\big)$$

and

$$\Upsilon\big(f_{-m, \varepsilon \eta q^m p z, z} \big) = g_z(\,.\,;\, p, m, \varepsilon, \eta)$$

for all $z \in J(p, m, \varepsilon, \eta)$.

If $p \in q^{\mathbb{Z}}$, $m \in \mathbb{Z}$ and $\varepsilon, \eta \in \{-, +\}$, we will denote the restriction of Υ to $\mathcal{K}(p, m, \varepsilon, \eta)$ by $\Upsilon_{p,m}^{\varepsilon,\eta}$.

Theorem 6.2. *Let $p \in q^{\mathbb{Z}}$, $m \in \mathbb{Z}$ and $\varepsilon, \eta \in \{-, +\}$. Then, Ω maps $\mathcal{K}(p, m, \varepsilon, \eta) \cap D(\Omega)$ into $\mathcal{K}(p, m, \varepsilon, \eta)$ and*

$$\Upsilon_{p,m}^{\varepsilon,\eta}\, \Omega \restriction_{\mathcal{K}(p,m,\varepsilon,\eta)} \big(\Upsilon_{p,m}^{\varepsilon,\eta}\big)^*$$

is the multiplication, with the identity function, operator on $L^2\big(I(p, m, \varepsilon, \eta)\big)$.

In general, $\mathcal{K}(p, m, \varepsilon, \eta)$ is not contained in the domain of Ω so we use the convention that the operator $\Omega\restriction_{\mathcal{K}(p,m,\varepsilon,\eta)}$ is defined to have $\mathcal{K}(p, m, \varepsilon, \eta) \cap D(\Omega)$ as its domain. The same remark applies to other operators in this paper.

We also use the following notation. Let $p \in q^{\mathbb{Z}}$, $m \in \mathbb{Z}$, $\varepsilon, \eta \in \{+,-\}$ and $x = \mu(y)$ for some $y \in -q^{-\mathbb{N}} \cup q^{-\mathbb{N}}$. If $x \in \sigma_d(p,m,\varepsilon,\eta)$ (which is equivalent to saying that Ω has an eigenvector of eigenvalue x inside $\mathcal{K}(p,m,\varepsilon,\eta)$) we define

$$\xi(x;\, p,m,\varepsilon,\eta) = (\Upsilon^{\varepsilon,\eta}_{p,m})^*(\delta_x) = \sum_{z\in J(p,m,\varepsilon,\eta)} g_z(x;\, p,m,\eta,\varepsilon)\ f_{-m,\varepsilon\eta q^m pz,z},$$

so that $\xi(x;\, p,m,\varepsilon,\eta)$ is a unit eigenvector of Ω inside $\mathcal{K}(p,m,\varepsilon,\eta)$. This unit eigenvector is uniquely determined by the fact that the number

$$\langle \xi(x;\, p,m,\varepsilon,\eta),\, f_{-m,\varepsilon\eta q^m pz,z}\rangle\ (-\varepsilon\,\eta\, y)^{\chi(z)}$$

converges to a positive number $c(x;\, p,m,\varepsilon,\eta)$ (which is explicitly known) as $z \to 0$. We also set $\dot{\xi}(x;\, p,m,\varepsilon,\eta) = c(x;\, p,m,\varepsilon,\eta)^{-1}\ \xi(x;\, p,m,\varepsilon,\eta)$ which is the unique eigenvector of Ω inside $\mathcal{K}(p,m,\varepsilon,\eta)$ so that

$$\langle \dot{\xi}(x;\, p,m,\varepsilon,\eta),\, f_{-m,\varepsilon\eta q^m pz,z}\rangle\ (-\varepsilon\,\eta\, y)^{\chi(z)}$$

converges to 1 as $z \to 0$.

If $x \notin \sigma_d(p,m,\varepsilon,\eta)$, we set $\xi(x;\, p,m,\varepsilon,\eta) = \dot{\xi}(x;\, p,m,\varepsilon,\eta) = 0$.

The normalization for the functions $g_z(\,.\,;\, p,m,\varepsilon,\eta)$ that we introduced before Definition 6.1 is not necessary for the validity of the theorem above. However, this specific normalization is useful to get elegant expressions for $\Upsilon\, E\, \Upsilon^*$ (and similarly, for the extra generators of $\hat{M}$).

Proposition 6.3. *Let $p \in q^{\mathbb{Z}}$, $m \in \mathbb{Z}$ and $\varepsilon, \eta \in \{-,+\}$. Then,*

(1) *$\mathcal{K}(p,m,\varepsilon,\eta) \subseteq D(K)$, K leaves $\mathcal{K}(p,m,\varepsilon,\eta)$ invariant and*

$$\Upsilon^{\varepsilon,\eta}_{p,m}\ K\restriction_{\mathcal{K}(p,m,\varepsilon,\eta)}\ (\Upsilon^{\varepsilon,\eta}_{p,m})^* = q^m\ p^{\frac{1}{2}}\ 1.$$

(2) *E maps $\mathcal{K}(p,m,\varepsilon,\eta) \cap D(E)$ into $\mathcal{K}(p,m+1,\varepsilon,\eta)$ and*

$$\Upsilon^{\varepsilon,\eta}_{p,m+1}\ E\restriction_{\mathcal{K}(p,m,\varepsilon,\eta)}\ (\Upsilon^{\varepsilon,\eta}_{p,m})^*$$

is the multiplication operator with the function

$$x \mapsto q^m\ \sqrt{qp}\ \sqrt{1 + 2p^{-1}\, q^{1-2m}\, x + p^{-2} q^{2-4m}}$$

A word of caution is in order. The space $L^2(I(p,m,\varepsilon,\eta))$ can be different from $L^2(I(p,m+1,\varepsilon,\eta))$ because $\sigma_d(p,m,\varepsilon,\eta)$ does not have to be equal to $\sigma_d(p,m+1,\varepsilon,\eta)$. Therefore, the last statement of the previous proposition should be interpreted in the following way. For f in the domain of $\Upsilon^{\varepsilon,\eta}_{p,m+1}\ E\restriction_{\mathcal{K}(p,m,\varepsilon,\eta)}\ (\Upsilon^{\varepsilon,\eta}_{p,m})^*$ and $x \in \sigma_d(p,m+1,\varepsilon,\eta) \setminus \sigma_d(p,m,\varepsilon,\eta)$, we have that

$$\big(\Upsilon^{\varepsilon,\eta}_{p,m+1}\ E\restriction_{\mathcal{K}(p,m,\varepsilon,\eta)}\ (\Upsilon^{\varepsilon,\eta}_{p,m})^*\big)(f)(x) = 0,$$

which is consistent with the convention (2) of the introduction.

7 The extra generators for the dual

The description of $\mathcal{K}$ based on the spectral decomposition of the Casimir operator in the previous section is essential to describe the dual $\hat{M}$ in a more natural way. As a first application we will introduce extra generators, other than E and K, for $\hat{M}$ based on this new description.

Definition 7.1. Consider $s, t \in \{+, -\}$ and $n \in \mathbb{Z}$. We define the operator $U_n^{s,t} \in B(\mathcal{K})$ so that for $p \in q^{\mathbb{Z}}$, $m \in \mathbb{Z}$ and $\varepsilon, \eta \in \{-, +\}$, the space $\mathcal{K}(p, m, \varepsilon, \eta)$ is mapped into $\mathcal{K}(p, m+n, s\,\varepsilon, t\,\eta)$ by $U_n^{s,t}$ and

$$\big(\Upsilon_{p,m+n}^{s\,\varepsilon, t\,\eta}\; U_n^{s,t}\; (\Upsilon_{p,m}^{\varepsilon,\eta})^*\big)(f)(x) = \varepsilon^{\frac{1}{2}(1-s)}\; \eta^{\frac{1}{2}(1-t)+n}\; f(s\,t\,x)$$

for $f \in \mathcal{L}^2(I(p, m, \varepsilon, \eta))$ and almost all $x \in I(p, m+n, s\varepsilon, t\eta)$.

The same words of caution apply as after Proposition 6.3 to the fact that $I(p, m + n, s\,\varepsilon, t\,\eta)$ can be different from $I(p, m, \varepsilon, \eta)$. For the eigenvectors of Ω, we have that

$$U_n^{s,t}\; \xi(x; p, m, \varepsilon, \eta) = \varepsilon^{\frac{1}{2}(1-s)}\; \eta^{\frac{1}{2}(1-t)+n}\; \xi(s\,t\,x; p, m, s\,\varepsilon, t\,\eta)$$

for $p \in q^{\mathbb{Z}}, m \in \mathbb{Z}, \varepsilon, \eta \in \{+, -\}$ and $x \in \sigma_d(p, m, \varepsilon, \eta)$. So $U_n^{s,t}\; \xi(x; p, m, \varepsilon, \eta) \neq 0$ if and only if $s\,t\,x \in \sigma_d(p, m+n, s\,\varepsilon, t\,\eta)$.

These operators are defined in such a way that

Proposition 7.2. *If $s, t \in \{+, -\}$ and $n \in \mathbb{Z}$, the operator $U_n^{s,t}$ belongs to $\hat{M}$.*

Furthermore, this family provides the necessary extra generators.

Theorem 7.3. *The von Neumann algebra $\hat{M}$ is generated by E, K, U_0^{+-} and U_0^{-+}.*

In the rest of this section we explain the basic ideas behind these results by adding some further detail. By definition of $\hat{M}$, the operators $(\omega_{f_{m_1,p_1,t}, f_{m_2,p_2,t}} \otimes \iota)(W^*)$ generate $\hat{M}$. The following result is an easy consequence of Proposition 4.3.

Lemma 7.4. *Consider $p \in q^{\mathbb{Z}}$, $m \in \mathbb{Z}$, $\varepsilon, \eta \in \{-, +\}$ and $v \in \mathcal{K}(p, m, \varepsilon, \eta)$. Consider also $m_1, m_2 \in \mathbb{Z}$ and $p_1, p_2, r \in I_q$ and set $m' = m + m_2 - m_1$, $\varepsilon' = \operatorname{sgn}(p_1)\,\varepsilon$ and $\eta' = \operatorname{sgn}(p_2)\,\eta$. Then, the vector*

$$w := (\omega_{f_{m_1,p_1,r}, f_{m_2,p_2,r}} \otimes \iota)(W^*)\; v$$

belongs to $\mathcal{K}(p, m', \varepsilon', \eta')$. If $q^{2m}\, p = q^{m_1 - m_2}\, |p_2/p_1|$, then $\langle w, f_{-m',\varepsilon'\,\eta'\,q^{m'}\,p\,x,x}\rangle$ equals

$$(-1)^{m'}\; (\varepsilon'\eta')^{\chi(x)}\; (\eta')^{\chi(p)+m'}\; \frac{|p_1 p_2|}{q^m\, p}\; \frac{1}{|x|}$$

$$\times \sum_{y \in J(p,m,\varepsilon,\eta)} \frac{1}{|y|}\; a_{p_1}(y, x)\, a_{p_2}(\varepsilon\,\eta\,q^m\,p\,y, \varepsilon'\,\eta'\,q^{m'}\,p\,x)\; \langle v, f_{-m,\varepsilon\,\eta\,q^m\,p\,y,y}\rangle$$

for all $x \in J(p, m', \varepsilon', \eta')$ (where the sum is absolutely convergent). If $q^{2m}\, p \neq q^{m_1-m_2}\, |p_2/p_1|$, then $w = 0$.

This expression for this kind of slice of W^* is to involved to work with. We will now indicate how a more useful formula can be deduced from it. For $p_1, p_2 \in I_q$, $n \in \mathbb{Z}$ and $a \in \mathbb{C} \setminus \{0\}$, we define the complex number $c(a, p_1, p_2, n)$ as

$$c_q^2\, q^n\, |p_1 p_2| \sqrt{(-\kappa(p_1), -\kappa(p_2); q^2)_\infty}$$
$$\times \sum_{\substack{t \in -q^{\mathbb{Z}} \cup q^{\mathbb{Z}} \\ \operatorname{sgn}(t) = \operatorname{sgn}(p_1)}} \frac{\big(-\operatorname{sgn}(p_1 p_2)\, a\big)^{-\chi(t)}}{|t|}\, \nu(p_1/t)\, \nu(p_2 q^n/t)$$
$$\times\ {}_1\varphi_1\left(\begin{matrix} -q^2/\kappa(p_1) \\ 0 \end{matrix}\ ;\ q^2\kappa(t)\right)\ {}_1\varphi_1\left(\begin{matrix} -q^2/\kappa(p_2) \\ 0 \end{matrix}\ ;\ q^2\kappa(\operatorname{sgn}(p_1 p_2) q^{-n} t)\right),$$

where the ${}_1\varphi_1$-functions are in base q^2. For most of the analysis concerning quantum $\widetilde{\mathrm{SU}}(1,1)$ we do not need to know what this value is. However, at one point we need to know that this expression is non-zero in order to prove that the corepresentations in the principal unitary series are irreducible. Thus, a more explicit expression for this sum is required. It turns out that this sum is known in the literature in the case $p_1, p_2 > 0$ (see [11]) and can be evaluated. In [7], we use some other method to evaluate this sum (based on some basic but clever manipulations) in order to be able to incorporate all cases where $p_1 > 0$ or $p_2 > 0$. Relying on the triple product identity also the case $p_1, p_2 < 0$ can be treated. In all cases, $c(a, p_1, p_2, n)$ can be simply expressed in terms of a ${}_2\varphi_1$-function.

Let us now explain the basic idea behind the computation of the above slice of W^* by looking at the point spectrum of Ω, which is the simplest case.

So take $p \in q^{\mathbb{Z}}$, $m \in \mathbb{Z}$, $\varepsilon, \eta \in \{-, +\}$ and set $m' = m + m_2 - m_1$, $\varepsilon' = \operatorname{sgn}(p_1)\,\varepsilon$ and $\eta' = \operatorname{sgn}(p_2)\,\eta$. Let $x = \mu(y) \in \sigma_d(p, m, \varepsilon, \eta)$ for some $y \in -q^{-\mathbb{N}} \cup q^{-\mathbb{N}}$. Assume that $q^{2m}\, p = q^{m_1-m_2}\, |p_2/p_1|$.

Set $v = \dot{\xi}(x; p, m, \varepsilon, \eta)$ and $w = (\omega_{f_{m_1,p_1,r}, f_{m_2,p_2,r}} \otimes \iota)(W^*)\, \dot{\xi}(x; p, m, \varepsilon, \eta)$. Thus, v is the eigenvector of Ω inside $\mathcal{K}(p, m, \varepsilon, \eta)$ determined by the fact that $\langle v, f_{-m, \varepsilon\eta q^m p z, z}\rangle\, (-\varepsilon\,\eta\, y)^{\chi(z)} \to 1$ as $z \to 0$.

Lemma 7.4 shows that w belongs to $\mathcal{K}(p, m', \varepsilon', \eta')$ whilst Proposition 5.8 and Lemma 5.7 guarantee that $w = 0$ or w is an eigenvector of Ω with eigenvalue $\operatorname{sgn}(p_1 p_2)\, x$.

Lemma 7.4 also allows to prove that $\langle w, f_{-m', \varepsilon'\eta' q^{m'} p z, z}\rangle\, (-\varepsilon\,\eta\, y)^{\chi(z)}$ converges to

$$(\eta')^{m_1+m_2}\, s(\varepsilon, \varepsilon')\, s(\eta, \eta')\ c(y, p_1, p_2, m_2 - m_1)$$

as $z \to 0$. It should be noted that in order to prove this asymptotic behavior, we only need to know the asymptotic behavior of $\langle v, f_{-m, \varepsilon\eta q^m p z, z}\rangle$ as $z \to 0$.

Thus, we conclude that $(\omega_{f_{m_1,p_1,r},f_{m_2,p_2,r}} \otimes \iota)(W^*)\, \dot{\xi}(x; p, m, \varepsilon, \eta)$ equals

$$(\eta')^{m_1+m_2}\, s(\varepsilon, \varepsilon')\, s(\eta, \eta')\;\; c(y, p_1, p_2, m_2 - m_1)\; \dot{\xi}(x; p, m', \varepsilon', \eta'),$$

where it is important to note that $c(y, p_1, p_2, m_2 - m_1)$ only depends on p_1,p_2,m_1,m_2 and x.

The above principle can be refined to obtain similar results on the continuous spectrum of Ω. In the end, this results in the following generic form for $(\omega_{f_{m_1,p_1,r},f_{m_2,p_2,r}} \otimes \iota)(W^*)$.

Proposition 7.5. *For every $p_1, p_2 \in I_q$ and $n \in \mathbb{Z}$ there exists a continuous function $h(\,.\,; p_1, p_2, n) : [-1, 1] \cup \mu(-q^{-\mathbb{N}} \cup q^{-\mathbb{N}}) \to \mathbb{C}$ so that the following holds.*

Consider $r \in I_q$, $m_1, m_2 \in \mathbb{Z}$ and set $p = |p_2/p_1|\, q^{m_1-m_2}$, $s = \operatorname{sgn}(p_1)$, $t = \operatorname{sgn}(p_2)$. If $m \in \mathbb{Z}$ and $\varepsilon, \eta \in \{-, +\}$, then

$$(\omega_{f_{m_1,p_1,r},f_{m_2,p_2,r}} \otimes \iota)(W^*)\;\; \mathcal{K}(q^{-2m} p, m, \varepsilon, \eta) \subseteq \mathcal{K}(q^{-2m} p, m + m_2 - m_1, s\,\varepsilon, t\,\eta)$$

and for $f \in \mathcal{L}^2\big(I(q^{-2m} p, m, \varepsilon, \eta)\big)$,

$$\begin{aligned}\Big(\Upsilon^{s\,\varepsilon, t\,\eta}_{q^{-2m} p, m+m_2-m_1}\, (\omega_{f_{m_1,p_1,r},f_{m_2,p_2,r}} \otimes \iota)(W^*)\, (\Upsilon^{\varepsilon,\eta}_{q^{-2m} p,m})^*\Big)(f)(x) \\ = \varepsilon^{\frac{1}{2}(1-s)}\; \eta^{\frac{1}{2}(1-t)+m_2-m_1}\;\; h(s\,t\,x; p_1, p_2, m_2 - m_1)\;\; f(s\,t\,x)\end{aligned}$$

for almost all $x \in I(q^{-2m} p, m + m_2 - m_1, s\,\varepsilon, t\,\eta)$.

Also, $(\omega_{f_{m_1,p_1,r},f_{m_2,p_2,r}} \otimes \iota)(W^*)$ is 0 on the orthogonal complement of

$$\oplus_{m\in\mathbb{Z},\varepsilon,\eta\in\{-,+\}}\; \mathcal{K}(q^{-2m} p, m, \varepsilon, \eta).$$

The function $h(\,.\,; p_1, p_2, n)$ in the above proposition is uniquely determined on $[-1, 1]$ and on some subset $\Sigma(p_1, p_2, n)$ of $\mu(-q^{-\mathbb{N}} \cup q^{-\mathbb{N}})$ and, as should be clear from the above exposition, can be given in terms of $c(\,.\,; p_1, p_2, m_2 - m_1)$. Explicit formulas for $h(\,.\,; p_1, p_2, n)$ in terms of ${}_2\varphi_1$-functions can be found in [7].

In any case, if P denotes the spectral projection of K corresponding to the eigenvalue $p^{\frac{1}{2}}$, the above proposition says that

$$(\omega_{f_{m_1,p_1,r},f_{m_2,p_2,r}} \otimes \iota)(W^*) = U^{s,t}_{m_2-m_1}\; P\, h(\Omega; p_1, p_2, m_2 - m_1).$$

Since $U^{s,t}_{m_2-m_1}$ can be written in terms of U_0^{+-}, U_0^{-+} and E and Ω is defined via E and K, this result implies that $\hat{M}$ is contained in the von Neumann algebra generated by E, K, U_0^{+-} and U_0^{-+}.

Using the above proposition (without having to know the precise expression for $h(\,.\,; p_1, p_2, n)$), the bicommutant theorem $\hat{M}'' = \hat{M}$ and the fact that $\hat{M}' = \hat{J}\,\hat{M}\,\hat{J}$ one can also prove the validity of Proposition 7.2 and thereby establishing Theorem 7.3 in the end.

8 Unitary corepresentations: the discrete series

To each value in the spectrum of the Casimir operator Ω, one can associate a unitary corepresentation via the multiplicative unitary W. In this section we look at the procedure of doing so for the point spectrum of Ω, which is the simplest case.

Fix $p \in q^{\mathbb{Z}}$ and $x \in \mu(-q^{-\mathbb{N}} \cup q^{-\mathbb{N}})$. Write $x = \mu(y)$ for $y \in -q^{-\mathbb{N}} \cup q^{-\mathbb{N}}$. For every $m \in \mathbb{Z}$ and $\varepsilon, \eta \in \{-,+\}$, we will use the shorthand notation

$$e_m^{\varepsilon,\eta} = \xi(\varepsilon\,\eta\,x;\, p, m, \varepsilon, \eta),$$

where you should remember that $e_m^{\varepsilon,\eta} \neq 0$ if and only if $\varepsilon\,\eta\,x \in \sigma_d(p, m, \varepsilon, \eta)$.

Definition 8.1. Assume there exist $m \in \mathbb{Z}$ and $\varepsilon, \eta \in \{-,+\}$ so that $\varepsilon\,\eta\,x \in \sigma_d(p, m, \varepsilon, \eta)$. We define the non-zero closed subspace $\mathcal{L}_{p,x}$ of $\mathcal{K}$ as

$$\mathcal{L}_{p,x} = [\, e_m^{\varepsilon,\eta} \mid m \in \mathbb{Z}, \varepsilon, \eta \in \{-,+\}\,].$$

The space $\mathcal{L}_{p,x}$ is an invariant subspace of the corepresentation W of (M, Δ) and the element $W_{p,x} := W\lceil_{\mathcal{K}\otimes\mathcal{L}_{p,x}}$ is a unitary corepresentation of (M, Δ) on $\mathcal{L}_{p,x}$. In this case, we say that (p, x) determines a corepresentation of (M, Δ).

Notice that the invariance of $\mathcal{L}_{p,x}$ follows from Lemma 7.4 and the fact that

$$\operatorname{sgn}(p_1 p_2)\,(\omega_{f_{m_1,p_1,t_1}, f_{m_2,p_2,t_2}} \otimes \iota)(W^*)\,\Omega \subseteq \Omega\,(\omega_{f_{m_1,p_1,t_1}, f_{m_2,p_2,t_2}} \otimes \iota)(W^*).$$

For the rest of this section we assume that (p, x) determines a corepresentation of (M, Δ). Then, $|y| \in p\,q^{1+2\mathbb{Z}}$ (if $y \in p\,q^{1+2\mathbb{Z}}$, then (p, x) determines a corepresentation of (M, Δ)).

Proposition 8.2. *The corepresentation $W_{p,x}$ is irreducible.*

This irreducibility result will not hold for the continuous spectrum of Ω but there we will be able to describe the decomposition into irreducible components.

In terms of these eigenvectors of Ω, Proposition 6.3 takes on the following form. Consider $m \in \mathbb{Z}$ and $\varepsilon, \eta \in \{-,+\}$ such that $x \in \sigma_d(p, m, \varepsilon, \eta)$. Then, the vector $e_m^{\varepsilon,\eta}$ belongs to the domain of Ω, K and E and

$$\begin{aligned}
\Omega\, e_m^{\varepsilon,\eta} &= \varepsilon\,\eta\,x\, e_m^{\varepsilon,\eta},\\
K\, e_m^{\varepsilon,\eta} &= q^m\, p^{\frac{1}{2}}\, e_m^{\varepsilon,\eta},\\
E\, e_m^{\varepsilon,\eta} &= q^m\, \sqrt{qp}\,\sqrt{1 + 2\,\varepsilon\,\eta\, p^{-1} q^{1-2m}\, x + p^{-2} q^{2-4m}}\; e_{m+1}^{\varepsilon,\eta},\\
U_n^{s,t}\, e_m^{\varepsilon,\eta} &= \varepsilon^{\frac{1}{2}(1-s)}\,\eta^{\frac{1}{2}(1-t)+n}\, e_{m+n}^{s\,\varepsilon,t\,\eta}, \qquad (n \in \mathbb{Z}, s, t \in \{+,-\})
\end{aligned} \tag{8.1}$$

where one should remember that $e_{m+1}^{\varepsilon,\eta} = 0$ iff $\varepsilon\,\eta\,x \notin \sigma_d(p, m+1, \varepsilon, \eta)$ and $e_{m+n}^{s\,\varepsilon,t\,\eta} = 0$ iff $s\,t\,\varepsilon\,\eta\,x \notin \sigma_d(p, m+n, s\,\varepsilon, t\,\eta)$.

Now, write $p = q^t$ en $|y| = q^{t+1+2l}$, where $t, l \in \mathbb{Z}$ and $l < -(t+l)$. Then, the Hilbert space $\mathcal{L}_{p,x}$ has the following orthonormal basis.

(i) If $x > 0$,

$$\{ e_m^{++} \mid m \in \mathbb{Z} \} \cup \{ e_m^{-+} \mid m \in \mathbb{Z},\ m \leq l \} \cup \{ e_m^{+-} \mid m \in \mathbb{Z},\ m \geq -(t+l) \},$$

(ii) If $x < 0$, $l \geq 0$ and $l + t < 0$,

$$\{ e_m^{-+} \mid m \in \mathbb{Z} \} \cup \{ e_m^{++} \mid m \in \mathbb{Z},\ m \leq l \} \cup \{ e_m^{--} \mid m \in \mathbb{Z},\ m \geq -(t+l) \},$$

(iii) If $x < 0$, $l < 0$ and $l + t \geq 0$,

$$\{ e_m^{+-} \mid m \in \mathbb{Z} \} \cup \{ e_m^{--} \mid m \in \mathbb{Z},\ m \leq l \} \cup \{ e_m^{++} \mid p \in \mathbb{Z},\ m \geq -(t+l) \}.$$

If we combine these facts with the action of Ω, K and E described in Eqs. (8.1), we see that in each case the corepresentation $W_{p,x}$ is obtained by exponentiating a positive discrete series, a negative discrete series and a strange series *-representations of $\mathcal{U}_q(\mathfrak{su}(1,1))$ (see Eq. (1,1), Eq. (1.10), Prop. 4 and the comments after this proposition in [17]) and combining them via U_0^{+-}, U_0^{-+}.

9 Unitary corepresentations: the principal unitary series

In this section, we associate to every value in $(-1, 1)$ a unitary corepresentation via the multiplicative unitary. These corepresentations are reducible and we will describe their irreducible components. Fix $p \in \mathbb{Z}$ and $x \in (-1, 1)$. Write $x = \cos\theta$, where $\theta \in (0, \pi)$.

We define the Hilbert space

$$\mathcal{L} = \ell^2_{++}(\mathbb{Z}) \oplus \ell^2_{+-}(\mathbb{Z}) \oplus \ell^2_{-+}(\mathbb{Z}) \oplus \ell^2_{--}(\mathbb{Z}),$$

where $\ell^2_{\varepsilon,\eta}(\mathbb{Z})$ is a copy of $\ell^2(\mathbb{Z})$ and the standard basis of $\ell^2_{\varepsilon,\eta}(\mathbb{Z})$ is denoted by $\left(e_m^{\varepsilon,\eta}\right)_{m\in\mathbb{Z}}$ $(\varepsilon, \eta \in \{+,-\})$.

The starting point is Proposition (7.5). Recall that the function $h(\,.\,; p_1, p_2, n)$ in this proposition is uniquely determined on $[-1, 1]$ and that it can be written down explicitly in terms of ${}_2\varphi_1$-functions.

Proposition 9.1. *There exists a unique unitary corepresentation $W_{p,x} \in M \otimes B(\mathcal{L})$ so that for $p_1, t_1 \in I_q$, $m_1 \in \mathbb{Z}$ and $m \in \mathbb{Z}$, $\varepsilon, \eta \in \{+,-\}$,*

$$W_{p,x}^* \left(f_{m_1,p_1,t_1} \otimes e_m^{\varepsilon,\eta}\right) = \sum_{p_2 \in I_q} \varepsilon^{\frac{1}{2}(1-\operatorname{sgn}(p_1))}\, \eta^{\frac{1}{2}(1-\operatorname{sgn}(p_2))+\chi(p_1 p_2 p)+2m}$$

$$h(\varepsilon\,\eta\,x\,;\, p_1, p_2, \chi(p_2/p_1 p) - 2m)\ f_{m_1+\chi(p_2/p_1 p)-2m,\,p_2,\,t_1} \otimes e_{\chi(p_2/p_1 p)-m}^{\operatorname{sgn}(p_1)\,\varepsilon,\,\operatorname{sgn}(p_2)\,\eta}.$$

The proof of this proposition is based on the following basic principle. The unitarity and multiplicativity of W imply certain properties of the functions $h(\,.\,; p_1, p_2, n)$. In turn, these properties allow us to prove that $W_{p,x}$ is a unitary and a corepresentation.

As mentioned before, the corepresentation $W_{p,x}$ is not irreducible. Recall that the invariant subspaces of $W_{p,x}$ are the invariant subspaces of the von Neumann algebra $\hat{M}_{p,x}$ that is the σ-weak closure of $\{ (\omega \otimes \iota)(W^*_{p,x}) \mid \omega \in M_* \}$. By definition,

$$(\omega_{f_{m_1,p_1,t_1}, f_{m_2,p_2,t_2}} \otimes \iota)(W_{p,x})\, e_m^{\varepsilon,\eta} = \delta_{t_1,t_2}\, \delta_{2m,m_1-m_2+\chi(p_2/p_1 p)} \tag{9.1}$$

$$\varepsilon^{\frac{1}{2}(1-\mathrm{sgn}(p_1))}\, \eta^{\frac{1}{2}(1-\mathrm{sgn}(p_2))+m_2-m_1}\, h(\varepsilon\,\eta\, x;\, p_1, p_2, m_2 - m_1)\, e_{m+m_2-m_1}^{\mathrm{sgn}(p_1)\,\varepsilon, \mathrm{sgn}(p_2)\,\eta}$$

for $p_1, p_2, t_1, t_2 \in I_q$ and $m_1, m_2 \in \mathbb{Z}$ and $m \in \mathbb{Z}$, $\varepsilon, \eta \in \{-, +\}$.

Using the operators in Eq. (9.1), it is not difficult to check that $\mathcal{L}$ decomposes in the following invariant subspaces.

Proposition 9.2. (1) *Suppose $x \neq 0$. Then $\mathcal{L} = \mathcal{L}_0 \oplus \mathcal{L}_1$ is an orthogonal decomposition of $\mathcal{L}$ into irreducible invariant subspaces of $W_{p,x}$, where*

$$\mathcal{L}_j = [\, e_m^{++} + (-1)^{m+j}\, e_m^{--}, e_m^{+-} + (-1)^{m+j+1}\, e_m^{-+} \mid m \in \mathbb{Z}\,] \qquad (j = 0, 1).$$

(2) *Suppose $x = 0$. Then $\mathcal{L} = \mathcal{L}_{0,0} \oplus \mathcal{L}_{0,1} \oplus \mathcal{L}_{1,0} \oplus \mathcal{L}_{1,1}$ is an orthogonal decomposition of $\mathcal{L}$ into irreducible invariant subspaces of $W_{p,x}$, where*

$$\mathcal{L}_{j,k} = [\, e_m^{++} + (-1)^{m+j}\, e_m^{--} + (-1)^k\, i\, \big((-1)^{m+j+1}\, e_m^{+-} + e_m^{-+}\big) \mid m \in \mathbb{Z}\,]$$

for $j = 0, 1$, $k = 0, 1$

The irreducibility is obvious if $x = 0$ but requires some explanation if $x \neq 0$.

Define $E_{p,x}$ and $K_{p,x}$ as the minimal closed linear operators in $\mathcal{L}$ so that for $m \in \mathbb{Z}$ and $\varepsilon, \eta \in \{+, -\}$, the vector $e_m^{\varepsilon,\eta}$ belongs to the domain of $E_{p,x}$ and $K_{p,x}$ and

$$E_{p,x}\, e_m^{\varepsilon,\eta} = q^m \sqrt{qp}\, \sqrt{1 + 2\,\varepsilon\,\eta\, p^{-1} q^{1-2m}\, x + p^{-2} q^{2-4m}}\; e_{m+1}^{\varepsilon,\eta}$$

and

$$K_{p,x}\, e_m^{\varepsilon,\eta} = q^m\, p^{\frac{1}{2}}\, e_m^{\varepsilon,\eta}.$$

Using the properties of the functions $h(\,.\,; p_1, p_2, n)$ one shows that $E_{p,x}$ and $K_{p,x}$ are affiliated to $\hat{M}_{p,x}$.

Define the operator $\Omega_{p,x} \in B(\mathcal{L})$ so that $\Omega_{p,x}\, e_m^{\varepsilon,\eta} = \varepsilon\,\eta\, x\, e_m^{\varepsilon,\eta}$. Then, $2\,\Omega_{p,x}$ extends

$$(q - q^{-1})^2\, E^*_{p,x} E_{p,x} - q^{-1}\, K^{-2}_{p,x} - q\, K^2_{p,x},$$

implying that $\Omega_{p,x}$ belongs to $\hat{M}_{p,x}$.

Combining this with the operators in (9.1) one also proves that the operators $U^{+-}_{p,x}, U^{-+}_{p,x} \in B(\mathcal{L})$ defined by

$$U^{-+}_{p,x}\, e_m^{\varepsilon,\eta} = \varepsilon\, e_m^{-\varepsilon,\eta} \quad \text{and} \quad U^{+-}_{p,x}\, e_m^{\varepsilon,\eta} = \eta\, e_m^{\varepsilon,-\eta}$$

belong to $\hat{M}_{p,x}$.

With these operators inside $\hat{M}_{p,x}$, one easily checks the irreducibility of the spaces $\mathcal{L}_0$ and $\mathcal{L}_1$ with respect to $W_{p,x}$.

Similar to the discussion in the previous section, the corepresentations $W_{p,x}\restriction_{\mathcal{K}\otimes\mathcal{L}_j}$ $(j = 0, 1)$ are obtained by exponentiating two principal unitary series $*$-representations of $\mathcal{U}_q(\mathfrak{su}(1, 1))$ (see [17]) and combining them via $U_{p,x}^{+-}$ and $U_{p,x}^{-+}$.

For the discrete series corepresentations we could have followed the same method as here to define $\hat{W}_{p,x}$ but it is obviously much easier to define it in the discrete case by restriction (and also obtain the operators in $\hat{M}_{p,x}$ by restriction). The difference in irreducibility results between the two kinds of corepresentations stems from the fact that in the discrete case, there exist $m \in \mathbb{Z}$ and $\varepsilon, \eta \in \{+, -\}$ for which $e_m^{\varepsilon,\eta} \neq 0$, but $e_m^{s\,\varepsilon,t\,\eta} = 0$ if $s = -$ or $t = -$.

Notice that by Proposition 9.1 the matrix elements of $W_{p,x}$ are given by

$$
\begin{aligned}
(\iota \otimes \omega_{e_m^{\varepsilon,\eta}, e_n^{\varepsilon',\eta'}})(W_{p,x})\, f_{m_1,p_1,t_1} \;=\; & \delta_{\mathrm{sgn}(p_1),\varepsilon\,\varepsilon'}\; \varepsilon^{\frac{1}{2}(1-\varepsilon\,\varepsilon')}\, \eta^{\frac{1}{2}(1-\eta\,\eta')+m+n} \\
& h(\varepsilon\,\eta\, x;\, p_1, \varepsilon\,\varepsilon'\,\eta\,\eta'\, p_1\, q^{m+n}\, p, n-m)\; f_{m_1+n-m,\varepsilon\,\varepsilon'\,\eta\,\eta'\,p_1\,q^{m+n}\,p,t_1}
\end{aligned}
$$

and recall that $h(\varepsilon\,\eta\, x;\, p_1, \varepsilon\,\varepsilon'\,\eta\,\eta'\, p_1\, q^{m+n}\, p, n-m)$ can be expressed in terms of ${}_2\varphi_1$-functions. It will be shown in [7] that in the case $p = 1$ and $\varepsilon = \varepsilon' = \eta = \eta' = +$, this corresponds to Proposition 3 of [17].

In the last two sections we introduced all irreducible unitary corepresentations of (M, Δ) appearing in the direct integral decomposition of the multiplicative unitary W. It is however to be expected (as in the classical case) that these are not all irreducible unitary corepresentations of (M, Δ) since none of the complementary series corepresentations introduced in [17] have been considered. We hope to fill this void in the near term future.

References

[1] W. A. Al-Salam and L. Carlitz, Some orthogonal q-polynomials, *Math. Nachr.* 30 (1965), 47–61.

[2] N. Ciccoli, E. Koelink and T. Koornwinder, q-Laguerre polynomials and big q-Bessel functions and their orthogonality relations, *Methods Appl. Anal.* 6 (1999), 109–127.

[3] M. Enock and J.-M. Schwartz, *Kac algebras and duality of locally compact groups*, Springer-Verlag, Berlin 1992.

[4] G. Gasper and M. Rahman, *Basic hypergeometric series*, Cambridge University Press, Cambridge 1990.

[5] E. Koelink, Spectral theory and special functions, in *Lecture notes for the 2000 Laredo Summer school of the SIAM Activity Group* 2000, `http://aw.twi.tudelft.nl/~koelink/laredo.html`.

[6] E. Koelink and J. Kustermans, A locally compact quantum group analogue of the normalizer of $SU(1, 1)$ in $\mathrm{SL}(2, \mathbb{C})$, to appear in *Comm. Math. Phys.*, preprint `math.QA/0105117` (2001).

[7] E. Koelink and J. Kustermans, The Pontryagin dual of quantum $\widetilde{SU}(1, 1)$, in preparation (2002).

[8] E. Koelink and J. V. Stokman, The big q-Jacobi function transform, to appear in *Constr. Approx.*, preprint `math.CA/9904111` (2002).

[9] E. Koelink and J. V. Stokman, with an appendix by M. Rahman, Fourier transforms on the quantum $SU(1, 1)$ group, *Publ. Res. Inst. Math. Sci.* 37 (2001), 621–715.

[10] E. Koelink and J. V. Stokman, The Askey-Wilson function transform scheme, *Special Functions 2000: Current Perspective and Future Directions*, NATO Sci. Ser. C Math. Phys. Sci. 30, Kluwer, Dordrecht 2001, 221–241.

[11] T. H. Koornwinder and R. F. Swarttouw, On q-analogues of the Fourier and Hankel transforms, *Trans. Amer. Math. Soc.* 333 (1992), 445–461

[12] L. I. Korogodskii, Quantum Group $SU(1, 1) \rtimes \mathbb{Z}_2$ and super tensor products, *Comm. Math. Phys.* 163 (1994), 433–460.

[13] L. I. Korogodskii and L. L. Vaksman, Spherical functions on the quantum group $SU(1, 1)$ and the q-analogue of the Mehler-Fock formula, *Funktsional. Anal. i Prilozhen.* 25 (1) (1991), 60–62; English translation in *Funct. Anal. Appl.* 25 (1991), 48–49.

[14] J. Kustermans and S. Vaes, Locally compact quantum groups, *Ann. Sci. École Norm. Sup.* (4) 33 (2000), 837–934.

[15] J. Kustermans and S. Vaes, Locally compact quantum groups in the von Neumann algebraic setting, to appear in *Math. Scand.*, preprint `math.QA/0005219` (2000).

[16] T. Masuda, K. Mimachi, Y. Nakagami, M. Noumi, Y. Saburi and K. Ueno, Unitary representations of the quantum group $SU_q(1, 1)$: structure of the dual space of $\mathcal{U}_q(\mathfrak{sl}(2))$, *Lett. Math. Phys.* 19 (1990), 187–194.

[17] T. Masuda, K. Mimachi, Y. Nakagami, M. Noumi, Y. Saburi and K. Ueno, Unitary representations of the quantum group $SU_q(1, 1)$: II - matrix elements of unitary representations and the basic hypergeometric functions, *Lett. Math. Phys.* 19 (1990), 195–204.

[18] G. K. Pedersen and M. Takesaki, The Radon-Nikodym theorem for von Neumann algebras, *Acta Math.* 130 (1973), 53–87.

[19] K. Schmüdgen, *Unbounded operator algebras and representation theory*, Oper. Theory Adv. Appl. 37, Birkhäuser, Basel 1990.

[20] S. Vaes, A Radon-Nikodym theorem for von Neumann algebras, *J. Operator Theory* 46 (2001), 477–489.

[21] S. L. Woronowicz, Unbounded elements affiliated with C*-algebras and non-compact quantum groups, *Commun. Math. Phys.* 136 (1991), 399–432.

[22] S. L. Woronowicz, Extended $SU(1, 1)$ quantum group, Hilbert space level, preprint KMMF, in preparation.

[23] S. L. Woronowicz and S. Zakrzewski, Quantum $ax + b$ group, preprint KMMF (1999).

Morita base change in quantum groupoids

Peter Schauenburg

Mathematisches Institut der Universität München
Theresienstr. 39, 80333 München, Germany
email: `schauen@mathematik.uni-muenchen.de`

Abstract. Let L be a quantum semigroupoid, more precisely a $\times_R$-bialgebra in the sense of Takeuchi. We describe a procedure replacing the algebra R by any Morita equivalent, or in fact more generally any $\sqrt{\text{Morita}}$ equivalent (in the sense of Takeuchi) algebra S to obtain a $\times_S$-bialgebra $\widetilde{H}$ with the same monoidal representation category.

1 Introduction

Quantum groupoids (or Hopf algebroids) are algebraic structures designed to be the analogs of (the function algebras of) groupoids in the realm of noncommutative geometry. A groupoid consists of a set G of arrows, and a set V of vertices. Thus a quantum groupoid consists of an algebra L (the function algebra on the noncommutative space of arrows) and an algebra R (the function algebra on the noncommutative space of vertices. The assignment to an arrow of its source and target vertices defines two maps $G \rightrightarrows V$. Thus the definition of a quantum groupoid involves two maps $R \rightrightarrows L$; it turns out to be the right choice to assume one of these to be an algebra, the other an anti-algebra map, and to assume that the images of the two commute. Since multiplication in G is an only partially defined map, comultiplication in L maps from L to some tensor product $L \otimes_R L$; one has to make the right choice of module structures to define the tensor product, and one needs to assume that comultiplication actually maps to a certain subspace of $L \otimes_R L$ to be able to state that comultiplication is assumed to be an algebra map.

The first version of a quantum (semi)groupoid or bialgebroid or Hopf algebroid was considered by Takeuchi [27], following work of Sweedler [25] in which R is by assumption commutative. Actually Takeuchi invents his $\times_R$-bialgebras from different motivations, involving generalizations of Brauer groups, and does not seem to be thinking of groupoids at all. Lu [14] and Xu [30] reinvent his notion, now with the motivation by noncommutative-geometric groupoids in mind. (Actually most of Lu's or Xu's definition is the very same as Takeuchi's up to changes in notation, at least as far as comultiplication is concerned. For a detailed translation, and the removal of any doubt about the notion of counit, consult Brzeziński and Militaru [3].)

Of the other possible definitions of a quantum groupoid we should mention the weak Hopf algebras of Böhm and Szlachányi [2], see also the recent survey [17] by Nikshych and Vainerman and the literature cited there, and the notion of a face algebra due to Hayashi [8, 10]. Face algebras were shown to be precisely the $\times_R$-bialgebras in which R is commutative and separable in [21]. Also, face algebras are precisely the weak bialgebras whose target counital subalgebras are commutative. Etingof and Nikshych [5] have shown that weak Hopf algebras are $\times_R$-bialgebras. In fact weak bialgebras are precisely those $\times_R$-bialgebras in which R is Frobenius-separable (for example semisimple over the complex numbers) [24].

In the present paper we will discuss a construction that allows us to replace the algebra R in any $\times_R$-bialgebra L by a Morita-equivalent algebra S to obtain a $\times_S$-bialgebra that has the same representation theory, more precisely a monoidal category of representations equivalent to that of L. In fact we can, more generally, replace R by any $\sqrt{\text{Morita}}$ equivalent algebra S. The notion of $\sqrt{\text{Morita}}$ equivalence is due to Takeuchi [28]. Two algebras R, S are by definition $\sqrt{\text{Morita}}$ equivalent if we have an equivalence of k-linear monoidal categories ${}_R\mathcal{M}_R \cong {}_S\mathcal{M}_S$. The definition is already at the heart of our application: A $\times_R$-bialgebra can be characterized as having a monoidal category of representations with tensor product based on the tensor product in ${}_R\mathcal{M}_R$. However, for some purposes it does seem that Morita base change (replacing R by a Morita equivalent algebra S) is more well-behaved than the more general $\sqrt{\text{Morita}}$ base change (replacing R by a $\sqrt{\text{Morita}}$ equivalent algebra S): We will show that Morita base change respects duality.

Morita (or $\sqrt{\text{Morita}}$) base change can serve two immediate purposes: One is to produce new examples of quantum groupoids. The other, and perhaps more useful one, is to on the contrary reduce the supply of essentially different examples – we can consider two $\times_R$-bialgebras to be not very essentially different if they are obtained from each other by $\sqrt{\text{Morita}}$ base change. Note that the equivalence relation thus imposed on $\times_R$-bialgebras is weaker than the natural relation that would consider two $\times_R$-bialgebras to be equivalent if their monoidal categories of representations are equivalent. In fact this latter equivalence relation is known to be important and nontrivial also in the realm of ordinary bialgebras, where Morita base change is meaningless. Thus Morita base change presents a possibility of relating different $\times_R$-bialgebras very closely, in a way that cannot occur between ordinary bialgebras.

Let us state very briefly two ways in which $\sqrt{\text{Morita}}$ base change reduces the supply of examples: If R is an Azumaya k-algebra, then any $\times_R$-bialgebra is, up to $\sqrt{\text{Morita}}$ base change, an ordinary bialgebra. Over the field of complex numbers, every weak bialgebra is, up to Morita base change, a face algebra. Of course, in neither case our results show that certain $\times_R$-bialgebras are entirely superfluous, since examples may occur in natural situations that come with a specific choice of R.

The plan of the paper is as follows: After recalling some definitions in Section 2 and Section 3 we present the general $\sqrt{\text{Morita}}$ base change procedure in Section 4. More detailed information on Morita base change will be given in Section 5. In Section 6 we discuss the canonical Tannaka duality of Hayashi [11, 10]; this construction assigns a face algebra F to any finite split semisimple k-linear monoidal category. For example, it assigns such a face algebra to the category of representations of a split semisimple (quasi)Hopf algebra H. At first sight, there is no apparent relation between the original H and Hayashi's F (beyond, of course, the fact that their monoidal representation categories are equivalent). We show that F can be obtained from H in two steps: First, one applies a kind of smash product construction that builds from H a $\times_H$-bialgebra isomorphic to $H \otimes H \otimes H^*$ as a vector space. Next, applying Morita base change to replace the base H by the Morita equivalent product of copies of the field, one obtains a face algebra – which turns out to be Hayashi's face algebra F. In Section 7 we compute the dimension of the face algebra obtained by Morita base change from a certain weak Hopf algebra constructed by Nikshych and Vainerman from a subfactor of a type II_1 factor. It turns out that Morita base change reduces the dimension from 122 to 24 without affecting the monoidal category of representations.

Acknowledgements. The author is indebted to Leonid Vainerman for interesting discussions, and in particular for some help in understanding the example underlying Section 7.

2 Hopf algebroids

In this section we will very briefly recall the necessary definitions and notations on $\times_R$-bialgebras. For more details we refer to [25, 27, 20].

Throughout the paper, k denotes a commutative base ring, and all modules, algebras, unadorned tensor products etc. are understood to be over k

Let R be a k-algebra. We denote the opposite algebra by $\overline{R}$, we let $R \ni r \mapsto \overline{r} \in \overline{R}$ denote the obvious k-algebra antiisomorphism, and abbreviate the enveloping algebra $R^e := R \otimes \overline{R}$. We write $r\overline{s} := r \otimes \overline{s} \in R \otimes \overline{R}$ for $r, s \in R$.

For our purposes, a handy characterization of $\times_R$-bialgebras is the following [20, Thm. 5.1]: A $\times_R$-bialgebra L is an R^e-ring (that is, a k-algebra equipped with a k-algebra map $R^e \to L$, which we write $r \otimes \overline{s} \mapsto r\overline{s}$) for which the category ${}_L\mathcal{M}$ is equipped with a monoidal category structure such that the "underlying" functor ${}_L\mathcal{M} \to {}_{R^e}\mathcal{M}$ is a strict monoidal functor. Here, the monoidal category structure on ${}_{R^e}\mathcal{M}$ is induced via the identification with the category ${}_R\mathcal{M}_R$ of bimodules; we denote tensor product in ${}_{R^e}\mathcal{M}$ by $\diamond_R$, or $\diamond$ if no confusion is likely.

Thus, for two L-modules M, N, there is an L-module structure on $M \diamond_R N$, and this tensor product of L-modules defines a monoidal category structure on ${}_L\mathcal{M}$. The connection with the original definition in [27] is that the module structure on $M \diamond N$ can be described in terms of a certain comultiplication on L, which, however, has a more intricate definition than in the ordinary bialgebra case. First of all, the comultiplication is an algebra map $L \to L \times_R L$ into a certain subset $L \times_R L \subset L \diamond_R L$ which has an algebra structure induced by that of $L \otimes L$, and whose definition we shall now recall.

The notations $\int_r := \int_{r\in R}$ and $\int^r := \int^{r\in R}$, which we will introduce only by example, are due to MacLane, see [25, 27]. For $M, N \in {}_{R^e}\mathcal{M}_{R^e}$ we let

$$\int_r {}_{\bar r}M \otimes {}_rN := M \otimes N \big/ \langle \bar r m \otimes n - m \otimes rn \mid r \in R, m \in M, n \in N\rangle$$

and we let $\int^r M_{\bar r} \otimes N_r \subset M \otimes N$ denote the k-submodule consisting of all elements $\sum m_i \otimes n_i \in M \otimes N$ satisfying $\sum m_i \bar r \otimes n_i = \sum m_i \otimes n_i r$ for all $r \in R$. Note $\int_r {}_{\bar r}M \otimes {}_rN = M \diamond_R N$ for $M, N \in {}_{R^e}\mathcal{M}$.

For two R^e-bimodules M and N we abbreviate

$$M \times_R N := \int^s \int_r {}_{\bar r}M_{\bar s} \otimes {}_rN_s.$$

If M, N are R^e-rings, then so is $M \times_R N$, with R^e-ring structure

$$R^e \ni r \otimes \bar s \mapsto r \otimes \bar s \in M \times_R N,$$

and multiplication given by $\left(\sum m_i \otimes n_i\right)\left(\sum m'_j \otimes n'_j\right) = \sum m_i m'_j \otimes n_i n'_j$.

For $M, N, P \in {}_{R^e}\mathcal{M}_{R^e}$ one defines

$$M \times_R P \times_R N := \int^{s,u} \int_{r,t} {}_{\bar r}M_{\bar s} \otimes {}_{r,\bar t}P_{s,\bar u} \otimes {}_tN_u$$

(where $\int^{s,u} := \int^s \int^u = \int^u \int^s$). There are associativity maps

$$(M \times_R P) \times_R N \xrightarrow{\alpha} M \times_R P \times_R N$$

$$M \times_R (P \times_R N) \xrightarrow{\alpha'} M \times_R P \times_R N$$

given on elements by the obvious formulas (doing nothing), but which need not be isomorphisms. If M, N and P are R^e-rings, so is $M \times_R N \times_R P$, and α, α' are R^e-ring maps.

An R^e-ring structure on the algebra $\mathrm{E} = \mathrm{End}(R)$ is given by $r \otimes \bar s \mapsto (t \mapsto rts)$. We have, for any $M \in {}_{R^e}\mathcal{M}_{R^e}$, two R^e-bimodule maps

$$\theta : M \times_R \mathrm{End}(R) \to M; \qquad m \otimes f \mapsto \overline{f(1)}m$$

$$\theta' : \mathrm{End}(R) \times_R M \to M; \qquad f \otimes m \mapsto f(1)m.$$

which are R^e-ring homomorphisms if M is an R^e-ring.

Now we are prepared to write down the definition of a $\times_R$-bialgebra L. This is by definition an R^e-ring equipped with a comultiplication, a map $\Delta\colon L \to L \times_R L$ of R^e-rings over R^e, and a counit, a map $\varepsilon : L \to \mathrm{E}$ of R^e-rings, such that

$$\alpha(\Delta \times_R L)\Delta = \alpha'(L \times_R \Delta)\Delta\colon L \to L \times_R L \times_R L \tag{2.1}$$

$$\theta(L \times_R \varepsilon)\Delta = \mathrm{id}_L = \theta(\varepsilon \times_R L)\Delta. \tag{2.2}$$

For $\times_R$-bialgebras we will make use of the usual Sweedler notation, writing $\Delta(\ell) =: \ell_{(1)} \otimes \ell_{(2)} \in L \times_R L$.

If L is a $\times_R$-bialgebra, then the module structure on the tensor product $M \diamond_R N$ of $M, N \in {}_L\mathcal{M}$ can be described in terms of the comultiplication of L by the usual formula $\ell(m \otimes n) = \ell_{(1)}m \otimes \ell_{(2)}n$.

As we mentioned above, $\times_R$-bialgebra structures on an R^e-ring L are in bijection with monoidal category structures on ${}_L\mathcal{M}$ for which the underlying functor ${}_L\mathcal{M} \to {}_{R^e}\mathcal{M}$ is strictly monoidal [20, Thm. 5.1]. It is easy to see that any strict monoidal category equivalence ${}_L\mathcal{M} \cong {}_H\mathcal{M}$ for which

$$\begin{array}{ccc} {}_L\mathcal{M} & \longrightarrow & {}_H\mathcal{M} \\ & \searrow \quad \swarrow & \\ & {}_{R^e}\mathcal{M} & \end{array}$$

is a commutative diagram of (strict) monoidal functors is induced by an isomorphism $H \cong L$ of $\times_R$-bialgebras.

The assumptions for the reconstruction of a $\times_R$-bialgebra structure on L from a monoidal category structure on ${}_L\mathcal{M}$ can easily be relaxed as follows [20, Thm. 5.3]: Whenever $({}_L\mathcal{M}, \odot)$ is a monoidal category, and the underlying functor ${}_L\mathcal{M} \to {}_{R^e}\mathcal{M}$ is given the structure of a monoidal functor, we can find a unique $\times_R$-bialgebra structure on L such that

$$\begin{array}{ccc} ({}_L\mathcal{M}, \odot) & \xrightarrow{\mathcal{Id}} & ({}_L\mathcal{M}, \diamond) \\ & \searrow \quad \swarrow & \\ & {}_{R^e}\mathcal{M} & \end{array}$$

is a commutative diagram of monoidal functors, where the left slanted arrow is the given monoidal underlying functor, the right one is the strict monoidal underlying functor, and the monoidal functor structure on the top arrow is determined by commutativity of the diagram. This is done by endowing the tensor product $V \diamond W$ of $V, W \in {}_L\mathcal{M}$ with that L-module structure that makes the monoidal functor structure $V \odot W \cong V \diamond W$ an L-module map. In the same spirit, one can relax the assumptions for the reconstruction of a $\times_R$-bialgebra isomorphism from a monoidal category equivalence:

Lemma 2.1. *Let L, H be two $\times_R$-bialgebras, and assume given a monoidal category equivalence $(\mathcal{T}, \xi)\colon {}_L\mathcal{M} \to {}_H\mathcal{M}$ such that the diagram*

$$\begin{array}{ccc} {}_L\mathcal{M} & \xrightarrow{\mathcal{T}} & {}_H\mathcal{M} \\ & \mathcal{U}\searrow \quad \swarrow\mathcal{U}' & \\ & {}_{R^e}\mathcal{M} & \end{array}$$

of monoidal functors commutes up to monoidal isomorphism. Then $H \cong L$ as $\times_R$-bialgebras.

Proof. We only have to replace $(\mathcal{T}, \xi)$ by an isomorphic strict monoidal functor $\mathcal{T}'$ such that the analogous triangle of strict monoidal functors commutes on the nose. Given a monoidal isomorphism $\phi\colon \mathcal{U} \to \mathcal{U}'\mathcal{T}$, we achieve this by letting $\mathcal{T}'(V)$ be the R^e-module $\mathcal{U}(V)$, equipped with that H-module structure for which $\phi\colon \mathcal{U}(V) \to \mathcal{U}'\mathcal{T}(V)$ is H-linear. Monoidality of ϕ means that the diagrams

$$\begin{array}{ccc}
\mathcal{U}(V \diamond W) & = & \mathcal{U}(V) \diamond \mathcal{U}(W) \\
\downarrow{\scriptstyle \phi} & & \downarrow{\scriptstyle \phi\diamond\phi} \\
\mathcal{U}'\mathcal{F}(V \diamond W) & \xrightarrow{\mathcal{U}'\xi} & \mathcal{U}'\mathcal{T}(V) \diamond \mathcal{U}'\mathcal{T}(W)
\end{array}$$

commute, which tells us that $\mathcal{U}(V \diamond W) = \mathcal{U}(V) \diamond \mathcal{U}(W)$ as H-modules, i.e. $\mathcal{T}'(V \diamond W) = \mathcal{T}'(V) \diamond \mathcal{T}'(W)$. Hence $\mathcal{T}'$ is strict monoidal. □

The suitable definition of comodules over a $\times_R$-bialgebra L is as follows: A left L-comodule is an R-bimodule M together with a map $\lambda\colon M \to L \times_R M$ of R-bimodules such that

$$\alpha'(L \times_R \lambda)\lambda = \alpha(\Delta \times_R M)\lambda\colon M \to L \times_R L \times_R M$$

and $\theta'(\varepsilon \times_R M)\lambda = \mathrm{id}_M$ hold. We will denote by ${}^L\mathcal{M}$ the category of left L-comodules. We will use Sweedler notation in the form $\lambda(m) = m_{(-1)} \otimes m_{(0)}$ and $\alpha(\Delta \times_R M)(m) = m_{(-2)} \otimes m_{(-1)} \otimes m_{(0)}$ for L-comodules.

The category ${}^L\mathcal{M}$ of left L-comodules over a $\times_R$-bialgebra is monoidal. The tensor product of $M, N \in {}^L\mathcal{M}$ is their tensor product $M \otimes_R N$ over R, equipped with the comodule structure

$$\begin{aligned}
M \underset{R}{\otimes} N &\to L \times_R (M \underset{R}{\otimes} N) \\
m \otimes n &\mapsto m_{(-1)}n_{(-1)} \otimes m_{(0)} \otimes n_{(0)}
\end{aligned}$$

In [22, Thm. and Def. 3.5] we have introduced a notion of $\times_R$-Hopf algebra. It is rather different from that of a Hopf algebroid given by Lu [14], although Lu's bialgebroids are the same as $\times_R$-bialgebras. By definition, a $\times_R$-bialgebra is a $\times_R$-Hopf algebra if and only if the map

$$\beta\colon L \underset{\overline{R}}{\otimes} L \ni \ell \otimes m \mapsto \ell_{(1)} \otimes \ell_{(2)} m \in L \diamond L$$

is a bijection. Note that this is equivalent to a well-known characterization of Hopf algebras among ordinary bialgebras. By [22] it is equivalent to saying that the underlying functor ${}_L\mathcal{M} \to {}_{R^e}\mathcal{M}$ preserves inner hom-functors. More precisely, for each $M \in {}_L\mathcal{M}$ the functor ${}_L\mathcal{M} \ni N \mapsto N \diamond M \in {}_L\mathcal{M}$ has a right adjoint $\underline{\mathrm{hom}}(M, -)$. The $\times_R$-bialgebra is a $\times_R$-Hopf algebra if and only if a canonically defined map $\underline{\mathrm{hom}}(M, N) \to \mathrm{Hom}_{\overline{R}-}(M, N)$ is a bijection for all $M, N \in {}_L\mathcal{M}$.

Let us finally recall two special cases of the notion of a $\times_R$-bialgebra. Weak bialgebras and weak Hopf algebras were introduced by Böhm and Szlachányi [2]. We refer to the survey [17] by Nikshych and Vainerman and the literature cited there. It was shown in [5] that weak Hopf algebras are $\times_R$-bialgebras. More details and a converse are in [24]. By definition, a weak bialgebra H is a k-coalgebra and k-algebra such that comultiplication is multiplicative, but not necessarily unit-preserving (and neither is multiplication assumed to be comultiplicative). There are specific axioms replacing the "missing" compatibility axioms for a bialgebra, namely, for $f, g, h \in H$:

$$\varepsilon(fgh) = \varepsilon(fg_{(1)})\varepsilon(g_{(2)}h) = \varepsilon(fg_{(2)})\varepsilon(g_{(1)}h),$$
$$1_{(1)} \otimes 1_{(2)} \otimes 1_{(3)} = 1_{(1)} \otimes 1_{(2)}1'_{(1)} \otimes 1'_{(2)} = 1_{(1)} \otimes 1'_{(1)}1_{(2)} \otimes 1'_{(2)}.$$

If H is a weak bialgebra, then the target counital subalgebra H_t consists by definition of all elements of the form $\varepsilon(1_{(1)}h)1_{(2)}$ with $h \in H$. The source counital subalgebra H_s is the target counital subalgebra in the coopposite of H. It turns out that H_t is a subalgebra which is Frobenius-separable (i.e. a multi-matrix algebra when $k = \mathbb{C}$ is the field of complex numbers), anti-isomorphic to H_s, and that H has the structure of a $\times_R$-bialgebra for $R = H_t$, in which H_s is the image of $\overline{R}$ in H. Moreover, any $\times_R$-bialgebra in which R is Frobenius-separable can be obtained in this way from a weak bialgebra. A weak Hopf algebra is by definition a weak bialgebra H with an antipode, which in turn is an anti-automorphism of H whose axioms we shall not recall. The antipode maps H_t isomorphically onto H_s and vice versa. We have shown in [24] that a weak bialgebra has an antipode if and only if the associated $\times_R$-bialgebra is a $\times_R$-Hopf algebra.

The face algebras introduced earlier by Hayashi [8, 10] are recovered as a yet more special case of weak bialgebras, namely that where the target (and source) counital subalgebra is commutative. In particular, as shown in [21], a face algebra H the same thing as a $\times_R$-bialgebra in which R is commutative and separable. We will only be using the case where the base field is the field of complex numbers, so that R is a direct product of copies of the field. In particular, the images of the minimal idempotents of R in H form a distinguished family of idempotents in H, which feature prominently in Hayashi's original definition (along, of course, with the images of the corresponding idempotents in $\overline{R}$). We shall refer to them as the face idempotents of H; their number, or the dimension of R, is an important structure element of H.

3 Morita- and $\sqrt{\text{Morita}}$-equivalence

Two k-algebras R and S are said to be Morita equivalent (we shall write $R \overset{\mathrm{M}}{\sim} S$ for short) if the categories ${}_R\mathcal{M}$ and ${}_S\mathcal{M}$ of left modules are equivalent as k-linear abelian categories. Let us recall the explicit description of such an equivalence in terms of a strict Morita context. By definition, a Morita context (R, S, P, Q, f, g) consists of two algebras R, S, two bimodules $P \in {}_S\mathcal{M}_R$, $Q \in {}_R\mathcal{M}_S$, and two homomorphisms $f\colon P \otimes_R Q \to S$ of S-bimodules and $g\colon Q \otimes_S P \to R$ of R-bimodules, which we shall write $f(p \otimes q) = pq$, and $g(q \otimes p) = qp$; the data are required to fulfill the additional associativity axioms $(qp)q' = q(pq')$ and $(pq)p' = p(qp')$ for $p, p' \in P$ and $q, q' \in Q$. If $R \overset{\mathrm{M}}{\sim} S$ are Morita equivalent k-algebras, then there is a strict (meaning f, g are isomorphisms) Morita context (R, S, P, Q, f, g) so that the equivalence ${}_R\mathcal{M} \cong {}_S\mathcal{M}$ assumed to exist by definition of Morita equivalence is isomorphic to $\mathcal{F}\colon {}_R\mathcal{M} \ni M \mapsto P \otimes_R M \in {}_S\mathcal{M}$; a quasiinverse $\mathcal{F}^{-1}$ for $\mathcal{F}$ can be given by $\mathcal{F}^{-1}(N) = Q \otimes_S N$ for $N \in {}_S\mathcal{M}$, the relevant isomorphisms $\mathcal{F}\mathcal{F}^{-1} \cong \mathcal{Id}$ and $\mathcal{F}^{-1}\mathcal{F} \cong \mathcal{Id}$ are $P \otimes_R Q \otimes_S N \xrightarrow{f \otimes_S N} N$ for $N \in {}_S\mathcal{M}$ and $Q \otimes_S P \otimes_R M \xrightarrow{g \otimes_R M} M$ for $M \in {}_R\mathcal{M}$.

Whenever $R \overset{\mathrm{M}}{\sim} S$, we also get an equivalence $({}_R\mathcal{M}_R, \otimes_R) \cong ({}_S\mathcal{M}_S, \otimes_S)$ of k-linear monoidal categories. This can be seen by applying Watts' theorem [29], which says that the monoidal category ${}_R\mathcal{M}_R$ can be viewed as the category of right exact k-linear endofunctors of ${}_R\mathcal{M}$. It will be somewhat more useful to describe the monoidal equivalence more explicitly in terms of a Morita context (R, S, P, Q, f, g) as above; we get an equivalence of bimodule categories

$$(\hat{\mathcal{F}}, \xi)\colon ({}_R\mathcal{M}_R, \underset{R}{\otimes}) \to ({}_S\mathcal{M}_S, \underset{S}{\otimes})$$

as follows: We set $\hat{\mathcal{F}}(M) = P \otimes_R M \otimes_R Q$ as S-bimodules, and we define the monoidal functor structure

$$\xi\colon \mathcal{F}(M) \underset{S}{\otimes} \mathcal{F}(N) \to \mathcal{F}(M \underset{R}{\otimes} N)$$

as the composition

$$\begin{aligned} P \underset{R}{\otimes} M \underset{R}{\otimes} Q \underset{S}{\otimes} P \underset{R}{\otimes} N \underset{R}{\otimes} Q \xrightarrow{P \otimes M \otimes g \otimes N \otimes Q} & P \underset{R}{\otimes} M \underset{R}{\otimes} R \underset{R}{\otimes} N \underset{R}{\otimes} Q \\ & \cong P \underset{R}{\otimes} M \underset{R}{\otimes} N \underset{R}{\otimes} Q \end{aligned}$$

It is useful to know that the equivalence $\mathcal{F}$ and the monoidal equivalence $\hat{\mathcal{F}}$ are compatible in the following sense: The category ${}_R\mathcal{M}$ is in a natural way a left ${}_R\mathcal{M}_R$-category in the sense of Pareigis [19], that is, a category on which ${}_R\mathcal{M}_R$ acts (by tensor product). The compatibility says that the following diagram commutes up to coherent natural isomorphisms:

$$\begin{array}{ccc} {}_R\mathcal{M}_R \times {}_R\mathcal{M} & \xrightarrow{\otimes_R} & {}_R\mathcal{M} \\ \downarrow{\scriptstyle \hat{\mathcal{F}} \times \mathcal{F}} & & \downarrow{\scriptstyle \mathcal{F}} \\ {}_S\mathcal{M}_S \times {}_S\mathcal{M} & \xrightarrow{\otimes_S} & {}_S\mathcal{M} \end{array}$$

Takeuchi [28] has introduced and investigated the notion of $\sqrt{\text{Morita}}$-equivalence of k-algebras; by his definition, two k-algebras R, S are $\sqrt{\text{Morita}}$-equivalent, written $R \overset{\sqrt{\mathrm{M}}}{\sim} S$, if there is an equivalence of k-linear monoidal categories ${}_R\mathcal{M}_R \cong {}_S\mathcal{M}_S$. By the above, Morita equivalence clearly implies $\sqrt{\text{Morita}}$-equivalence. On the other hand, since ${}_R\mathcal{M}_R \cong {}_{R^e}\mathcal{M}$, the enveloping algebras of $\sqrt{\text{Morita}}$-equvialent algebras are Morita equivalent, so that

$$R \overset{\mathrm{M}}{\sim} S \quad \Rightarrow \quad R \overset{\sqrt{\mathrm{M}}}{\sim} S \quad \Rightarrow \quad R^e \overset{\mathrm{M}}{\sim} S^e.$$

Neither of the reverse implications holds.

Note that a bimodule $M \in {}_R\mathcal{M}_R$ has a left dual object in the monoidal category $({}_R\mathcal{M}_R, \otimes_R)$ if and only if it is finitely generated projective as a right R-module. It follows that any equivalence of monoidal categories ${}_R\mathcal{M}_R \cong {}_S\mathcal{M}_S$ maps left (or right) finitely generated projective modules to left (or right) finitely generated projective modules. This reduces to a standard fact on projective modules, if the equivalence comes from a Morita equivalence $R \overset{\mathrm{M}}{\sim} S$, for then it maps $M \in {}_R\mathcal{M}_R$ to $P \otimes_R M \otimes_R Q$, which is finitely generated projective as left S-module if ${}_RM$ is finitely generated projective, for ${}_SP$ and ${}_RQ$ are finitely generated projective.

4 A $\sqrt{\text{Morita}}$-base change principle

An A-ring L for a k-algebra A is an algebra in the monoidal category of A-bimodules. As an immediate consequence, if $A \overset{\mathrm{M}}{\sim} B$, then an A-ring is essentially the same as a B-ring. Moreover, if $\widetilde{L}$ is the B-ring corresponding to the A-ring L, then $\widetilde{L}$-modules are essentially the same as L-modules, since the actions of ${}_B\mathcal{M}_B$ on ${}_B\mathcal{M}$ and of ${}_A\mathcal{M}_A$ on ${}_A\mathcal{M}$ are compatible with the equivalences. Thus we have:

Lemma 4.1. *Let L be an A-ring over the k-algebra A. Assume given a k-algebra B and a strict Morita context (A, B, C, D, ϕ, ψ). Then there is a B-ring $\widetilde{L}$, and a category equivalence $\mathcal{G}\colon {}_L\mathcal{M} \to {}_{\widetilde{L}}\mathcal{M}$ lifting the equivalence $\mathcal{F}\colon {}_A\mathcal{M} \to {}_B\mathcal{M}$ given by tensoring with C, as in the following diagram:*

$$\begin{array}{ccc} {}_L\mathcal{M} & \xrightarrow{\mathcal{G}} & {}_{\widetilde{L}}\mathcal{M} \\ \downarrow & & \downarrow \\ {}_A\mathcal{M} & \xrightarrow{\mathcal{F}} & {}_B\mathcal{M} \end{array}$$

in which the vertical arrows are the underlying functors induced by the A-ring structure of L, and the B-ring structure of $\widetilde{L}$, respectively.

Explicitly, the B-ring $\widetilde{L}$ is given by $\widetilde{L} := C \otimes_A L \otimes_A D$ with unit map

$$B \cong C \underset{A}{\otimes} D \xrightarrow{C \otimes_A \eta \otimes_A D} C \underset{A}{\otimes} L \underset{A}{\otimes} D$$

in which η is the unit map of the A-ring L, and multiplication map

$$\begin{aligned}\widetilde{L}\underset{B}{\otimes}\widetilde{L}&=C\underset{A}{\otimes}L\underset{A}{\otimes}D\underset{B}{\otimes}C\underset{A}{\otimes}L\underset{A}{\otimes}D\\&\cong C\underset{A}{\otimes}L\underset{A}{\otimes}L\underset{A}{\otimes}D\xrightarrow{C\otimes_A\nabla\otimes_A D}C\underset{A}{\otimes}L\underset{A}{\otimes}D.\end{aligned}$$

When M is a L-module, then the $\widetilde{L}$-module $\mathcal{F}(M)$ is $C\otimes_A M$ equipped with the $\widetilde{L}$-module structure

$$\widetilde{L}\underset{B}{\otimes}\mathcal{F}(M)=C\underset{A}{\otimes}L\underset{A}{\otimes}D\underset{B}{\otimes}C\underset{A}{\otimes}M=C\underset{A}{\otimes}L\underset{A}{\otimes}M\xrightarrow{C\otimes_A\mu_M}C\underset{A}{\otimes}M=\mathcal{F}(M).$$

Remark 4.2. Assume that the algebras involved in the situation above are multi-matrix algebras over a field k.

Recall that the inclusion $A\subset L$ can then be described in terms of its inclusion matrix, which records, for any simple L-module and simple A-module, the number of times that the latter occurs in the decomposition of the former as an A-module. It should be obvious from the construction in the lemma that the inclusion matrix of $B\subset\widetilde{L}$ is the same as that of $A\subset L$.

The Bratteli diagram is a convenient way of graphically depicting the information contained in the inclusion matrix. The Bratteli diagram for $A\subset L$ is a graph with vertices on two (top and bottom) levels called its floors: The top (resp. bottom) level has a vertex for each full matrix ring in the decomposition of L (resp. A); we label it with the rank of that component. The numbers of edges or links between the vertices on the two floors are the relevant entries in the inclusion matrix. Note that the number on a top floor vertex has to be the sum of the numbers on the bottom floor vertices linked to it, weighted by the respective number of links.

We see that the Bratteli diagram of the inclusion $B\subset\widetilde{L}$ is the same as that of the inclusion $A\subset L$, except that the ranks of the components of A have to be replaced by the ranks of the components of B, and the top floor representing $\widetilde{L}$ has to be adjusted accordingly.

For more details on the theory of inclusion matrices and Bratteli diagrams we refer the reader to [6]. We shall use Bratteli diagrams in Section 7, and note here already that they can be conveniently stacked to describe towers of multi-matrix algebras; we will then refer to the Bratteli diagram describing one of the inclusions as a story of the total diagram describing the tower.

Theorem 4.3. *Let L be a $\times_R$-bialgebra for a k-algebra R. Let S be a k-algebra which is $\sqrt{\text{Morita}}$-equivalent to R. Then there is a $\times_S$-bialgebra $\widetilde{L}$ whose module category is equivalent to that of L, as a monoidal category.*

More precisely, assume given a monoidal category equivalence $\mathcal{F}:{}_{R^e}\mathcal{M}\to{}_{S^e}\mathcal{M}$. Then there is a $\times_S$-bialgebra $\widetilde{L}$ and a monoidal category equivalence $\mathcal{G}:{}_L\mathcal{M}\to{}_{\widetilde{L}}\mathcal{M}$

making

$$\begin{array}{ccc} {}_L\mathcal{M} & \xrightarrow{\mathcal{G}} & {}_{\widetilde{L}}\mathcal{M} \\ \Big\downarrow{\scriptstyle \mathcal{U}} & & \Big\downarrow{\scriptstyle \widetilde{\mathcal{U}}} \\ {}_{R^e}\mathcal{M} & \xrightarrow{\mathcal{F}} & {}_{S^e}\mathcal{M} \end{array} \tag{4.1}$$

a commutative diagram of monoidal functors (in which the vertical arrows are underlying functors). The $\times_S$*-bialgebra* $\widetilde{L}$ *is unique up to isomorphism.*

Proof. We know already that there is an S^e-ring $\widetilde{L}$ and a category equivalence $\mathcal{G}$ making the square in the theorem a commutative diagram of k-linear functors. We can endow ${}_{\widetilde{L}}\mathcal{M}$ with a monoidal category structure such that $\mathcal{G}$ is a monoidal functor. Then the underlying functor $\widetilde{\mathcal{U}}$ is monoidal as well, since it can be written as the composition $\widetilde{\mathcal{U}} = \mathcal{F}\mathcal{U}\mathcal{G}^{-1}$. Now [20, Thm. 5.1, Rem. 5.3] imply that there exists a unique $\times_S$-bialgebra structure on $\widetilde{L}$ inducing the given monoidal category structure on ${}_{\widetilde{L}}\mathcal{M}$. □

Definition 4.4. Let L be a $\times_R$-bialgebra, and S a k-algebra $\sqrt{\text{Morita}}$-equivalent to R. We will say that the $\times_S$-bialgebra $\widetilde{L}$ obtained from L as in the proof of Theorem 4.3 is obtained from L by a $\sqrt{\text{Morita}}$ base change.

Thus a $\times_S$-bialgebra obtained from a $\times_R$-bialgebra L by a $\sqrt{\text{Morita}}$-base change has the same monoidal representation category as L itself. The difference is "merely" in a change of the base algebra.

We will be somewhat sloppy in our terminology: Given a $\times_R$-bialgebra L and $S \overset{\sqrt{\text{M}}}{\sim} R$, we will speak of *the* $\times_S$-bialgebra $\widetilde{L}$ obtained from L by $\sqrt{\text{Morita}}$ base change. This suppresses the choice of a monoidal category equivalence ${}_R\mathcal{M}_R \cong {}_S\mathcal{M}_S$, by which (and not by S alone) $\widetilde{L}$ is determined.

Corollary 4.5. *Let* L *be a* $\times_R$*-bialgebra, where* R *is an Azumaya* k*-algebra. Then* L *can be obtained by* $\sqrt{\text{Morita}}$*-base change from an ordinary bialgebra* H.

Proof. Takeuchi [28, Ex. 2.4] has observed that the algebra R is Azumaya if and only if $R \overset{\sqrt{\text{M}}}{\sim} k$. □

Takeuchi also gives the following description of the monoidal category equivalence ${}_R\mathcal{M}_R \to \mathcal{M}_k$: It maps M to the centralizer M^R of R in M, whereas its inverse maps $V \in \mathcal{M}_k$ to $V \otimes R$ with the obvious R-bimodule structure. Thus, the ordinary k-bialgebra associated to a $\times_R$-bialgebra L is

$$H = \{\ell \in L \mid \forall r \in R\colon r\ell = \overline{r}\ell \wedge \ell r = \ell\overline{r}\}$$

whereas L can be obtained from H by merely tensoring with two copies of R, one of which gives the left R^e-module structure, and the other one the right R^e-module structure of $R \otimes L \otimes R$.

In a sense our corollary says that examples of $\times_R$-bialgebras in which R is Azumaya are irrelevant; they are just versions of ordinary bialgebras in which the base ring is enlarged, without affecting the representation theory. That would also apply to examples like that considered by Kadison in [12,Thm. 5.2]. In fact in the example of a $\times_R$-bialgebra T there, R is Azumaya over its center Z. Moreover, the two algebra maps $Z \to T$ coming from the maps $R \to T$ and $\overline{R} \to H$ coincide by construction and have central image. Thus T can be considered as a $\times_R$-bialgebra for the Z-algebra R; since this is Azumaya, Corollary 4.5 applies, so that T can be obtained by $\sqrt{\text{Morita}}$ base change from a Z-bialgebra. However, we should rush to concede that Corollary 4.5 does of course not rule out that interesting examples of $\times_R$-bialgebras over Azumaya k-algebras arise naturally. In fact Kadison's example is constructed from a natural situation that comes with a natural choice of base R. Moreover, the example gives us the opportunity to point out a certain subtlety about $\sqrt{\text{Morita}}$ base change: The $\times_R$-bialgebra T in [12, Sec. 4] occurs in duality with another $\times_R$-bialgebra S, which can also be considered as a $\times_R$-bialgebra for the Z-algebra R. Thus, if R is Azumaya over Z, then S can be reduced by $\sqrt{\text{Morita}}$ base change to a Z-bialgebra S', while T can be replaced by a Z-bialgebra T'. However, we do not have any indication that S' and T' are still dual to each other. It is conceivable that the duality only shows over the ring R. We will show below that Morita base change is compatible with duality.

Closing the section, let us show that $\sqrt{\text{Morita}}$ base change preserves the property of a $\times_R$-bialgebra of being a $\times_R$-Hopf algebra in the sense of [22]:

Proposition 4.6. *Let L be a $\times_R$-bialgebra, $S \overset{\sqrt{M}}{\sim} R$, and let $\widetilde{L}$ be the $\times_S$-bialgebra obtained from L by $\sqrt{\text{Morita}}$ base change. Then $\widetilde{L}$ is a $\times_S$-Hopf algebra if and only if L is a $\times_R$-Hopf algebra.*

Proof. In the diagram (4.1), the horizontal functors are monoidal equivalences, hence preserve inner hom-functors. Thus the left hand vertical functor preserves inner hom-functors if and only the right hand one does. □

5 Morita base change

Morita equivalence implies $\sqrt{\text{Morita}}$ equivalence. Thus, given a $\times_R$-bialgebra L and a k-algebra S Morita equivalent to R, we can apply $\sqrt{\text{Morita}}$ base change (which, of course, we shall call Morita base change in this case) to L to obtain a $\times_S$-bialgebra $\widetilde{L}$ with equivalent monoidal module category.

5.1 Morita base change – explicitly

In order to find out what the result looks like more explicitly, we fix a Morita context (R, S, P, Q, f, g). We will write $f(p \otimes q) = pq$, $g(q \otimes p) = qp$, $f^{-1}(1_S) =$

$p_i \otimes q^i \in P \otimes_R Q$ and $g^{-1}(1_R) = q_i \otimes p^i \in Q \otimes_S P$ (with a summation over upper and lower indices understood). Write $\overline{P} \in {}_{\overline{R}}\mathcal{M}_{\overline{S}}$ for the bimodule opposite to P, and $\overline{p}$ with $p \in P$ for a typical element; similarly for $\overline{Q} \in {}_{\overline{S}}\mathcal{M}_{\overline{R}}$. Somewhat dangerously we write $P^e := P \otimes \overline{Q} \in {}_{S^e}\mathcal{M}_{R^e}$ and $Q^e := Q \otimes \overline{P} \in {}_{R^e}\mathcal{M}_{S^e}$, so that the bimodules P^e and Q^e induce the equivalence ${}_{R^e}\mathcal{M} \cong {}_{S^e}\mathcal{M}$ underlying the $\sqrt{\text{Morita}}$ equivalence between R and S induced by the Morita equivalence between R and S.

To keep our formulas a manageable size, we will write $p\overline{q} := p \otimes \overline{q} \in P \otimes \overline{Q} = P^e$, and similarly for the typical elements of Q^e.

Now let L be a $\times_R$-bialgebra. The $\times_S$-bialgebra $\widetilde{L}$ obtained from L by Morita base change has underlying S^e-bimodule $P^e \otimes_{R^e} L \otimes_{R^e} Q^e$. The equivalence ${}_L\mathcal{M} \cong {}_{\widetilde{L}}\mathcal{M}$ sends $M \in {}_L\mathcal{M}$ to $P^e \otimes_{R^e} M$, with the $\widetilde{L}$-module structure given by

$$(p_1\overline{q_1} \otimes \ell \otimes q_2\overline{p_2})(p_3\overline{q_3} \otimes m) = p_1\overline{q_1} \otimes \ell(q_2p_3)(\overline{q_3p_2})m$$

for $p_1, p_2, p_3 \in P$ and $q_1, q_2, q_3 \in Q$. The monoidal functor structure of the equivalence is given by

$$\begin{aligned}
\xi \colon (P^e \underset{R^e}{\otimes} M) \underset{S}{\diamond} (P^e \underset{R^e}{\otimes} N) &\rightleftarrows P^e \underset{R^e}{\otimes} (M \underset{R}{\diamond} N) \\
p_1\overline{q_1} \otimes m \otimes p_2\overline{q_2} \otimes n &\mapsto p_1\overline{q_2} \otimes m \otimes (q_1p_2)n \\
&= p_1\overline{q_{(2)}} \otimes \overline{q_1p_2}m \otimes n \\
p\overline{q_i} \otimes m \otimes p^i\overline{q} &\leftarrow p\overline{q} \otimes m \otimes n
\end{aligned}$$

for $M, N \in {}_L\mathcal{M}, m \in M, n \in N, p_1, p_2 \in P, q_1, q_2 \in Q$.

It follows that the $\widetilde{L}$-module structure of the tensor product of two $\widetilde{L}$-modules coming via the equivalence from L-modules M, N can be computed as the composition

$$\begin{aligned}
&(P^e \underset{R^e}{\otimes} L \underset{R^e}{\otimes} Q^e) \underset{S^e}{\otimes} ((P^e \underset{R^e}{\otimes} M) \underset{S}{\diamond} (P^e \overset{e}{\underset{R}{\otimes}} N)) \\
&\xrightarrow{\mathrm{id}\otimes\xi} (P^e \underset{R^e}{\otimes} L \underset{R^e}{\otimes} Q^e) \underset{S^e}{\otimes} (P^e \underset{R^e}{\otimes} (M \underset{R}{\diamond} N)) \\
&\xrightarrow{\mu} P^e \underset{R^e}{\otimes} (M \underset{R}{\diamond} N) \xrightarrow{\xi^{-1}} (P^e \underset{R^e}{\otimes} M) \underset{S}{\diamond} (P^e \underset{R^e}{\otimes} N)
\end{aligned}$$

hence is given by

$$\begin{aligned}
&(p_1\overline{q_1} \otimes \ell \otimes q_2\overline{p_2})(p_3\overline{q_3} \otimes m \otimes p_4\overline{q_4} \otimes n) \\
&= \xi^{-1}((p_1\overline{q_1} \otimes \ell \otimes q_2\overline{p_2})(p_3\overline{q_4} \otimes m \otimes (q_3p_4)n) \\
&= \xi^{-1}(p_1\overline{q_1} \otimes \ell(q_2p_3)(\overline{q_4p_2})(m \otimes (q_3p_4)n)) \\
&= \xi^{-1}(p_1\overline{q_1} \otimes \ell_{(1)}(q_2p_3)m \otimes \ell_2(\overline{q_4p_2})(q_3p_4)n) \\
&= p_1\overline{q_i} \otimes \ell_{(1)}(q_2p_3)m \otimes p^i\overline{q_1} \otimes \ell_2(\overline{q_4p_2})(q_3p_4)n =
\end{aligned}$$

$$= p_1\overline{q_i} \otimes \ell_{(1)}(q_2p_3)m \otimes p^i\overline{q_1} \otimes \ell_2(q_3p_j)(q^jp_4)(\overline{q_4p_2})n$$
$$= p_1\overline{q_i} \otimes \ell_{(1)}(q_2p_3)(\overline{q_3p_j})m \otimes p^i\overline{q_1} \otimes \ell_{(2)}(q^jp_4)(\overline{q_4p_2})n$$
$$= (p_1\overline{q_i} \otimes \ell_{(1)} \otimes q_2\overline{p_j})(p_3\overline{q_3} \otimes m) \otimes (p^i\overline{q_1} \otimes \ell_{(2)} \otimes q^j\overline{p_2})(p_4\overline{q_4} \otimes n)$$

for $p_1, \dots, p_4 \in P, q_1, \dots, q_4 \in Q, \ell \in L, m \in M$, and $n \in N$. this proves that the comultiplication in $\widetilde{L}$ is given by the formula

$$\Delta(p_1\overline{q_1} \otimes \ell \otimes q_2\overline{p_2}) = (p_1\overline{q_i} \otimes \ell_{(1)} \otimes q_2\overline{p_j}) \otimes (p^i\overline{q_1} \otimes \ell_{(2)} \otimes q^j\overline{p_2})$$

for $p_1, p_2 \in P$ and $q_1, q_2 \in Q$.

5.2 Weak bialgebras versus face algebras

Let us be yet more concrete for the case that R is a multi-matrix algebra $R = \bigoplus_{\alpha=1}^n M_{d_\alpha}(k)$, and $S = k^n$. A Morita context (R, S, P, Q, f, g) can be given as follows: P is generated as a right R-module by one element p which is a sum $p = \sum_{\alpha=1}^n E_{11}^{(\alpha)}$ of minimal idempotents (where we have denoted the matrix units in the α-th component by $E_{ij}^{(\alpha)}$). Q is generated as left R-module by the same element p. both maps f, g are given by matrix multiplication. We have $f^{-1}(1_S) = p \otimes p$, and $g^{-1}(1_R) = \sum_{\alpha=1}^n \sum_{i=1}^{d_\alpha} E_{i1}^{(\alpha)} \otimes E_{1i}^{(\alpha)}$. Let L be a $\times_R$-bialgebra, and $\widetilde{L}$ the $\times_S$-bialgebra obtained from it by Morita base change. Then $\widetilde{L} = p\overline{p}Lp\overline{p} \subset L$, with multiplication given by multiplication in L, unit $p\overline{p}$, and comultiplication

$$\Delta(p\overline{p}\ell p\overline{p}) = p\overline{E_{i1}^{(\alpha)}}\ell_{(1)}p\overline{p} \otimes E_{1i}^{(\alpha)}\overline{p}\ell_{(2)}p\overline{p}.$$

Now let $k = \mathbb{C}$ be the field of complex numbers. Then a $\times_R$-bialgebra for a multi-matrix algebra R is the same as a weak bialgebra in the sense of Böhm and Szlachányi [2, 1]. If R is commutative, then this is in turn the same thing as a face algebra in the sense of Hayashi [10]. Thus Morita base change says that Hayashi's face algebras are a sufficiently general case of weak bialgebras, at least as long as we are interested in the respective module categories:

Corollary 5.1. *Let H be a weak bialgebra over the field of complex numbers.*

Then H can be obtained by Morita base change from a weak bialgebra whose source counital subalgebra is commutative.

In particular, there is a face algebra F and a monoidal category equivalence ${}_H\mathcal{M} \cong {}_F\mathcal{M}$.

H is a weak Hopf algebra if and only if F is a face Hopf algebra.

Remark 5.2. For the case of semisimple H, it follows in fact from Hayashi's canonical Tannaka duality [10, 11] that there is a face algebra F and a monoidal category equivalence ${}_H\mathcal{M} \cong {}_F\mathcal{M}$. The corollary above shows the same for non-semisimple H, but it is also a different result in the semisimple case: A peculiar feature of Hayashi's canonical Tannaka duality (on which we will give more details in Section 6) is that it yields semisimple face algebras with the same number of face idempotents and irreducible representations. This clearly needs not be the case for the face algebras obtained by Morita base change. A trivial example is the trivial Morita base change applied to an ordinary Hopf algebra, which leaves us with the same Hopf algebra, or only one face idempotent.

Hayashi's canonical Tannaka duality proceeds in two steps: Given a semisimple monoidal category $\mathcal{C}$ with n simples, like ${}_H\mathcal{M}$ above, one first constructs a faithful k-linear exact monoidal functor $\mathcal{C} \to {}_R\mathcal{M}_R$ for $R = k^n$, and from this fiber functor one reconstructs a face algebra.

Ostrik [18] describes a more general way to construct monoidal functors $\mathcal{C} \to {}_R\mathcal{M}_R$ from a semisimple category $\mathcal{C}$ to a bimodule category over a multi-matrix algebra, based on actions of $\mathcal{C}$ on the module category ${}_R\mathcal{M}$. If one can show that $\mathcal{C} \to {}_R\mathcal{M}_R$ factors over an equivalence $\mathcal{C} \cong {}_A\mathcal{M}$ for some algebra A, a lemma of Szláchanyi [26] then shows that A is a weak bialgebra (alternatively, one may of course use [20, Thm. 5.1] again to obtain a bialgebroid structure on A); this accounts for the general case of Morita base change for semisimple weak Hopf algebras.

5.3 Duality

There is a well-behaved notion of duality for $\times_R$-bialgebras, developed in [22], and shown to be compatible with the duality for weak bialgebras in [24]. The main difficulty in the definitions is to sort out how the four module structures in a $\times_R$-bialgebra should be translated through the duality, and to check that the formulas defining the dual structures are well defined with respect to the various tensor products over R. A specialty is that one can define the $\times_R$-bialgebra analog of the opposite or coopposite of the dual of an ordinary bialgebra, but not the direct analog of the dual (unless one wants to allow two versions of "left" and "right" bialgebroids like Kadison and Szlachányi [13]). More generally, one defines [22, Def. 5.1] a skew pairing between two $\times_R$-bialgebras Λ and L to be a k-linear map $\tau\colon \Lambda \otimes L \to R$ satisfying

$$\tau((r \otimes \overline{s})\xi(t \otimes \overline{u})|\ell)v = r\tau(\xi|(t \otimes \overline{v})\ell(u \otimes \overline{s})), \tag{5.1}$$

$$\tau(\xi|\ell m) = \tau(\overline{\tau(\xi_{(2)}|m)}\xi_{(1)}|\ell), \qquad \tau(\xi|1) = \varepsilon(\xi)(1), \tag{5.2}$$

$$\tau(\xi\zeta|\ell) = \tau(\xi|\tau(\zeta|\ell_{(1)})\ell_{(2)}), \qquad \tau(1|\ell) = \varepsilon(\ell)(1) \tag{5.3}$$

for all $\xi, \zeta \in \Lambda$, $\ell, m \in L$, $r, s, t, u, v \in R$.

Proposition 5.3. *Let $\tau\colon \Lambda \otimes L \to R$ be a skew pairing between $\times_R$-bialgebras Λ and L. Let $S \overset{M}{\sim} R$, and let $\widetilde{\Lambda}, \widetilde{L}$ be the $\times_S$-bialgebras obtained from Λ and L by Morita base change.*

Then a skew pairing $\widetilde{\tau}\colon \widetilde{\Lambda}\otimes\widetilde{L}\to S$ can be defined by

$$\widetilde{\tau}(p_1\otimes\overline{q_1}\otimes\xi\otimes q_2\otimes\overline{p_2}|p_3\otimes\overline{q_3}\otimes\ell\otimes q_4\otimes\overline{p_4}) = p_1\tau(\overline{q_1p_4}\xi(q_2p_3)(\overline{q_4p_2})|\ell)q_3$$
$$= p_1\tau(\xi|(q_2p_3)\ell(q_4p_2)(\overline{q_1p_4}))q_3$$

for $p_1,\dots,p_4\in P$, $q_1,\dots,q_4\in Q$, $\xi\in\Lambda$, and $\ell\in L$.

Proof. We omit checking (5.1) for $\widetilde{\tau}$. Now let $p_1,\dots,p_6\in P$, $q_1,\dots,q_6\in Q$, $\xi\in\Lambda$, and $\ell,m\in L$. Then

$$\begin{aligned}
&\widetilde{\tau}((p_1\overline{q_1}\otimes\xi\otimes q_2\overline{p_2})_{(1)}|p_3\overline{q_3}\otimes\ell\otimes q_4\overline{p_4}\overline{\widetilde{\tau}((p_1\overline{q_1}\otimes\xi\otimes q_2\overline{p_2})_{(2)}|p_5\overline{q_5}\otimes m\otimes q_6\overline{p_6})})\\
&\quad=\widetilde{\tau}(p_1\overline{q_i}\otimes\xi_{(1)}\otimes q_2\overline{p_j}|p_3\overline{q_3}\otimes\ell\otimes q_4\overline{p_4}\overline{\widetilde{\tau}(p^i\overline{q_1}\otimes\xi_{(2)}\otimes q^j\overline{p_2}|p_5\overline{q_5}\otimes m\otimes q_6\overline{p_6})})\\
&\quad=\widetilde{\tau}(p_1\overline{q_i}\otimes\xi_{(1)}\otimes q_2\overline{p_j}|p_3\overline{q_3}\otimes\ell\otimes q_4\overline{p_4}\overline{p^i\tau(\overline{q_1p_6}\xi_{(2)}(q^jp_5)(\overline{q_6p_2})|m)q_5})\\
&\quad=\widetilde{\tau}(p_1\overline{q_i}\xi_{(1)}q_2\overline{p_5}|p_3\overline{q_3}\ell(\overline{q_5p_4})\overline{\tau(\overline{q_1p_6}\xi_{(2)}(\overline{q_6p_2})|m)}q_4\overline{p^i})\\
&\quad=p_1\tau(\overline{q_ip^i}\xi_{(1)}(q_2p_3)(\overline{q_4p_5})|\ell(\overline{q_5p_4})\overline{\tau(\overline{q_1p_6}\xi_{(2)}\overline{q_6p_2}|m)})q_3\\
&\quad==p_1\tau(\xi_{(1)}(q_2p_3)|\ell(q_4p_5)\overline{\tau(\overline{q_1p_6}\xi_{(2)}\overline{q_6p_2}|\overline{q_5p_4}m)})q_3\\
&\quad=p_1\tau(\overline{q_1p_6}\xi(q_2p_3)(\overline{q_6p_2})|\ell(q_4p_5)(\overline{q_5p_4})m)q_3\\
&\quad==\widetilde{\tau}(p_1\overline{q_1}\otimes\xi\otimes q_2\overline{p_2}|p_3\overline{q_3}\otimes\ell(q_4p_5)(\overline{q_5p_4})m\otimes q_6\overline{p_6})\\
&\quad=\widetilde{\tau}(p_1\overline{q_1}\otimes\xi\otimes q_2\overline{p_2}|(p_3\overline{q_3}\otimes\ell\otimes q_3\overline{p_4})(p_5\overline{q_5}\otimes m\otimes q_6\overline{p_6}))
\end{aligned}$$

proves (5.2) for $\widetilde{\tau}$ (we omit treating the second part).

The proof for (5.3) is similar. □

A skew pairing $\tau\colon\Lambda\otimes L\to R$ induces a map $\phi\colon\Lambda\to\operatorname{Hom}_{\overline{R}-}(L,R)$. There is an R^e-ring structure [22, Lem.5.5] on $L^\vee:=\operatorname{Hom}_{\overline{R}-}(L,R)$ for which ϕ is a morphism of R^e-rings. In particular, the induced R^e-bimodule structure [22, Def.5.4] satisfies $(r\overline{s}\xi t\overline{u})(\ell)=r\xi(t\ell u\overline{s})$ for $r,s,t,u\in R$, $\xi\in L^\vee$, and $\ell\in L$.

If L is finitely generated projective as left $\overline{R}$-module, then $L^\vee$ has a $\times_R$-bialgebra structure [22, Thm.5.12] such that evaluation defines a skew pairing between $L^\vee$ and L. We call this $\times_R$-bialgebra the left dual of L.

Proposition 5.4. *Let L be a $\times_R$-bialgebra that is finitely generated projective as left $\overline{R}$-module. Let $S\overset{M}{\sim}R$, and let $\widetilde{L}$ be the $\times_S$-bialgebra obtained from L by Morita base change. Then $\widetilde{L}$ is finitely generated projective as left $\overline{S}$-module, and its left dual $\times_S$-bialgebra $\widetilde{L}^\vee$ is isomorphic to the $\times_S$-bialgebra $\widetilde{L^\vee}$ that is obtained from the left dual $L^\vee$ of L by Morita base change.*

Proof. $\widetilde{L}=P^e\otimes_{R^e}L\otimes_{R^e}Q^e$ is finitely generated projective as left $\overline{S}$-module since the modules P_R, ${}_{\overline{S}}\overline{Q}$, ${}_{\overline{R}}L$, and ${}_{R^e}Q^e$ are finitely generated projective.

Since the R-modules P and Q are finitely generated projective and each other's dual, we have $\operatorname{Hom}(P\otimes_R M,V)\cong\operatorname{Hom}(M,V)\otimes_R Q$ for any $M\in{}_R\mathcal{M}$ and k-module V, and similarly $\operatorname{Hom}(N\otimes_R Q,V)\cong P\otimes_R\operatorname{Hom}(N,V)$. We use this three times in the second isomorphism in the following calculation. The first isomorphism uses the category equivalence ${}_{\overline{R}}\mathcal{M} \cong {}_{\overline{S}}\mathcal{M}$ given by tensoring with $\overline{Q}$. The fourth isomorphism is an instance of the general isomorphism $\mathrm{Hom}_{R-}(M,P) \cong \mathrm{Hom}_{R-}(M,R) \otimes_R P$, in the third we have used that $\int^r$ commutes with tensor products by flat (in particular by projective) modules. The last step merely replaces the rightmost P by $\overline{P}$.

$$\begin{aligned}
\mathrm{Hom}_{\overline{S}-}(\widetilde{L}, S) &= \mathrm{Hom}_{\overline{S}-}(P^e \underset{R^e}{\otimes} L \underset{R^e}{\otimes} Q^e, P \underset{R}{\otimes} Q)\\
&\cong \int^r \mathrm{Hom}(P \underset{R}{\otimes} {}_{\overline{r}}L \underset{R^e}{\otimes} Q^e, P_r)\\
&\cong \int^r \int_{s,t,u} P_s \otimes \overline{Q}_{\overline{t}} \otimes \mathrm{Hom}({}_{u\overline{r}}L_{s\overline{t}}, P_r) \otimes {}_u Q\\
&\cong \int_{s,t,u} P_s \otimes \overline{Q}_{\overline{t}} \otimes \int^r \mathrm{Hom}({}_{u\overline{r}}L_{s\overline{t}}, P_r) \otimes {}_u Q\\
&\cong \int_{s,t,u,v} P_s \otimes \overline{Q}_{\overline{t}} \otimes \mathrm{Hom}_{\overline{R}}({}_uL_{s\overline{t}}, {}_vR) \otimes {}_u Q \otimes {}_v P\\
&\cong P^e \underset{R^e}{\otimes} L^\vee \underset{R^e}{\otimes} Q^e.
\end{aligned}$$

Let $F\colon P^e \otimes_{R^e} L^\vee \otimes_{R^e} Q^e \to \widetilde{L}^\vee$ denote the resulting isomorphism. It is easy to check that

$$F(p_1\overline{q_1} \otimes \xi \otimes q_2\overline{p_2})(p_3\overline{q_3} \otimes \ell \otimes q_4\overline{p_4}) = p_1\xi((q_2p_3)\ell(q_4p_2)(\overline{q_1p_4}))q_3$$

for $p_1,\dots,p_4 \in P$, $q_1,\dots,q_4 \in Q$, $\xi \in L^\vee$ and $\ell \in L$. In other words, the evaluation of $\widetilde{L}^\vee$ on $\widetilde{L}$ can be written as the composition

$$\widetilde{L}^\vee \otimes \widetilde{L} \xrightarrow{F\otimes\widetilde{L}} \widetilde{L^\vee} \otimes \widetilde{L} \xrightarrow{\widetilde{\tau}} S,$$

where $\widetilde{\tau}$ is the skew pairing of $\times_S$-bialgebras induced by the evaluation $\tau : L^\vee \otimes L \to R$. It follows that F is an isomorphism of $\times_S$-bialgebras. □

Remark 5.5. Let R, S, L be as above. Since ${}^L\mathcal{M} \cong {}_{L^\vee}\mathcal{M}$ as monoidal categories by [22, Cor. 5.15], we can conclude that ${}^L\mathcal{M} \cong {}^{\widetilde{L}}\mathcal{M}$ as monoidal categories. Moreover, the equivalence is induced by the monoidal category equivalence ${}_R\mathcal{M}_R \cong {}_S\mathcal{M}_S$. Explicitly, it asssigns to $M \in {}^L\mathcal{M}$ the S-bimodule $P \otimes_R M \otimes_R Q$ endowed with the left $\widetilde{L}$-comodule structure

$$\begin{aligned}
P \underset{R}{\otimes} M \underset{R}{\otimes} Q &\to (P^e \underset{R^e}{\otimes} L \underset{R^e}{\otimes} Q^e) \times_S (P \underset{R}{\otimes} M \underset{R}{\otimes} Q)\\
p \otimes m \otimes q &\mapsto (p \otimes \overline{q_i} \otimes m_{(-1)} \otimes q \otimes \overline{p_j}) \otimes (p^i \otimes \otimes m_{(0)} \otimes q^j)
\end{aligned}$$

(where $M \ni m \mapsto m_{(-1)} \otimes m_{(0)} \in L \times_R M$ denotes the comodule structure on M).

We conjecture that the same formula can be used to define a category equivalence ${}^L\mathcal{M} \cong {}^{\widetilde{L}}\mathcal{M}$ when L is not finitely generated projective as left $\overline{R}$-module.

Remark 5.6. Let L be a $\times_R$-bialgebra which is finitely generated projective as left $\overline{R}$-module, and let $R \overset{\sqrt{\mathrm{M}}}{\sim} S$. Let the monoidal equivalence ${}_{R^e}\mathcal{M} \cong {}_{S^e}\mathcal{M}$ be induced by a Morita context involving the modules $C \in {}_{S^e}\mathcal{M}_{R^e}$ and $D \in {}_{R^e}\mathcal{M}_{S^e}$. By the remarks closing Section 3, we know that $C \otimes_{R^e} L$ is a finitely generated projective left $\overline{S}$-module. Since D is a finitely generated projective left R^e-module, we can conclude that $\widetilde{L} = C \otimes_{R^e} L \otimes_{R^e} D$ is a finitely generated projective left $\overline{S}$-module.

However, we do not know in this situation whether $\widetilde{L}^\vee$ and $\widetilde{L^\vee}$ are isomorphic $\times_S$-bialgebras.

Recall that the left dual $L^\vee$ is finitely generated projective as left R-module. For $\times_R$-bialgebras H such that ${}_R H$ is finitely generated projective, one can define a right dual $\times_R$-bialgebra ${}^\vee H$ in such a way that ${}^\vee(L^\vee) \cong L$.

Now let $\hat{L} := {}^\vee\widetilde{L^\vee}$ be the right dual $\times_S$-bialgebra of the $\times_S$-bialgebra obtained by $\sqrt{\text{Morita}}$ base change from the right dual of L (note that $\widetilde{L^\vee}$ is a finitely generated projective left S-module by reasoning similar to that used for ${}_{\overline{S}}\widetilde{L}$). Then we have equivalences of monoidal categories

$$^L\mathcal{M} \cong {}_{L^\vee}\mathcal{M} \cong {}_{\widetilde{L^\vee}}\mathcal{M} \cong {}^{\hat{L}}\mathcal{M}.$$

If our $\sqrt{\text{Morita}}$ equivalence comes from a Morita equivalence, then $\hat{L} \cong \widetilde{L}$. Otherwise, it seems that we have another version of $\sqrt{\text{Morita}}$ base change, suitable for comodules instead of modules.

More conjecturally, such a dual version of $\sqrt{\text{Morita}}$ base change should also be possible if L is not assumed to be finitely generated projective as left $\overline{R}$-module.

6 Canonical Tannaka duality

In this section we let k be a field. Let $\mathcal{C}$ be a semisimple k-linear tensor category with a finite number of simple objects whose endomorphism rings are isomorphic to k. Hayashi [11, 10] has proved that $\mathcal{C}$ is equivalent to the category of modules over a finite dimensional face algebra F. The construction can of course be applied to ${}_H\mathcal{M}$ where H is a split semisimple quasi-Hopf algebra, though it is not so clear how F is related to H.

In this section we will describe a connection between the "given" H and the "canonical" F. This proceeds in two steps. First, one uses a generalized smash product construction that produces a $\times_H$-bialgebra L isomorphic to $H \otimes H \otimes H^*$ as a vector space, and with ${}_L\mathcal{M} \cong {}_H\mathcal{M}$ as monoidal categories. In a second step, we use Morita base change to replace H by the Morita equivalent product of copies of the base field. The result is a face algebra $\widetilde{L}$, and we shall show that $\widetilde{L} \cong F$.

Let us first recall some elements of Hayashi's construction. An important step is the construction of a monoidal functor $\Omega_0 \colon \mathcal{C} \to {}_R\mathcal{M}_R$, where $R = k^n$ and n is the number of isomorphism classes of simple objects in $\mathcal{C}$. We will not go into details on the second important step, which is the construction of a unique face algebra F with n face idempotents such that Ω_0 factors over a monoidal equivalence $\Omega\colon \mathcal{C} \to {}_F\mathcal{M}$. (In fact Hayashi uses modules rather than comodules, which is of no importance since the face algebra he constructs is finite dimensional.)

Let Λ be the set of isomorphism classes of simple objects in $\mathcal{C}$, and pick a representative L_λ in each class $\lambda \in \Lambda$. For simplicity we let $R \cong k^n$ have the set Λ as its canonical basis of idempotents.

Hayashi's canonical functor Ω_0 sends $X \in \mathcal{C}$ to the R-bimodule $\Omega_0(X)$ with $\mu\Omega_0(X)\lambda = \mathrm{Hom}_{H-}(L_\mu, X \otimes L_\lambda)$, where, compared with Hayashi's convention, we have switched the sides in R-bimodules and replaced tensor product in $\mathcal{C}$ by its opposite. The monoidal functor structure ω of Ω_0 is the map

$$\Omega_0(X) \underset{R}{\otimes} \Omega_0(Y) \xrightarrow{\omega} \Omega_0(X \otimes Y)$$

sending $f \otimes g \in \mu\Omega_0(X)\rho \otimes \rho\Omega_0(Y)\lambda$ to the composite

$$L_\mu \xrightarrow{f} X \otimes L_\rho \xrightarrow{X\otimes g} X \otimes (Y \otimes L_\lambda) \xrightarrow{\Phi^{-1}} (X \otimes Y) \otimes L_\lambda,$$

where $\mu, \rho, \lambda \in \Lambda$, and Φ is the associator isomorphism in the category $\mathcal{C}$.

Now consider a split semisimple quasi-Hopf algebra H. We will apply Hayashi's constructions to $\mathcal{C} = {}_H\mathcal{M}$, and investigate the relation of F to H.

The first step is a construction suggested by Hausser and Nill (see [7], Proposition 3.11 and the remarks following the proof): They have defined a category ${}_H\mathcal{M}_H^H$ of Hopf modules over H, which is monoidal in such a way that the underlying functor $\mathcal{U}\colon {}_H\mathcal{M}_H^H \to {}_H\mathcal{M}_H$ is a strict monoidal functor, where ${}_H\mathcal{M}_H$ is a monoidal category by tensor product over H. They show ${}_H\mathcal{M}_H^H$ to be equivalent as a monoidal category to ${}_H\mathcal{M}$ via a monoidal functor

$$(\mathcal{R}, \xi)\colon {}_H\mathcal{M} \ni V \to V \otimes H \in {}_H\mathcal{M}_H^H.$$

Now by translating the coaction of H on a Hopf module into an action of the dual H^*, one can describe Hopf modules in ${}_H\mathcal{M}_H^H$ equivalently as modules over a certain generalized smash product $L := (H \otimes H^{\mathrm{op}})\#H^*$. This kind of classification of Hopf modules by modules over an algebra which is a product of several copies of H and its dual goes back to Cibils and Rosso [4]. We refer the reader to [23, Ex.4.12] for details on the construction of L. Since the underlying functor

$${}_{(H\otimes H^{\mathrm{op}})\#H^*}\mathcal{M} \cong {}_H\mathcal{M}_H^H \to {}_H\mathcal{M}_H$$

is strictly monoidal, it follows from [20, Thm.5.1] that L has the structure of a $\times_H$-bialgebra such that ${}_L\mathcal{M} \cong {}_H\mathcal{M}_H^H \cong {}_H\mathcal{M}$ are equivalences of monoidal categories.

H being split semisimple, it is Morita equivalent to a direct product of copies of k. We claim that Hayashi's F results from applying the appropriate Morita base change to L.

Let $\mathcal{F}$ denote the monoidal functor given by the Morita equivalence between R and H. By Lemma 2.1 and the definition of Morita base change we only have to verify that the diagram of monoidal functors

$$\begin{array}{ccc} {}_H\mathcal{M} & \xrightarrow{\mathcal{R}} & {}_H\mathcal{M}_H^H \\ \downarrow{\scriptstyle \Omega_0} & & \downarrow{\scriptstyle \mathcal{U}} \\ {}_R\mathcal{M}_R & \xleftarrow[\mathcal{F}]{} & {}_H\mathcal{M}_H \end{array}$$

commutes up to isomorphism of monoidal functors.

For $\lambda \in \Lambda$, now the set of simple modules in ${}_H\mathcal{M} = \mathcal{C}$, fix a minimal idempotent $e_\lambda \in H$ such that $L_\lambda := He_\lambda \in \Lambda$.

The functor $\mathcal{U}\mathcal{R}$ maps $X \in {}_H\mathcal{M}$ to $\mathcal{U}\mathcal{R}(X) = {\bullet}X \otimes {\bullet}H{\bullet} \in {}_H\mathcal{M}_H$; here the dots indicate that $X \otimes H$ is equipped with the diagonal left H-module structure and the right module structure induced by that of the right tensor factor. The functor $\mathcal{F}$ maps $M \in {}_H\mathcal{M}_H$ to the R-bimodule defined by $\mu\mathcal{F}(M)\lambda = e_\mu M e_\lambda$, so that $\mathcal{F}\mathcal{U}\mathcal{R}(X)$ satisfies

$$\begin{aligned}\mu\mathcal{F}\mathcal{U}\mathcal{R}(X)\lambda &= e_\mu \mathcal{U}\mathcal{R}(X)e_\lambda = e_\mu(X \otimes H)e_\lambda \\ &\cong \mathrm{Hom}_{H-}(He_\mu, X \otimes He_\lambda) = \mathrm{Hom}_{H-}(L_\mu, X \otimes L_\lambda) = \mu\Omega_0(X)\lambda.\end{aligned}$$

Note that the isomorphism $\psi\colon \Omega_0(X) \cong \mathcal{F}(X \otimes H)$ we have found maps $f \in \mathrm{Hom}_{H-}(L_\mu, X \otimes L_\lambda)$ to $\psi(f) = f(e_\mu) \in e_\mu(X \otimes H)e_\lambda$.

We still have to show that $\Omega_0 \cong \mathcal{F}\mathcal{U}\mathcal{R}$ as *monoidal* functors.

The monoidal functor structure ξ of $\mathcal{R}$ is the isomorphism

$$(X \otimes H) \underset{H}{\otimes} (Y \otimes H) = X \otimes (Y \otimes H) \xrightarrow{\Phi^{-1}} (X \otimes Y) \otimes H$$

in which the first equality is the canonical identification.

For $M, N \in {}_H\mathcal{M}_H$ we can identify $\mathcal{F}(M \otimes_H N) \subset M \otimes_H N$ with $\mathcal{F}(M) \otimes_R \mathcal{F}(N)$, which makes $\mathcal{F}$ a strict monoidal functor. In particular, the monoidal functor structure of $\mathcal{F}\mathcal{U}\mathcal{R}$ is the restriction of that of $\mathcal{R}$; we shall denote it by ξ again.

We need to show that the diagrams

$$\begin{array}{ccc} \Omega_0(X) \otimes_R \Omega_0(Y) & \xrightarrow{\omega} & \Omega_0(X \otimes Y) \\ \downarrow{\scriptstyle \psi\otimes\psi} & & \downarrow{\scriptstyle \psi} \\ \mathcal{F}\mathcal{U}\mathcal{R}(X) \otimes_R \mathcal{F}\mathcal{U}\mathcal{R}(Y) & \xrightarrow{\xi} & \mathcal{F}\mathcal{U}\mathcal{R}(X \otimes Y) \end{array}$$

commute for $X, Y \in {}_H\mathcal{M}$. Let $f \in \mathrm{Hom}_H(L_\mu, X \otimes L_\rho)$ and $g \in \mathrm{Hom}_H(L_\rho, Y \otimes L_\lambda)$. Then $\psi\omega(f \otimes g) = \omega(f \otimes g)(e_\mu) = \Phi^{-1}(X \otimes g)f(e_\mu)$. On the other hand, write $f(e_\mu) = \sum x_i \otimes h_i$ with $x_i \in X$ and $h_i \in L_\rho$. Then we have

$$\begin{aligned}\xi(\psi \otimes \psi)(f \otimes g) &= \xi(f(e_\mu) \otimes g(e_\rho)) = \Phi^{-1}\Big(\sum x_i \otimes h_i g(e_\rho)\Big) \\ &= \Phi^{-1}\Big(\sum x_i \otimes g(h_i e_\rho)\Big) = \Phi^{-1}\Big(\sum x_i \otimes g(h_i)\Big) = \Phi^{-1}(X \otimes g)f(e_\mu).\end{aligned}$$

7 An example from subfactor theory

Nikshych and Vainerman have shown how to associate a weak Hopf algebra to a subfactor of finite depth of a von Neumann algebra factor [15, 16]. The case of a subfactor $N \subset M$ of a type II_1 factor of index $\beta = 4\cos^2 \frac{\pi}{n+3}$ with $n \geq 2$ is treated in more detail in [15, 17]. The associated weak Hopf algebra can be described as follows (we summarize the beginning of [17, sec. 2.7]): Let $\mathcal{A}_{\beta,k}$ be the Temperley-Lieb algebra as in [6], that is, the unital algebra freely generated by idempotents $e_1, \dots, e_{k-1}$ subject to the relations $\beta e_i e_j e_i = e_i$ for $|i-j| = 1$ and $e_i e_j = e_j e_i$ for $|i-j| \geq 2$. Then $\mathcal{A}_{\beta,k}$ is semisimple for $k \leq n+1$ by the choice of β (cf. [6, §2.8]). Define $A_{1,k}$ by $A_{1,k} = \mathcal{A}_{\beta,k+1}$ if $k \leq n+1$, and let $A_{1,k+1}$ be obtained by applying the Jones basic construction to the inclusion $A_{1,k-1} \subset A_{1,k}$ for $k \geq n+1$. Thus $H := A_{1,2n-1}$ is generated by idempotents $e_1, \dots, e_{2n-1}$, and contains $A_{1,n-1}$, generated by $e_1, \dots, e_{n-1}$, and $A_{n+1,2n-1}$, generated by $e_{n+1}, \dots, e_{2n-1}$, as subalgebras. Nikshych and Vainerman describe a weak Hopf algebra structure of H with target counital subalgebra $H_t = A_{1,n-1}$ and $H_s = A_{n+1,2n-1}$. For $n = 2$, $H_t \cong \mathbb{C} \oplus \mathbb{C}$ is commutative, and $H \cong M_2(\mathbb{C}) \oplus M_3(\mathbb{C})$ is a face algebra of dimension 13.

We shall examine the case $n = 3$. We take the Bratteli diagram for the inclusion of $A_{1,2} = \mathcal{A}_{\beta,3}$ into $A_{1,3} = \mathcal{A}_{\beta,4}$ from the picture on page 101 of [6]. The remaining stories in the Bratteli diagram of the tower $A_{1,2} \subset \cdots \subset A_{1,5}$ are obtained by two applications of Jones' basic construction. By [6, §2.4] this amounts each time to adding a story on top which is an upside down mirror image of the preceding story. As a result, we have the following diagram:

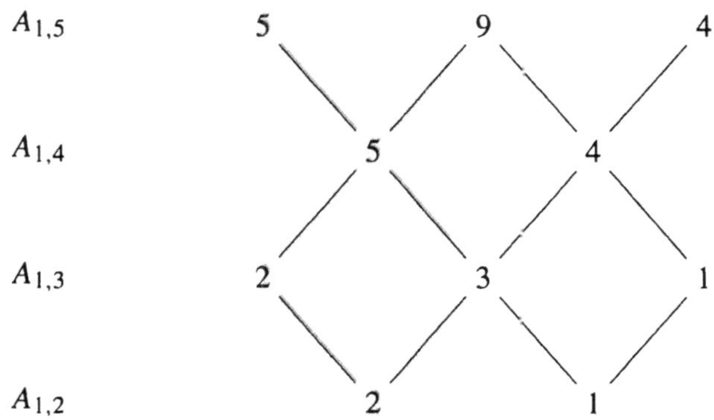

We see that $H = A_{1,5}$ has dimension $5^2 + 9^2 + 4^2 = 122$, and its counital subalgebras are isomorphic to $\mathbb{C} \oplus M_2(\mathbb{C})$, of dimension 5. We will apply Morita base change to this example, reducing the counital subalgebra to $\mathbb{C} \oplus \mathbb{C}$. We will not derive an explicit description of the resulting face algebra, but will be content with determining its algebra structure, hence its dimension.

The Bratteli diagram for the inclusion $H_t \subset H$ is obtained by composing the stories of the Bratteli diagram above:

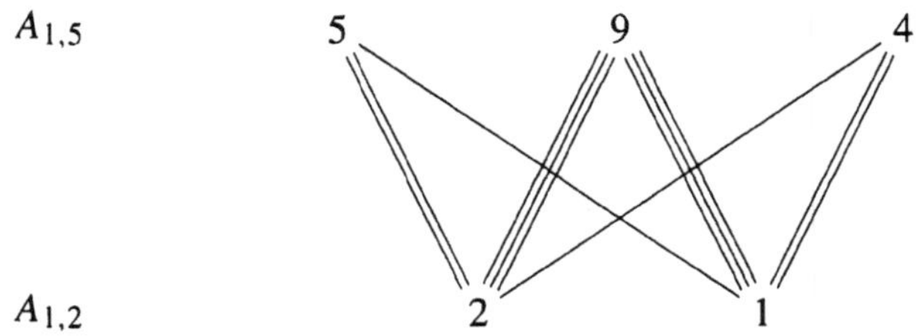

To apply Morita base change, we need the Bratteli diagram for the map $H_s \otimes H_t \to H$. Of course $H_s \otimes H_t \cong M_4(\mathbb{C}) \oplus M_2(\mathbb{C}) \oplus M_2(\mathbb{C}) \oplus \mathbb{C}$, and the antipode of H, which exchanges H_s and H_t, will switch the two copies of $M_2(\mathbb{C})$. The lower story of the following tower is the inclusion $H_t \subset H_t \otimes H_s$:

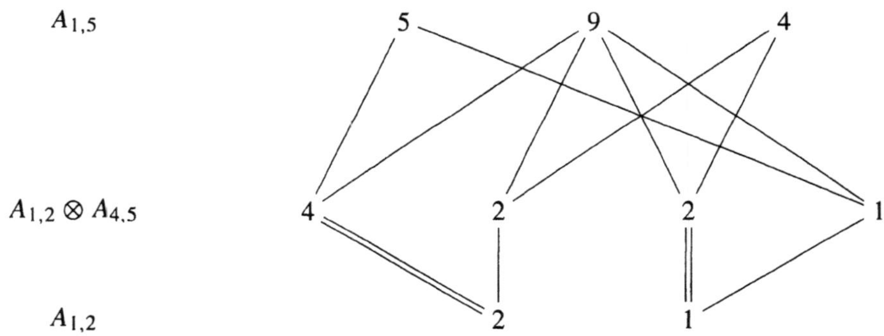

We claim that the upper story is the Bratteli diagram for the map $H_t \otimes H_s \to H$. It will help to know that, since the antipode switches the two vertices labelled 2 on the middle floor, any top floor vertex has the same number of edges to each of these vertices. In particular, there can be no edge from the 5 on the top floor to a 2 on the middle floor, since there is only one 1 on the bottom floor linked to 5. Also, there can be no more than one link from 5 to the 1 on the middle floor, since there should be only one link to the 1 on the bottom floor. This makes the two links from 5 to the middle floor inevitable as shown. There should be three links from the top 9 to the bottom 2. These cannot be accounted for by three links to the left 2 on the middle floor, since that would also entail three links to the right 2, and hence six links to the bottom 1. So there has to be one link to the 4 and one to the left 2, hence also the right 2 on the middle floor, which makes the one link to the 1 on the middle floor also inevitable. Finally the one link from the top 4 to the bottom 2 can only be accounted for by a single link from the top 4 to the left 2 on the middle floor, hence there also has to be one to the right 2, and there is no room for more.

Now we apply Morita base change to pass from H with counital subalgebra $H_t \cong \mathbb{C} \oplus M_2(\mathbb{C})$ to $\widetilde{H}$ with counital subalgebra $\widetilde{H}_t \cong \mathbb{C} \oplus \mathbb{C}$. The Bratteli diagram for the inclusion $\widetilde{H}_t \oplus \widetilde{H}_s \subset \widetilde{H}$ is the same as for $H_t \oplus H_s \subset H$, but with all ranks on the lower floor replaced by 1:

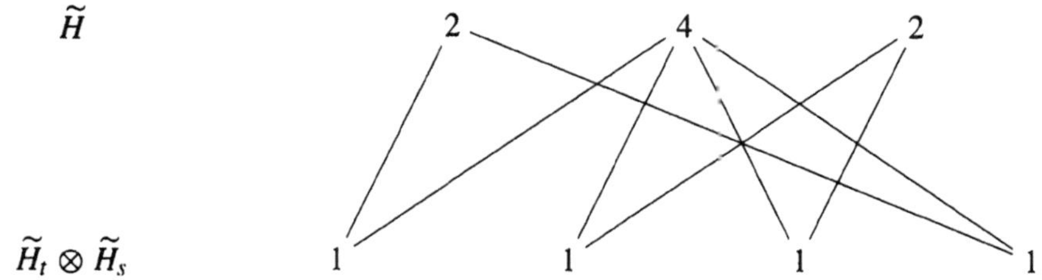

The resulting ranks on the upper floor show that $\dim \widetilde{H} = 2^2 + 4^2 + 2^2 = 24$. (Remember that $\dim H = 122$.)

Remark 7.1. By Hayashi's canonical Tannaka duality, there is a face algebra F with ${}_F\mathcal{M} \cong {}_H\mathcal{M}$ as monoidal categories, where F has three face idempotents (since H has three isomorphism classes of irreducible modules) whereas our $\widetilde{H}$ has two. Hayashi has also described another procedure to associate a face algebra to any subfactor of a II_1 factor [9], which will, however, also yield a face algebra that has as many faces as isomorphism classes of irreducible modules. By contrast, applying Morita base change to the weak Hopf algebra $A_{1,2n-1}$ of Nikshych and Vainerman will yield a face algebra with one face less than isomorphism classes of irreducible modules whenever n is odd.

References

[1] Böhm, G., Nill, F., and Szlachányi, K., Weak Hopf algebras I: Integral theory and C^*-structure, *J. Algebra* 221 (1999), 385–438.

[2] Böhm, G., and Szlachányi, K., A coassociative C^*-quantum group with nonintegral dimensions, *Lett. Math. Phys.* 38 (1996), 437–456.

[3] Brzeziński, T., and Militaru, G., Bialgebroids, $\times_A$-bialgebras and duality, *J. Algebra* 251 (2002), 279–294.

[4] Cibils, C., and Rosso, M., Hopf bimodules are modules, *J. Pure Appl. Algebra* 128 (1998), 225–231.

[5] Etingof, P., and Nikshych, D., Dynamical quantum groups at roots of 1, *Duke Math. J.* 108 (2001), 135–168.

[6] Goodman, F. M., de la Harpe, P., and Jones, V. F. R., *Coxeter graphs and towers of algebras*, Springer-Verlag, New York, 1989.

[7] Hausser, F., and Nill, F., Integral theory for quasi-Hopf algebras, preprint `math.QA/9904164`.

[8] Hayashi, T., Face algebras I–A generalization of quantum group theory, *J. Math. Soc. Japan* 50 (1998), 293–315.

[9] Hayashi, T., Galois quantum groups of II_1-subfactors, *Tohoku Math. J.* (2) 51 (1999), 365–389.

[10] Hayashi, T., A brief introduction to face algebras, in *New trends in Hopf algebra theory* (La Falda, 1999). Amer. Math. Soc., Providence, RI, 2000, 161–176.

[11] Hayashi, T., A canonical Tannaka duality for finite semisimple tensor categories, preprint `math.QA/9904073`.

[12] Kadison, L., Hopf algebroids and H-separable extensions, to appear in *Proc. Amer. Math. Soc.*, preprint `mathpreprints.com/math/Preprint/lkadison/20020130.1/2/` (2002).

[13] Kadison, L., and Szlachányi, K., Dual bialgebroids for depth two ring extensions, preprint `math.RA/0108067`.

[14] Lu, J.-H., Hopf algebroids and quantum groupoids, *Internat. J. Math.* 7 (1996), 47–70.

[15] Nikshych, D., and Vainerman, L., A characterization of depth 2 subfactors of II_1 factors, *J. Funct. Anal.* 171 (2000), 278–307.

[16] Nikshych, D., and Vainerman, L., A Galois correspondence for II_1 factors and quantum groupoids, *J. Funct. Anal.* 178 (2000), 113–142.

[17] Nikshych, D., and Vainerman, L., Finite quantum groupoids and their applications, in *New directions in Hopf algebras*, Math. Sci. Res. Inst. Publ. 43, Cambridge University Press, Cambridge 2002, 211–262.

[18] Ostrik, V., Monoidal categories, weak Hopf algebras and modular invariants, preprint `math.QA/0111139`.

[19] Pareigis, B., Non-additive ring and module theory II. $\mathcal{C}$-categories, $\mathcal{C}$-functors and $\mathcal{C}$-morphisms, *Publ. Math. Debrecen* 24 (1977), 351–361.

[20] Schauenburg, P., Bialgebras over noncommutative rings and a structure theorem for Hopf bimodules, *Appl. Categ. Structures* 6 (1998), 193–222.

[21] Schauenburg, P., Face algebras are $\times_R$-bialgebras, in *Rings, Hopf algebras, and Brauer groups* (S. Caenepeel and A. Verschoren, eds.), Lecture Notes in Pure and Appl. Math. 197, Marcel Dekker, Inc., New York, NY, 1998, 275–285. .

[22] Schauenburg, P., Duals and doubles of quantum groupoids ($\times_R$-bialgebras), in *New Trends in Hopf Algebra Theory* (N. Andruskiewitsch, W. R. Ferrer Santos, and H.-J. Schneider, eds.), Contemp. Math. 267, Amer. Math. Soc., Providence, RI, 273–299.

[23] Schauenburg, P., Actions of monoidal categories and generalized Hopf smash products, preprint (2001).

[24] Schauenburg, P., Weak Hopf algebras and quantum groupoids, preprint `math.QA/0204180`.

[25] Sweedler, M. E., Groups of simple algebras, *Inst. Hautes Études Sci. Publ. Math.* 44 (1974), 79–189.

[26] Szlachányi, K., Finite quantum groupoids and inclusions of finite type, preprint `math.QA/0011036`.

[27] Takeuchi, M., Groups of algebras over $A \otimes \overline{A}$, *J. Math. Soc. Japan* 29 (1977), 459–492.

[28] Takeuchi, M., $\sqrt{\text{Morita}}$ theory, *J. Math. Soc. Japan* 39 (1987), 301–336.

[29] Watts, C. E., Intrinsic characterizations of some additive functors, *Proc. Amer. Math. Soc.* 11 (1960), 5–8.

[30] Xu, P., Quantum groupoids, *Comm. Math. Phys.* 216 (2001), 539–581.

Galois actions by finite quantum groupoids

*Kornél Szlachányi**

Research Institute for Particle and Nuclear Physics
1525 Budapest, P.O.Box 49
email: `szlach@rmki.kfki.hu`

Abstract. Proposing a certain category of bialgebroid maps we show that the balanced depth 2 extensions appear as they were the finitary Galois extensions in the context of quantum groupoid actions, i.e., actions by finite bialgebroids, weak bialgebras or weak Hopf algebras. We comment on deformation of weak bialgebras, on half grouplike elements, on uniqueness of weak Hopf algebra reconstructions and discuss the example of separable field extensions

For extensions of rings, algebras or C^*-algebras the notion of depth 2, introduced originally for von Neumann factors by A. Ocneanu, has many features that makes it the analogue of Galois extension of fields. The extension $N \subset M$ of k-algebras is called of depth 2 if the canonical N-M bimodule $X = {}_N M_M$ and M-N bimodule $\bar{X} = {}_M M_N$ satisfy: $X \otimes \bar{X} \otimes X$ is a direct summand in a finite direct sum of copies of X and $\bar{X} \otimes X \otimes \bar{X}$ is a direct summand in a finite direct sum of $\bar{X}$'s. The right module M_N is called balanced if $\mathrm{End}_{\mathcal{E}}\, M \cong N$ where $\mathcal{E} = \mathrm{End}\, M_N$. For any balanced depth 2 extension the endomorphism ring $A = \mathrm{End}\ {}_N M_N$ carries a bialgebroid structure which is finite projective over the centralizer, or relative commutant, $R = C_M(N)$ both as a left and as a right module. Moreover the canonical action of A on M makes M to be a left A-module algebroid with invariant subalgebra equal to N [7]. This generalizes the result of Nikshych and Vainerman in subfactor theory [12] because finite index depth 2 extensions of von Neumann or C^*-algebras are always balanced depth 2 extensions. Moreover they are Frobenius extensions which causes the appearance of antipodes, hence leading to weak C^*-Hopf algebra structure [1] on A.

Our main purpose in this paper will be to study the uniqueness problem of the bialgebroid A. Using a kind of category for bialgebroids we show that A satisfies a universal property analogous to the one of the Galois group of a field extension. While the natural action of A on M generalizes Hopf–Galois extensions of Kreimer and Takeuchi [8], Hopf–Galois extensions do not have the universal property w.r.t the category of Hopf algebras as the example of Greither and Pareigis [4] has shown. Our proposal of a Galois bialgebroid $\mathrm{Gal}(M/N)$ of an algebra extension in Section 2 is

*Partially supported by Hungarian Scientific Research Fund, OTKA T-034512

close in spirit to Pareigis' Quantum Automorphism Group [14] but technically not a mature one.

In case of depth 2 Frobenius extensions we present an optimistic interpretation of the non-uniqueness of its associated weak Hopf algebra. The natural object to which a unique (measured) weak Hopf algebra can be associated is a Frobenius system [5], i.e., a Frobenius extension $N \subset M$ together with a Frobenius homomorphism $\phi\colon {}_N M_N \to {}_N N_N$.

Let me recall [2] for the definition of weak bialgebra (WBA) which is one of the main themes in this paper. Let K be a field. A finite dimensional K-space A together with a K-algebra structure $\langle A, m, u\rangle$ and a K-coalgebra structure $\langle A, \Delta, \varepsilon\rangle$ is called a *weak bialgebra* if

1. Δ is multiplicative / m is comultiplicative, i.e., as maps $A \otimes A \to A \otimes A$,
$$\Delta \circ m = (m \otimes m) \circ (\mathrm{id} \otimes \Sigma \otimes \mathrm{id}) \circ (\Delta \otimes \Delta)$$
where $\Sigma\colon A \otimes A \to A \otimes A$ denotes the flip map $x \otimes y \mapsto y \otimes x$,

2. ε is weakly multiplicative, i.e., as maps $A \otimes A \otimes A \to K$,
$$(\varepsilon \otimes \varepsilon) \circ (m \otimes m) \circ (\mathrm{id} \otimes \Delta \otimes \mathrm{id}) = \varepsilon \circ m \circ (m \otimes \mathrm{id})$$
$$(\varepsilon \otimes \varepsilon) \circ (m \otimes m) \circ (\mathrm{id} \otimes \Delta^{\mathrm{op}} \otimes \mathrm{id}) = \varepsilon \circ m \circ (m \otimes \mathrm{id})$$
where $\Delta^{\mathrm{op}} := \Sigma \circ \Delta$ is opposite comultiplication,

3. u is weakly comultiplicative, i.e., as maps $K \to A \otimes A \otimes A$,
$$(\mathrm{id} \otimes m \otimes \mathrm{id}) \circ (\Delta \otimes \Delta) \circ (u \otimes u) = (\Delta \otimes \mathrm{id}) \circ \Delta \circ u$$
$$(\mathrm{id} \otimes m^{\mathrm{op}} \otimes \mathrm{id}) \circ (\Delta \otimes \Delta) \circ (u \otimes u) = (\Delta \otimes \mathrm{id}) \circ \Delta \circ u$$
where $m^{\mathrm{op}} := m \circ \Sigma$ is opposite multiplication.

Weak bialgebras reduce to ordinary bialgebras iff Δ is unital. Weak bialgebras have canonical subalgebras A^L and A^R that are spanned by the right leg and left leg of $\Delta(1)$, respectively. A^L belongs to the relative commutant of A^R and there is a canonical antiisomorphism $A^L \to A^R$. The subalgebras A^L and A^R are separable K-algebras.

Takeuchi's $\times_R$-bialgebras [19] or, what is the same ([3, 22]), bialgebroids A [9], are defined over a ground ring R which is not supposed to be separable but plays the role of A^L (or of A^R). Indeed weak bialgebras are just the bialgebroids over separable base [16].

The weak bialgebras as well as the bialgebroids we speak about here are finite dimensional over the ground field and finitely generated projective over the base ring, respectively. Briefly saying they are finite quantum groupoids.

1 Weak bialgebras versus bialgebroids

A category of bialgebroids is introduced the arrows of which are called bialgebroid maps. They intend to be special arrows in a possible larger category. They have a special trend to point from bialgebras to bialgebroids but not vice versa. The category of maps of left/right bialgebroids will be denoted by Bia_l, Bia_r, respectively. Using the forgetful functor from weak bialgebras to bialgebroids we obtain weak morphisms of weak bialgebras as lifts of bialgebroid maps. We comment on deformed versions of WBA's and on half grouplike elements.

1.1 The category of bialgebroid maps

Let k be a commutative ring. All objects and maps below will be k-algebras and k-algebra maps, respectively. Thus our base category is k-Alg.

1.1.1 The objects. Following the original definition [19, 9], its reformulations in [22, 18], and the terminology of [7] we say that $\underline{A} = \langle A, R, s, t, \gamma, \pi\rangle$ is a *left bialgebroid* if

- $R \xrightarrow{s} A \xleftarrow{t} R^{\mathrm{op}}$ are k-algebra homomorphisms such that $s(r')t(r) = t(r)s(r')$ for $r, r' \in R$. Then A is made into an R-R-bimodule by setting $r \cdot a \cdot r' := s(r)t(r')a$.

- $\gamma\colon A \to A \otimes_R A$ and $\pi : A \to R$ are R-R-bimodule maps such that the triple $\langle A, \gamma, \pi\rangle$ is a comonoid in the category ${}_R\mathcal{M}_R$.

- γ is a ring homomorphism into the Takeuchi $\times_R$ -product $A \times_R A$, i.e.,

$$\gamma(a)(t(r) \otimes 1) = \gamma(a)(1 \otimes s(r)) \tag{1.1}$$
$$\gamma(a)\gamma(b) = \gamma(ab) \tag{1.2}$$
$$\gamma(1) = 1 \otimes 1 \tag{1.3}$$

  for all $a, b \in A$ and $r \in R$.

- π is compatible with the algebra structure, i.e.,

$$\pi(as(\pi(b))) = \pi(ab) = \pi(at(\pi(b))), \quad a, b \in A \tag{1.4}$$
$$\pi(1) = 1_R. \tag{1.5}$$

1.1.2 The arrows. For two left bialgebroids $\underline{A}$ and $\underline{B}$ a pair $\langle\varphi, \omega\rangle$ of algebra homomorphisms $\varphi\colon A \to B$ and $\omega\colon R_A \to R_B$ is called a *map of left bialgebroids* if

$$\begin{array}{ccc} A & \xrightarrow{\varphi} & B \\ {\scriptstyle s_A}\uparrow & & \uparrow{\scriptstyle s_B} \\ R_A & \xrightarrow{\omega} & R_B \end{array} \qquad \begin{array}{ccc} A & \xrightarrow{\varphi} & B \\ {\scriptstyle t_A}\uparrow & & \uparrow{\scriptstyle t_B} \\ R_A & \xrightarrow{\omega} & R_B \end{array} \tag{1.6}$$

are commutative diagrams in k-Alg and

$$\begin{array}{ccc} A & \xrightarrow{\varphi} & \Phi_\omega(B) \\ {\scriptstyle \pi_A}\downarrow & & \downarrow{\scriptstyle \pi_B} \\ R_A & \xrightarrow{\omega} & R_B \end{array} \tag{1.7}$$

$$\begin{array}{ccccc} A & & \xrightarrow{\quad\varphi\quad} & & \Phi_\omega(B) \\ {\scriptstyle \gamma_A}\downarrow & & & & \downarrow{\scriptstyle \Phi_\omega(\gamma_B)} \\ A \otimes_{R_A} A & & & & \Phi_\omega(B \otimes_{R_B} B) \\ & \searrow^{\varphi\otimes\varphi} & & \nearrow_{\tau^\omega_{B,B}} & \\ & & \Phi_\omega(B) \otimes_{R_A} \Phi_\omega(B) & & \end{array} \tag{1.8}$$

are commutative diagrams in ${}_{R_A}\mathcal{M}_{R_A}$. Here $\langle \Phi^\omega, \tau^\omega, \omega \rangle$ is the monoidal forgetful functor ${}_{R_B}\mathcal{M}_{R_B} \to {}_{R_A}\mathcal{M}_{R_A}$ associated to the algebra homomorphism ω. That is to say, with $R = R_A$ and $S = R_B$

$$\Phi_\omega \colon {}_S\mathcal{M}_S \to {}_R\mathcal{M}_R \quad \text{is the forgetful functor} \tag{1.9}$$

$$\tau^\omega_{X,Y} \colon \Phi_\omega(X) \otimes_R \Phi_\omega(Y) \to \Phi_\omega(X \otimes_S Y) \quad \text{is the canonical bimodule epimorphism} \tag{1.10}$$

$$\zeta_\omega \colon {}_R R_R \to {}_R S_R \quad \text{is } \omega \text{ as a bimodule map.} \tag{1.11}$$

Notice that ω is uniquely determined by φ as $\omega = \pi_B \circ \varphi \circ s_A$.

In order to see that the above properties of φ are preserved under composition of such maps, so we indeed have a category, one uses functoriality of $\omega \mapsto \Phi^\omega$, i.e., that in fact the monoidal forgetful functor is the arrow map of a functor $\Phi \colon k\text{-Alg}^{\text{op}} \to$ MonCat. The object map of this functor is

$$R \mapsto \langle {}_R\mathcal{M}_R, \otimes_R, {}_R R_R \rangle$$

Then functoriality means the identities

$$\Phi_\omega \circ \Phi_\sigma = \Phi_{\sigma\circ\omega}$$
$$(\Phi_\omega \circ \tau_\sigma) \bullet (\tau_\omega \circ (\Phi_\sigma \times \Phi_\sigma)) = \tau_{\sigma\circ\omega}$$
$$\Phi_\omega(\zeta_\sigma) \circ \zeta_\omega = \zeta_{\sigma\circ\omega}$$

for $R \xrightarrow{\omega} S \xrightarrow{\sigma} T$ in k-Alg.

In this way we have constructed a category Bia_l of left bialgebroids over the base category k-Alg, i.e., the objects are left bialgebroids *in* k-Alg. In Bia_l there is no fixed base ring and there are arrows between bialgebroids over different base rings. In particular there are bialgebroid maps from ordinary bialgebras to bialgebroids.

In a similar way one defines the category Bia_r of right bialgebroids and right bialgebroid maps the details of which we omit.

With the terminology "maps of bialgebroids" we intend to leave place for more general arrows between bialgebroids. Certain bimodules with a coproduct, so bimodule coalgebras, are natural candidates, they allow to formulate Morita equivalence [17] when only the forgetful functor $\mathcal{M}_A \to {}_R\mathcal{M}_R$ of the bialgebroid is considered as relevant. However, in the Galois problem of non-commutative rings maps of bialgebroids do play a role as we shall see in Section 2. In other words, Bia_l is large enough to contain maps from group algebras or bialgebras to bialgebroids but also small enough to contain only very restrictive isomorphisms.

1.2 From weak bialgebras to bialgebroids

Let $\underline{W} = \langle W, \Delta, \varepsilon \rangle$ be a WBA over K. Its left and right subalgebras are defined by

$$L = \{\varepsilon(1_{(1)}w)1_{(2)} \mid w \in W\} \tag{1.12}$$
$$R = \{1_{(1)}\varepsilon(w1_{(2)}) \mid w \in W\} \tag{1.13}$$

and are the images of the maps

$$\pi^L\colon W \to L, \quad w \mapsto \varepsilon(1_{(1)}w)1_{(2)} \tag{1.14}$$
$$\pi^R\colon W \to R, \quad w \mapsto 1_{(1)}\varepsilon(w1_{(2)}). \tag{1.15}$$

For the basic properties of these maps see [2].

Now we introduce data for bialgebroids as follows. Let

$$s^L\colon L \to W \tag{1.16}$$
$$s^R\colon R \to W \tag{1.17}$$

just the inclusion maps, hence algebra homomorphisms. Let

$$t^L\colon L^{\mathrm{op}} \to W, \quad l \mapsto 1_{(1)}\varepsilon(1_{(2)}l) \tag{1.18}$$
$$t^R\colon R^{\mathrm{op}} \to W, \quad r \mapsto \varepsilon(r1_{(1)})1_{(2)} \tag{1.19}$$

which are also algebra maps (if antipode exists they are the restrictions of S^{-1}). The ranges of s^L and t^L are L and R, so they commute. Similarly for s^R and t^R. This allows us to introduce bimodule structures ${}_L W_L$ and ${}_R W_R$, respectively, via the formulae

$$l \cdot w \cdot l' := s^L(l)t^L(l')w, \quad l, l' \in L, \ w \in W \tag{1.20}$$

$$r \cdot w \cdot r' := wt^R(r)s^R(r'), \quad r, r' \in R, \ w \in W. \tag{1.21}$$

Finally, let

$$\tau^L \colon W \otimes W \to W \otimes_L W \tag{1.22}$$

$$\tau^R \colon W \otimes W \to W \otimes_R W \tag{1.23}$$

be the canonical epimorphisms associated to the units $K \to L$, $K \to R$, respectively.

Lemma 1.1. *Let* $\underline{W} = \langle W, \Delta, \varepsilon \rangle$ *be a WBA/K. Then the maps*

$$\gamma^L := \tau^L \circ \Delta \colon W \to W \otimes_L W \tag{1.24}$$

$$\gamma^R := \tau^R \circ \Delta \colon W \to W \otimes_R W \tag{1.25}$$

are such that

$$\beta_l(\underline{W}) := \langle W, L, s^L, t^L, \gamma^L, \pi^L \rangle \text{ is a left bialgebroid} \tag{1.26}$$

and

$$\beta_r(\underline{W}) := \langle W, R, s^R, t^R, \gamma^R, \pi^R \rangle \text{ is a right bialgebroid.} \tag{1.27}$$

Proof. First notice that $\pi^L \circ s^L = \mathrm{id}_L$ and $\pi^L \circ t^L = \mathrm{id}_L$ by [2, (2.3a), (2.25a)]. Also using [2, Lemma 2.5]

$$\begin{aligned} \pi^L(l \cdot w \cdot l') &= \pi^L(lt^L(l')w) = l\pi^L(t^L(l')\pi^L(w)) = l\pi^L(w)\pi^L(t^L(l')) \\ &= l\pi^L(w)l' \end{aligned}$$

therefore π^L is an L-L bimodule map. Now $\langle {}_L W_L, \gamma^L, \pi^L \rangle$ is a comonoid in ${}_L\mathcal{M}_L$ because τ^L is natural and $\pi^L(w_{(1)})w_{(2)} = w$ and $t^L \circ \pi^L(w_{(2)})w_{(1)} = w$ are general WBA identities. Coassociativity follows using that τ^L extends to a natural transformation (1.10), namely, $\tau^L = \tau^{u_L}_{W,W}$ where $u_L \colon K \to L$ is the unit of the K-algebra L, and the latter satisfies the hexagon diagram of a (lax) monoidal functor. In order to show that the image of γ^L is in $W \times_L W$ it suffices to refer to the old WBA identity $1_{(1)} \otimes 1_{(2)}l = 1_{(1)}t^L(l) \otimes 1_{(2)}$ (cf. [2, (2.31a)]). Multiplicativity of γ^L then follows from multiplicativity of Δ. It remains to show the counit properties (1.4) but they are just the identities [2, (2.5a),(2.25a)]. Passing to the opposite-coopposite WBA one obtains the statement for β_r. □

The β_l and β_r defined by the lemma are expected to be the object maps of two functors

$$\mathsf{Bia}_l \xleftarrow{\ \beta_l\ } \mathrm{WBA} \xrightarrow{\ \beta_r\ } \mathsf{Bia}_r \tag{1.28}$$

the arrows of WBA, however, will be discussed later.

1.3 From bialgebroids over separable base to weak bialgebras

Since the left/right subalgebras of a WBA are always separable, we start from a left bialgebroid $\underline{B} = \langle B, R, s, t, \gamma, \pi \rangle$ in which R is a separable K-algebra. That is to say there exists an element $e = \sum_i e_i \otimes e^i \in R \otimes_K R$ such that

$$\sum_i e_i e^i = 1_R \quad \text{and} \quad \sum_i r e_i \otimes e^i = \sum_i e_i \otimes e^i r, \quad r \in R. \tag{1.29}$$

Such a separability idempotent provides a splitting map for the canonical epimorphism τ, namely

$$\sigma : B \otimes_R B \to B \otimes_K B, \quad b \otimes b' \mapsto \sum_i b \cdot e_i \otimes e^i \cdot b' \tag{1.30}$$

This formula is the same for right bialgebroids. For left bialgebroids we can write also

$$\sigma(b \otimes_R b') = \sum_i t(e_i) b \otimes s(e^i) b'. \tag{1.31}$$

But there is more than separability of R in a WBA. There is also a separability *structure* for R. For any weak bialgebra with left subalgebra R the restriction of the counit $\psi := \varepsilon|_R$ is a nondegenerate functional of index one. This means that ψ distinguishes a special separability idempotent, namely $e = S(1_{(1)}) \otimes 1_{(2)}$ which is the quasibasis of ψ. Comparing this with the above expression for σ one recognizes that σ is multiplication from the left by $\Delta(1)$ on any element from the inverse image $\tau^{-1}(\{b \otimes_R b'\})$.

Lemma 1.2. *Given a pair $\langle \underline{B}, \psi \rangle$, where $\underline{B}$ is a left or right bialgebroid over R and $\psi : R \to K$ is a nondegenerate functional of index 1, define*

$$\Delta := \sigma \circ \gamma : B \to B \otimes_K B \tag{1.32}$$

$$\varepsilon := \psi \circ \pi : B \to K \tag{1.33}$$

where σ is the splitting map of the canonical epimorphism $B \otimes_K B \to B \otimes_R B$ that is associated to the quasibasis e of ψ as in (1.30). *Then the triple $\langle B, \Delta, \varepsilon \rangle$ is a WBA over K.*

Proof. Mutatis mutandis, the proof has already been given in [7, Proposition 9.4] and in [16, Theorem 5.5]. □

The above lemma characterizes the fibres of the functor β_l in the following sense. The WBA's W with a fixed underlying (let's say left) bialgebroid $\beta_l(W) = \underline{B}$ are in one-to-one correspondence with separability structures $\langle R, \psi, e \rangle$ on R.

1.4 Strict and weak morphisms of weak bialgebras

Definition 1.3. Let $\langle W, \Delta, \varepsilon \rangle$ and $\langle W', \Delta', \varepsilon' \rangle$ be weak bialgebras over K. Then a K-linear map $f: W \to W'$ is called a

- *strict morphism* of weak bialgebras if f is an algebra map and a coalgebra map;
- *weak left morphism* of weak bialgebras if $f: \beta_l(W) \to \beta_l(W')$ is a map of the underlying left bialgebroids;
- *weak right morphism* of weak bialgebras if $f: \beta_r(W) \to \beta_r(W')$ is a map of the underlying right bialgebroids.

A strict morphism f not only preserves the left and right subalgebras, $f(L) \subset L'$ and $f(R) \subset R'$, but, because $(f \otimes f) \circ \Delta(1) = \Delta'(1')$, establishes isomorphisms $L \xrightarrow{\sim} L'$ and $R \xrightarrow{\sim} R'$. Therefore strict morphisms exist between two WBA's only if they have isomorphic left, resp. right subalgebras. This is definitely too strong since the original philosophy of [1] was to "blow up" Hopf algebras in order they could afford non-integral categorical dimensions, but the amount of the blowing up, i.e., the size of the left/right subalgebras should be considered as a gauge degree of freedom. Using weak morphisms we pursue this idea to some extent.

Since weak morphisms are just lifts of the rather involved bialgebroid maps into the WBA framework, they are useful only if they can be recognized directly without reference to bialgebroids. Therefore we make the

Proposition 1.4. *For weak bialgebras $\underline{W}$ and $\underline{W'}$ a K-linear map $f: W \to W'$ is a weak left morphism iff*

1. *f is a K-algebra map,*
2. $f(R) \subset R'$,
3. $\pi'^L \circ f = f \circ \pi^L$,
4. *and* $\Delta'(1') \cdot (f \otimes f)(\Delta(w)) = \Delta'(f(w))$, $w \in W$.

It is a weak right morphism iff

1. *f is a K-algebra map,*
2. $f(L) \subset L'$,
3. $\pi'^R \circ f = f \circ \pi^R$,
4. *and* $(f \otimes f)(\Delta(w)) \cdot \Delta'(1') = \Delta'(f(w))$, $w \in W$.

Proof. It suffices to prove the statement for left morphisms. Since s^L is just the injection $L \subset W$ and so is s'^L, the first diagram in (1.6) is equivalent to $f(L) \subset L'$, which in turn is a consequence of (1.7) which is nothing but condition 1.4 above. Having condition 1.4 anyway the second diagram of (1.6) is equivalent to the condition 1.4 because if 1.4 holds then $f \circ t^L = t'^L \circ \pi'^L \circ f \circ t^L = t'^L \circ f \circ \pi^L \circ t^L = t'^L \circ f$ and backwards is obvious. This proves that the three diagrams of (1.6) and (1.7) are equivalent to 1.4, 1.4 and 1.4. Assuming this we can equip W' with L-L bimodule structure $\hat{W}' = \Phi_\omega({}_{L'}W'_{L'})$ and lift f to an L-L bimodule map $\hat{f}$. Then (1.8) takes the form

$$\tau^\omega \circ (\hat{f} \otimes_L \hat{f}) \circ \gamma = \gamma' \circ \hat{f}. \tag{1.34}$$

Let τ' be the canonical epi for W' and σ' be its splitting map associated to the quasi basis e' of $\varepsilon'|_{L'}$. Let $\hat{\tau}$ be the canonical epi for $\hat{W}'$ and $\hat{\sigma}$ its splitting map that is associated to e, or to $(f \otimes f)(e)$ in some (bad) sense. Then we have $\tau^\omega \circ \hat{\tau} = \tau'$ and using the observation we made just before Lemma 1.2 we can write for all $w'_1 \otimes w'_2 \in \hat{W}' \otimes_L \hat{W}'$ that

$$\begin{aligned}\sigma' \circ \tau^\omega(w'_1 \otimes w'_2) &= \sigma' \circ \tau^\omega \circ \hat{\tau} \circ \hat{\sigma}(w'_1 \otimes w'_2) \\ &= 1'_{(1)} f(1_{(1)}) w'_1 \otimes 1'_{(2)} f(1_{(2)}) w'_2.\end{aligned}$$

Now acting by σ' on (1.34) we obtain 1.4 and acting by τ' on 1.4 we obtain (1.34). □

For weak Hopf algebras one defines weak left/right morphisms as those of its underlying weak bialgebra, disregarding whether they preserve antipodes or not.

The category of bialgebras, as well as the category of Hopf algebras, are full subcategories in each one of Bia_l, WBA and Bia_r.

Example 1.5. Let H be a Hopf algebra over K. Define its blowing up as the algebra $W := H \otimes M_n(K)$ with comultiplication $\Delta(h \otimes e_{ij}) := (h_{(1)} \otimes e_{ij}) \otimes (h_{(2)} \otimes e_{ij})$. Then W becomes a weak Hopf algebra. Its left and right subalgebras coincide and equal to the diagonal matrices with entries from K. The diagonal embedding of H, $f\colon H \to W$, $f(h) = h \otimes I_n$, is clearly an algebra map. It is not a coalgebra map however, but we have

$$\begin{aligned}\Delta(f(h)) &= \sum_i (h_{(1)} \otimes e_{ii}) \otimes (h_{(2)} \otimes e_{ii}) \\ &= \Delta(1_W)(f \otimes f)(\Delta_H(h)) = (f \otimes f)(\Delta_H(h))\Delta(1_W) \\ \pi^L(f(h)) &= f(1_H)\varepsilon_H(h) = \pi^R(f(h)).\end{aligned}$$

Now using the above proposition it is plain that f is weak left and right morphism of weak bialgebras and there is no strict morphism from H to W unless $n = 1$.

1.5 Weak automorphisms, twists and half grouplike elements

Let $f\colon W \to W'$ be a weak left morphism of WBA's. Then

$$\varepsilon'(f(w)) = \varepsilon'(\pi'^L(f(w))) = \varepsilon'(f(\pi^L(w)) = \varepsilon(1_{(1)}w)\varepsilon'(f(1_{(2)}))$$

therefore

$$\varepsilon'(f(w)) = \varepsilon(uw), \quad w \in W, \quad \text{where } u = \varepsilon'(f(1_{(1)}))1_{(2)} \in L \tag{1.35}$$

where, in order to get $u \in L$, we also made the $R \to L$ transformation

$$t^L(1_{(1)}\varepsilon'(f(1_{(2)}))) = \varepsilon'(f(\pi^L(1_{(1)})))1_{(2)} = u.$$

Especially we have $\varepsilon'(f(l)) = \varepsilon(ul)$ for $l \in L$. So assuming that $f|_L$ is an isomorphism onto L' we have u as a Radon–Nikodym derivative of a nondegenerate functional w.r.t another, hence invertible. Comparing their quasibases we obtain the equality

$$\omega^{-1}(\pi'^L(1'_{(1)})) \otimes \omega^{-1}(1'_{(2)}) = \pi^L(1_{(1)}) \otimes u^{-1}1_{(2)} \tag{1.36}$$

as elements of $L \otimes L$. Applying $f \otimes f$ and using that π^L restricts to an isomorphism $R \to L$ (the would-be antipode), we get

$$1'_{(1)} \otimes 1'_{(2)} = f(1_{(1)}) \otimes f(u^{-1})f(1_{(2)}). \tag{1.37}$$

Inserting this result to the 4th property of weak left morphisms in Proposition 1.4 one immediately arrives to the

Lemma 1.6. *Let $f\colon W \to W'$ be a weak left morphism of WBA's such that its restriction $\omega\colon L \to L'$ is an isomorphism. Then there is an invertible element $u \in L$ such that*

$$\Delta'(f(w)) = (f \otimes f)((1 \otimes u^{-1})\Delta(w)) \tag{1.38}$$
$$\varepsilon'(f(w)) = \varepsilon(uw) \tag{1.39}$$

for all $w \in W$.

This result holds in particular if f is an isomorphism. As a matter of fact Property 3 in Proposition 1.4 is invariant under changing f to f^{-1}. For completeness we remark that the forgetful functor $\mathsf{Bia}_l \to k\text{-}\mathsf{Alg}$ reflects isomorphisms. That is to say, if a weak left morphism is invertible as an algebra map then its inverse is a weak left morphism. So the lemma holds for $f = \mathrm{id}_W$. In this case u describes a deformation in the sense of [11, Remark 3.7]. Such (left) deformed WBA's have identical underlying left bialgebroids, so deformations should be interpreted as weak left automorphisms. Although the deformation changes the Nakayama automorphism of the counit, there may be no deformation at all which produces a tracial ε', unless the base L possesses a nondegenerate trace of index 1. For example if L is split semisimple then the only such trace is the regular trace. Since the Radon–Nikodym derivative of the regular trace w.r.t. $\varepsilon|_L$ is $1_{(2)}S(1_{(1)})$, tracial deformation exists iff $1_{(2)}S(1_{(1)})$ is invertible. In characteristic zero this is always the case, otherwise there are counter examples [21].

Now consider inner weak left automorphisms associated to left grouplike elements. For a bialgebroid $\langle B, L, s, t, \gamma, \pi\rangle$ an element $g \in B$ is grouplike if g is invertible and $\gamma(g) = g \otimes_L g$. For a WBA W define

$$\mathcal{G}^L(W) := \{g \in W \mid g \text{ is grouplike in } \beta_l(W)\} \tag{1.40}$$

$$\mathcal{G}^R(W) := \{g \in W \mid g \text{ is grouplike in } \beta_r(W)\} \tag{1.41}$$

the sets of left/right grouplike elements. This is of course equivalent to saying e.g. that g is left grouplike if it is invertible and $\Delta(g) = \Delta(1)(g \otimes g)$. In the next computations we assume that $g, h \in \mathcal{G}^L(W)$ and $w \in W$ is arbitrary.

$$\pi^L(g) = \varepsilon(1_{(1)}g)1_{(2)} = \varepsilon(g_{(1)})g_{(2)}g^{-1} = gg^{-1} = 1 \tag{1.42}$$

$$\begin{aligned}\Delta(g^{-1}) &= \Delta(g^{-1})\Delta(1) = \Delta(g^{-1})\Delta(g)(g^{-1} \otimes g^{-1})\\ &= \Delta(1)((g^{-1} \otimes g^{-1})\end{aligned} \tag{1.43}$$

$$\Delta(gh) = \Delta(g)\Delta(1)(h \otimes h) = \Delta(g)(h \otimes h) = \Delta(1)(gh \otimes gh) \tag{1.44}$$

$$\begin{aligned}\pi^L(gwg^{-1}) &= \pi^L(gw\pi^L(g^{-1})) = \pi^L(gw) = \varepsilon(1_{(1)}g\pi^L(w))1_{(2)}\\ &= \varepsilon(g_{(1)}\pi^L(w))g_{(2)}g^{-1} = g\pi^L(w)g^{-1}\end{aligned} \tag{1.45}$$

$$\begin{aligned}t^L \circ \pi^L(gwg^{-1}) &= 1_{(1)}\varepsilon(1_{(2)}gwg^{-1}) = g_{(1)}g^{-1}\varepsilon(g_{(2)}\pi^L(w))\\ &= g(t^L \circ \pi^L(w))g^{-1}\end{aligned} \tag{1.46}$$

$$\begin{aligned}\Delta(1)(gw_{(1)}g^{-1} \otimes gw_{(2)}g^{-1}) &= g_{(1)}w_{(1)}g^{-1} \otimes g_{(2)}w_{(2)}g^{-1}\\ &= g_{(1)}w_{(1)}g^{-1}_{(1)} \otimes g_{(2)}w_{(2)}g^{-1}_{(2)} = \Delta(gwg^{-1})\end{aligned} \tag{1.47}$$

This shows that $\mathcal{G}^L$ is a group and for all $g \in \mathcal{G}^L$ the inner automorphism $w \mapsto gwg^{-1}$ is a weak left automorphism $W \to W$. It does not leave the counit invariant but

$$\varepsilon(gwg^{-1}) = \varepsilon(uw) \quad w \in W, \quad \text{where } u = \varepsilon(g1_{(1)})1_{(2)} \tag{1.48}$$

implying for their quasibasis the relation

$$g^{-1}\pi^L(1_{(1)})g \otimes g^{-1}1_{(2)}g = \pi^L(1_{(1)}) \otimes u^{-1}1_{(2)}.$$

Applying $t^L \otimes \mathrm{id}$ and using that $\mathrm{Ad}_{g^{-1}}$ commutes with π^L and $t^L \circ \pi^L$ one obtains

$$g^{-1}1_{(1)}g \otimes g^{-1}1_{(2)}g = 1_{(2)} \otimes u^{-1}1_{(2)}$$

therefore

$$\Delta(g) = (g \otimes gu^{-1})\Delta(1) = (g\pi^R(u^{-1}) \otimes g)\Delta(1). \tag{1.49}$$

Since π^R on L is an algebra antiisomorphism and $\pi^R(u) = 1_{(1)}\varepsilon(gS^{-1}(1_{(2)})) = \pi^R(g)$, the above definition of $\mathcal{g}^L$ is equivalent to the one given by Vecsernyés [21].

2 Galois quantum groupoids

In this section we argue that the balanced depth 2 extensions [7] of rings or k-algebras are the proper analogues of the Galois extensions of fields (i.e., normal and separable field extensions) because they have finite quantum automorphism groups (cf. [14]) with invariant subalgebra just N and which are characterized by a universal property, hence unique. The role of finite groups are played by bialgebroids, i.e., $\times_R$-bialgebras, that are finitely generated projective over their base as a left and as a right module. They will be called finite bialgebroids. They are presumably also Hopf algebroids but the antipode raises several questions, so we skip their discussion altogether. The difference between Galois bialgebroid and Galois WBA will be found in the difference between depth 2 Frobenius extensions and depth 2 extensions with a Frobenius structure.

2.1 Quantum automorphisms

Recall the definition of left module algebroids over a left bialgebroid A in [7]. They are the monoids in the category of left A-modules.

Definition 2.1. Let $N \to M$ be an extension of k-algebras. Define the category $\mathrm{Aut}(M/N)$ as follows. Its objects are the pairs $\langle B, \alpha_B\rangle$ where B is a finite left bialgebroid in k-Alg and $\alpha_B\colon B \otimes M \to M$ is a left B-module algebroid action such that N is contained in the invariant subalgebra M^B. The arrows from $\langle B, \alpha_B\rangle$ to $\langle C, \alpha_C\rangle$ are the maps $f\colon B \to C$ of left bialgebroids such that

$$\alpha_C \circ (f \otimes \mathrm{id}_M) = \alpha_B.$$

A terminal object in $\mathrm{Aut}(M/N)$ is called a *universal action* on the extension M/N. The bialgebroid A, unique up to isomorphism in Bia_l, in a universal action is called the *Galois quantum groupoid* of the extension M/N and is denoted by $\mathrm{Gal}(M/N)$. If the invariant subalgebra of the universal action is equal to N the extension $N \to M$ is called a *Galois extension.*

If $F \subset E$ is a field extension then every finite group action on E which leaves F pointwise fixed factors uniquely through the Galois group $\mathrm{Gal}(E/F)$. This trivial fact is generalized by the above definition. Also, if the fixed points of $\mathrm{Gal}(E/F)$ coincide with the elements of F the extension is normal and separable, i.e., Galois, by Artin's Theorem.

Note that the real beauty of universal monoids of [14] has not been used in the above definition. One could consider much more general arrows $\alpha\colon B \otimes M \to M$ than just actions.

If H is a Hopf algebra, f.g.p. over k, then Kreimer and Takeuchi define an H-Galois extension to be a ring extension $N \subset M$ such that

- there is a left H-module algebra action $\alpha\colon H \otimes M \to M$,
- $N = M^H$, the invariant subring,
- M_N is finitely generated projective and
- the map

$$\phi\colon M \otimes H \to \operatorname{End} M_N, \quad m \otimes h \mapsto \{m' \mapsto m\alpha(h \otimes m')\} \tag{2.1}$$

  is an isomorphism.

(More precisely, this is a reformulation by Ulbrich [20].)

Hopf–Galois extensions in this sense, however, do not have the universal property with respect to the category of Hopf algebras. As it was pointed out by Greither and Pareigis in [4] there are separable field extensions which are H-Galois for two different Hopf algebras. We come back to this example in the last section.

2.2 Universal bialgebroid actions

The advantage of using bialgebroids is that there is a very general class of ring extensions for which a universal bialgebroid action exists. These are the depth 2 extensions $N \subset M$ for which M_N is balanced. They include all ring extensions that are H-Galois for some Hopf algebra, as it was shown by Kadison recently [6], but many more. The universal bialgebroid of a depth 2 balanced ring extension is a canonical structure on the endomorphism ring $A = \operatorname{End}_N M_N$ and it has been constructed in [7] although its universality was not formulated there. Below we shall give a proof for the special case of separable centralizer which leads us to weak bialgebra actions as follows.

If W is a weak bialgebra over K and $B = \beta_l(W)$ its underlying left bialgebroid then the category of left W-modules and the category of left B-modules are monoidally equivalent [16], Proposition 5.3, in fact isomorphic. Therefore these categories have the same monoids. Therefore a module algebra over W is the same as a module algebroid over B. This lends to a WBA action $\alpha\colon B \otimes M \to M$ the name *weak Galois action* if it has N as its invariant subalgebra and if it has the universal property w.r.t. weak left morphisms of WBA's.

Theorem 2.2. *Let K be a field and $N \subset M$ a K algebra extension such that*

- *$N \subset M$ is of depth 2,*
- *M_N is balanced,*
- *$R := C_M(N)$ is a separable K-algebra.*

Then the bialgebroid $A = \mathrm{End}_N M_N$ *constructed in* [7] *and acting on* M *in the natural way is the Galois bialgebroid* $\mathrm{Gal}(M/N)$. *That is to say, any weak bialgebra* $\langle A, \Delta_A, \varepsilon_A \rangle$ *with underlying left bialgebroid* A *has the following universal property. If* $\alpha_W\colon W \otimes M \to M$ *is a left module algebra action of a WBA* W *such that* $M^W \supset N$ *then there is a unique weak left morphism* $f\colon W \to A$ *of weak bialgebras such that* $\alpha \circ (f \otimes \mathrm{id}_M) = \alpha_W$.

Proof. That the invariant subalgebra is N was shown in [7]. To prove the universal property notice that for any weak bialgebra action on M

$$w \triangleright (nmn') = n(w \triangleright m)n', \quad w \in W,\ n, n' \in M^W,\ m \in M, \tag{2.2}$$

in particular for all $n, n' \in N$. Thus there is a unique algebra map $f\colon W \to A$ such that

$$f(w)(m) = w \triangleright m.$$

Already this implies uniqueness so we are left to show that f is a weak left morphism. We will use the criteria given in Proposition 1.4. At first compute the action of an $l \in W^L$.

$$l \triangleright m = (l_{(1)} \triangleright 1_M)(l_{(2)} \triangleright m) = 1 \triangleright ((l \triangleright 1_M)m) = (l \triangleright 1_M)m$$

Since $(l \triangleright 1_M)n = l \triangleright n = n(l \triangleright 1_M)$ for $n \in N$, $W^L \triangleright 1_M \subset C_M(N) = R$. Hence $f(W^L) \subset \lambda_M(R) = A^L$, where $\lambda_M(m)$ denotes left multiplication by m on M. For $r \in W^R$ we have

$$r \triangleright m = (r_{(1)} \triangleright m)(r_{(2)} \triangleright 1_M) = 1 \triangleright (m(r \triangleright 1_M)) = m(r \triangleright 1_M) = m(\pi^L_W(r) \triangleright 1_M)$$

Since $A^R = \rho_M(R)$ where $\rho_M(m)$ is right multiplication by m on M, we obtain $f(W^R) \subset \rho_M(R) = A^R$. Next recall that the counit of the left bialgebroid A is $\pi^L_A(a) = \lambda_M(a(1_M))$. Therefore

$$\pi^L_A(f(w)) = \lambda_M(f(w)(1_M)) = \lambda_M(w \triangleright 1_M) = \lambda_M(\pi^L_W \triangleright 1_M) = f(\pi^L_W(w))$$

Turning to the last condition of Proposition 1.4 we recall [7], Prop. 3.9 stating that

$$A \otimes_R A \xrightarrow{\sim} \mathrm{Hom}_{N-N}(M \otimes_N M, M), \quad (a \otimes a')(m \otimes m') = a(m)a'(m').$$

Now consider the composite K-linear maps

$$W \xrightarrow{\Delta} W \otimes W \xrightarrow{f \otimes f} A \otimes A \xrightarrow{\tau} A \otimes_R A$$
$$W \xrightarrow{f} A \xrightarrow{\gamma} A \otimes_R A$$

which are equal because their images act on $M \otimes_N M$ in the same way due to the module algebra property. So, composing them with the splitting map σ associated to $\Delta_A(1_A)$ and using $\sigma \circ \tau = \lambda_{A \otimes A}(\Delta_A(1_A))$ we obtain

$$\Delta_A(1_A) \cdot (f \otimes f)(\Delta_W(w)) = \Delta_A(f(w)), \quad w \in W. \qquad \square$$

2.3 Universal weak Hopf algebra actions

If we add to the conditions of Theorem 2.2 that $N \subset M$ is a Frobenius extension then it already implies that the WBA lift of its Galois bialgebroid is a WHA [7, Section 9]. This does not make the WHA unique, only up to weak left isomorphisms. This freedom of the WHA is precisely the freedom of choosing a Frobenius functional $\psi: R \to K$ of index 1. Therefore it is natural to associate WHA actions to Frobenius structures $\langle N \to M, \phi: {}_N M_N \to {}_N N_N \rangle$ rather than to just extensions $N \to M$.

If $\phi: {}_N M_N \to {}_N N_N$ is a Frobenius map, i.e., a bimodule map with quasibasis $\sum_i m_i \otimes m^i \in M \otimes_N M$, then its restriction to the centralizer $\phi|_R$ maps R into the center Z of N. Since R is not only part of M but belongs to the bialgebroid A as well, it is very natural to build $\phi|_R$ into the data of the WHA as $\varepsilon|_R$. Strictly speaking, this is possible only if the center of N is trivial. There is a tiny point here about the restriction. While in case of finite index C^*-algebra extensions one considers faithful conditional expectations ϕ which have faithful restrictions to the finite dimensional R, therefore $\phi|_R$ is a Frobenius map with invertible index, this is not automatic for general Frobenius algebra extensions.

Theorem 2.3. *Let $N \subset M$ be a depth 2 Frobenius extension of K-algebras with centralizer R a separable K-algebra and with* Center $N = K$. *Assume $\phi: M \to N$ is a Frobenius map with its restriction $\phi|_R$ being an index* 1 *Frobenius map. Then there exists a unique weak Hopf algebra A and a left module algebra action of A on M which satisfies the universal property of Theorem* 2.2 *and such that $\phi|_R = \varepsilon|_R$.*

Proof. The antipode of a WHA is unique therefore uniqueness of A follows if we show that its WBA structure is unique. The latter is uniquely determined by its underlying left bialgebroid $\beta_l(A)$ and by the restriction of its counit, $\varepsilon|_R$. The former is uniquely determined by the universal property as $\beta_l(A) = \mathrm{Gal}(M/N)$ by Theorem 2.2 and the latter by the requirement $\varepsilon|_R = \phi|_R$. This proves uniqueness. The existence part is an easy application of Theorem 9.5 of [7]. □

The question arises how to interpret ϕ if only its restriction to the centralizer matters. Since ϕ is an N-N bimodule map, it belongs to A as a nondegenerate left integral. Thus in fact the data of the theorem determine a measurable quantum groupoid, i.e., a WHA with a distinguished nondegenerate integral.

Generalizations to Center $N = Z$ a separable K-algebra is possible. It requires to use a slight generalization of the notion of a WHA. It requires WHA's not in k-Alg but in ${}_Z\mathcal{M}_Z$, cf. [18, Proposition 1.6].

In addition to the assumptions of Theorem 2.3 let us assume that $\phi(1_M)$ is invertible or only assume that M/N is split. Then M_N is balanced therefore M/N is a Galois extension in the sense of Definition 2.1.

3 Separable field extensions are weak Hopf Galois

Let $E|K$ be a separable field extension. Then the following results are standard.

1. E_K is finite dimensional, $\dim E_K = n < \infty$.
2. Let $\tau\colon E \to K$ be the trace associated to the regular E-module ${}_E E$. Then there exists $x_i, y_i \in E$, $i = 1, \ldots n$ such that
$$\sum_i x_i \tau(y_i x) = x, \quad \forall x \in E \tag{3.1}$$
$$\sum_i x_i y_i = 1. \tag{3.2}$$
3. The above set $\{x_i, y_i\}$, called a quasibasis for τ, satisfies
$$\sum_i x x_i \otimes y_i = \sum_i x_i \otimes y_i x, \quad \forall x \in E.$$
4. There exists a $\xi \in E$ which generates E as an K-algebra, i.e., $E = K(\xi)$. (Primitive Element Theorem)
5. The non-zero K-algebra endomorphisms of E are automorphisms. Their group G forms an K-linearly independent set in the K-algebra of K-linear endomorphisms $\operatorname{End} E_K$ of the K-module E_K. The G-invariants $F = E^G$ form a subfield of E and $E|F$ is (classically) Galois with Galois group G. Hence $|G| = \dim_F E = n/m$ where $m = \dim_K F$.

3.1 The universal weak Hopf algebra of E/K

Define A as the K-algebra $\operatorname{End} E_K$ and its weak Hopf algebra structure by
$$\Delta_A(a) = \sum_i \sum_j x_i \tau(x_j \_) \otimes y_j a(y_j \_), \tag{3.3}$$
$$\varepsilon_A(a) = \tau(a(1)), \tag{3.4}$$
$$S_A(a) = \sum_i x_i \tau(a(y_i)\_). \tag{3.5}$$

The WHA A is a very special one:

1. The left and right subalgebras coincide with E. As a matter of fact, identifying E with the subalgebra of A of (left) multiplications on E
$$\pi^L(a) = a(1), \quad \pi^R(a) = \sum_i x_i \tau(a(y_i))$$
thus $A^L = A^R = E$.

2. The antipode is involutive, $S_A^2 = \mathrm{id}_A$. What is more, it is transposition w.r.t. the nondegenerate bilinear form on $E \otimes_K E$ given by τ, i.e.,

$$\tau(xa(y)) = \tau(S_A(a)(x)y), \quad x, y \in E.$$

3. A is cocommutative, $a_{(1)} \otimes a_{(2)} = a_{(2)} \otimes a_{(1)}$ as elements of $A \otimes_K A$, holds for all $a \in A$ as a consequence of commutativity of E on the one hand and of the existence of an isomorphism $A \otimes_K A \cong \mathrm{End}_K(E \otimes_K E)$, i.e., finite dimensionality of E_K on the other hand.

4. Left integrals in A are the endomorphisms l of E_K such that $l(E) \subset K$. Their general form is $l = \tau(r\_)$ where $r \in E$. Normalized left integrals thus exist ($\Rightarrow$ A is a separable K-algebra) and the invariant subalgebra of E is K.

5. τ is a 2-sided nondegenerate integral. If n is invertible in K, especially in characteristic 0, then τ/n is a Haar integral in A.

6. The left grouplike elements of A are precisely the algebra automorphisms of E_K. Thus the number $|\mathcal{G}^L|$ of left grouplike elements is a divisor of n and is equal to n precisely if E/K is classically Galois. In the latter case A is a crossed product of E with the group algebra $K\mathcal{G}^L$.

7. The left A-module algebra A^L, i.e., the trivial A-module, coincides with E with its canonical left A-module structure. That is to say,

$$a(x) = ax(1) = a_{(1)}x(1\varepsilon_A(a_{(2)})) = a_{(1)}xS_A(a_{(2)})(1) = \pi^L(ax)(1) = \pi^L(ax)$$

for all $a \in A$ and $x \in E$.

8. The smash product $E \rtimes A$ is isomorphic to A as K-algebras via the canonical map $x \rtimes a \mapsto \{y \mapsto xa(y)\}$.

3.2 Weak Hopf Galois extensions

As an immediate generalization of Hopf–Galois extensions one can make the

Definition 3.1. Let W be a weak Hopf algebra over K. A finite field extension E/K is called W-Galois if there exists a weak Hopf module algebra action $\alpha\colon W \otimes_K E \to E$ such that the map

$$\Phi\colon E \otimes_L W \to \mathrm{End}_K E, \quad x \otimes w \mapsto \{y \mapsto x\alpha(w \otimes y)\},$$

where L stands for W^L, is an isomorphism.

Let E/K be a finite extension which is W-Galois. Below we shall write $w \triangleright x$ for $\alpha(w \otimes x)$, $w \in W$, $x \in E$.

1. The map Φ in the above Definition is an K-algebra isomorphism from the smash product $E \rtimes W$ to A. As a matter of fact, the underlying K-space of the smash product is the tensor product $E \otimes_L W$ of L modules where E_L is defined by $x \cdot l := x(l \triangleright 1)$. So the definition of W-Galois extension just claims that the smash product is isomorphic to A as an K-space. This map is an algebra map as it is obvious from the multiplication rule of the smash product. Let $\varphi\colon W \to A$ denote the restriction of this map.

2. The restriction of the K-algebra monomorphism $\varphi\colon W \to A$ to $L = W^L$ is $\varphi(l)\colon x \mapsto (l \triangleright 1)x$. Therefore $\varphi(L) \subset E$ and therefore φ identifies L with an intermediate field $K \subset L \subset E$.

3. Let $r \in W^R$. Then $r \triangleright x = x(r \triangleright 1) = (S_W(r) \triangleright 1)x$. Therefore $\varphi(r) = \varphi(S_W(r)) \in L$ which, using injectivity of φ, implies that $W^L = W^R = L$ and S_W acts as the identity on L.

4. It follows that $\Delta_W(1_W)$ is a separating idempotent for the separable algebra L over K. By commutativity of E it contains $\Delta_A(1_A)$ as a subprojection, i.e.,

$$(\varphi \otimes \varphi)(\Delta_W(1_W))\Delta_A(1_A) = \Delta_A(1_A)(\varphi \otimes \varphi)(\Delta_W(1_W)) = \Delta_A(1_A)$$

5. φ is a weak (two sided) morphism of weak bialgebras, i.e.,

$$\Delta_A(1_A)(\varphi \otimes \varphi)(\Delta_W(w)) = \Delta_A(\varphi(w)), \qquad w \in W. \tag{3.6}$$

This can be seen as follows. Upon identifying $A \otimes_K A$ with $\mathrm{End}_K(E \otimes_K E)$ the module algebra property of ${}_W E$ boils down to

$$\tau \circ (\varphi \otimes \varphi)(\Delta_W(w)) = \tau \circ \Delta_A(\varphi(w)).$$

Composing both hand sides with the section σ of τ which is given by $\sigma(1) = \Delta_A(1_A)$ we get precisely the required statement.

6. W is cocommutative. As a matter of fact, module algebra property of ${}_W E$ and commutativity of E immediately imply that $(w_{(1)} \triangleright x)(w_{(2)} \triangleright y) = (w_{(1)} \triangleright y)(w_{(2)} \triangleright x)$ for $x, y \in E$ and $w \in W$. Therefore $\Phi((w_{(1)} \triangleright y) \rtimes w_{(2)}) = \Phi((w_{(2)} \triangleright y) \rtimes w_{(1)})$ and Φ being mono we have equality of the arguments in the smash product. Now the arguments are images under $v_y \otimes \mathrm{id}_W$ of $w_{(1)} \otimes_L w_{(2)}$ and $w_{(2)} \otimes_L w_{(1)}$, respectively, where $v_y\colon W \to E$, $w \mapsto w \triangleright y$, is a right L-module map. Since the common kernel of maps $v_y \otimes \mathrm{id}_W\colon W \otimes_L W \to E \otimes_L W$, $y \in E$, is the kernel of $\varphi \otimes \mathrm{id}$, we conclude that

$$w_{(1)} \otimes_L w_{(2)} = w_{(2)} \otimes_L w_{(1)}, \qquad w \in W.$$

Now use separability of L/K and the isomorphism $W \otimes_L W \cong (W \otimes_K W)\Delta_W(1_W)$ to conclude that $w_{(1)} \otimes_K w_{(2)} = w_{(2)} \otimes_K w_{(1)}$, $w \in W$.

7. If $n \in {}_W E$ is an invariant then $\varphi(n)$ commutes with $\varphi(W)$ and since End E_K is generated by $\varphi(W)$ and the commutative E, it belongs to Center End $E_K = K$. This proves that $E^W = K$.

8. $\pi_A^L(\varphi(w)) = w \triangleright 1 = \pi_W^L(w) \triangleright 1 = \pi_W^L(w)$, the last equation identifying W^L with $L \subset E$. Therefore $\varepsilon_A(\varphi(w)) = \varepsilon_A(\pi_A^L(\varphi(w))) = \varepsilon_A(\pi_W^L(w)) = \tau(\pi_W^L(w))$. Since $\varepsilon_W|_L$ is nondegenerate there exist $u \in L_\times$ such that $\tau(l) = \varepsilon_W(ul)$ for $l \in L$. It follows that

$$\varepsilon_A(\varphi(w)) = \varepsilon_W(u\pi_W^L(w)) = \varepsilon_W(\pi_W^L(uw)) = \varepsilon_W(uw)$$

thus

$$\varepsilon_W(w) = \varepsilon_A(\varphi(u^{-1}w)), \quad w \in W. \tag{3.7}$$

In particular if E/K is H-Galois for some finite dimensional Hopf algebra H over K then H is embedded into A by a unique weak morphism of weak Hopf algebras, which is just the restriction of the Galois map. Moreover A is the crossed product of E with H.

Example 3.2 (Greither–Pareigis [4, 10]). Let $K = \mathbb{Q}$ the rational field and $E = \mathbb{Q}[x]/(x^4 - 2)$. Then E/K is separable but not normal. However it is H-Galois for two different Hopf algebras. One of these Hopf algebras is the commutative and co-commutative Hopf algebra $H = \mathbb{Q}[c, s]/(c^2 + s^2 - 1, cs)$ with comultiplication $\Delta(c) = c \otimes c - s \otimes s$, $\Delta(s) = c \otimes s + s \otimes c$, counit $\varepsilon(c) = 1$, $\varepsilon(s) = 0$ and antipode $S(c) = c$, $S(s) = -s$. The weak left embedding of H into A is clear from the presentation of A as

$$A = \mathbb{Q}\text{-alg}\langle c, s, x \mid c^2 + s^2 - 1, cs, sc, cx - xs, sx + xc, x^4 - 2\rangle \tag{3.8}$$

The general WHA structure of (3.3) can be cast into the form

$$\Delta(1) = \frac{1}{4} 1 \otimes 1 + \frac{1}{8} \sum_{k=1}^{3} x^k \otimes x^{4-k} \tag{3.9}$$

$$\Delta(x) = \frac{1}{4}(x \otimes 1 + 1 \otimes x) + \frac{1}{8}(x^3 \otimes x^2 + x^2 \otimes x^3) \tag{3.10}$$

$$\Delta(c) = \Delta(1)(c \otimes c - s \otimes s) \tag{3.11}$$

$$\Delta(s) = \Delta(1)(c \otimes s + s \otimes c) \tag{3.12}$$

$$\varepsilon(c) = 4 \quad \varepsilon(s) = 0 \quad \varepsilon(x) = 0 \tag{3.13}$$

$$S(c) = c \quad S(s) = -s \quad S(x) = x \tag{3.14}$$

3.3 Galois connection

Let E/K be separable and let A be its universal weak Hopf algebra. Let $\mathrm{Sub}_{\mathsf{Alg}/K}(E)$ be the set of subobjects of E in the category of K-algebras. Also, let $\mathrm{Sub}_{\mathsf{WHA}/K}(A)$

be the sub-WHA's of A. The latter means any K-subalgebra of A which is closed under comultiplication. That is to say we restrict ourselves to strict embeddings of WHA's in the sense of Definition 1.3.

We can define two order reversing functions (contravariant functors between preorders)

$$\mathsf{Sub}_{\mathsf{WHA}/K}(A) \xrightarrow{\text{Fix}} \mathsf{Sub}_{\mathsf{Alg}/K}(E) \xrightarrow{\text{Gal}} \mathsf{Sub}_{\mathsf{WHA}/K}(A)$$

as follows.

$$\text{Fix}(W) := \{ x \in E \mid a(x) = \pi^L(a)x, \ \forall a \in W \} \tag{3.15}$$

$$\text{Gal}(F) := \{ a \in A \mid a(xy) = a(x)y, \ \forall x \in E, \ y \in F \} \tag{3.16}$$

Then we have the adjointness relations

$$W \subset \text{Gal}(F) \Leftrightarrow F \subset \text{Fix}(W) \tag{3.17}$$

Therefore the pair $\langle \text{Fix}, \text{Gal} \rangle$ is a Galois connection between sub-WHA's of $A = \text{End}\, E_K$ and intermediate fields $E \supset F \supset K$. Moreover this is a half Galois correspondence since every intermediate field occurs as a fixed field of a sub-WHA,

$$F = \text{Fix}(\text{Gal}(F)), \quad \forall F \in \mathsf{Sub}_{\mathsf{Alg}/K}(E). \tag{3.18}$$

A full Galois correspondence would require further analysis of weak Hopf subalgebras and coideal subalgebras like in [13].

References

[1] G. Böhm, K. Szlachányi, A coassociative C^*-quantum group with nonintegral dimensions, *Lett. Math. Phys.* 35 (1996), 437–456.

[2] G. Böhm, F. Nill and K. Szlachányi, Weak Hopf algebras, I. Integral theory and C^*-structure, *J. Algebra* 221 (1999), 385–438.

[3] T. Brzezinski and G. Militaru, Bialgebroids, $\times_A$-bialgebras and duality, to appear in *J. Algebra*, preprint QA/0012164.

[4] C. Greither, B. Pareigis, Hopf Galois theory for separable field extensions, *J. Algebra* 106 (1987) 239–258

[5] L. Kadison, *New examples of Frobenius extensions*, Univ. Lecture Ser. 14, Amer. Math. Soc., Providence, RI, 1999.

[6] L. Kadison, A Hopf algebroid associated to a Galois extension, preprint www.mathpreprints.com.

[7] L. Kadison, K. Szlachányi, *Dual bialgebroids for depth 2 ring extensions*, to appear in *Adv. in Math.*, math.RA/0108067.

[8] H. F. Kreimer, M. Takeuchi, Hopf algebras and Galois extensions of an algebra, *Indiana Univ. Math. J.* 30 (1981) 675–692

[9] J.-H. Lu, Hopf algebroids and quantum groupoids, *Internat. J. Math.* 7 (1996), 47–70.

[10] S. Montgomery, *Hopf Algebras and Their Actions on Rings*, CBMS Regional Conf. Ser. Math. 82, Amer. Math. Soc., Providence, RI, 1993.

[11] D. Nikshych, On the structure of weak Hopf algebras, preprint `math.QA/0106010`.

[12] D. Nikshych and L. Vainerman, A characterization of depth 2 subfactors of II_1 factors, *J. Funct. Anal.* 171 (2000), 278–307.

[13] D. Nikshych, L. Vainerman, A Galois correspondence for actions of quantum groupoids on II_1-factors, *J. Funct. Anal.* 178 (2000), 113–142.

[14] B. Pareigis, Quantum Groups – The functorial side, preprint 2000

[15] P. Schauenburg, Duals and doubles of quantum groupoids ($\times_R$-Hopf algebras), in *New trends in Hopf algebra theory* (N. Andruskiewitsch et al., eds.), Contemp. Math. 267, Amer. Math. Soc., Providence, RI, 2000, 273–299

[16] P. Schauenburg, Weak Hopf algebras and quantum groupoids, preprint `math.QA/0204180` (2002).

[17] P. Schauenburg, Morita base change in quantum groupoids, preprint `math.QA/0204338` (2002).

[18] K. Szlachányi, Finite quantum groupoids and inclusions of finite type, in *Mathematical physics in mathematics and physics. Quantum and operator algebraic aspects* (R. Longo, ed.), Fields Inst. Commun. 30, Amer. Math. Soc., Providence, RI, 2001, 393–407

[19] M. Takeuchi, Groups of algebras over $A \otimes \overline{A}$, *J. Math. Soc. Japan* 29 (1977), 459–492.

[20] K.-H. Ulbrich, Galoiserweiterungen von nicht-kommutativen Ringen, *Comm. Algebra* 10 (1982) 655–672

[21] P. Vecsernyés, Larson-Sweedler theorem, grouplike elments and invertible modules in weak Hopf algebras, preprint `math.QA/0111045` (2001).

[22] P. Xu, Quantum groupoids, preprint `QA/9905192` (1999).

On low-dimensional locally compact quantum groups

Stefaan Vaes and Leonid Vainerman

Institut de Mathématiques de Jussieu
Algèbres d'Opérateurs, Plateau 7E
175, rue du Chevaleret, 75013 Paris, France
email: vaes@math.jussieu.fr

Département de Mathématiques et Méchanique
Université de Caen
Campus II – Boulevard de Maréchal Juin
B.P. 5186, 14032 Caen Cedex, France
email: leonid.vainerman@math.unicaen.fr

Abstract. Continuing our research on extensions of locally compact quantum groups, we give a classification of all cocycle matched pairs of Lie algebras in small dimensions and prove that all of them can be exponentiated to cocycle matched pairs of Lie groups. Hence, all of them give rise to locally compact quantum groups by the cocycle bicrossed product construction. We also clarify the notion of an extension of locally compact quantum groups by relating it to the concept of a closed normal quantum subgroup and the quotient construction. Finally, we describe the infinitesimal objects of locally compact quantum groups with 2 and 3 generators – Hopf $*$-algebras and Lie bialgebras.

1 Introduction

In this paper we continue the research on extensions of locally compact (l.c.) quantum groups, initiated in [51]. The first wide class of l.c. quantum groups, namely Kac algebras, was introduced in the early sixties (see [18]) in order to explain in a symmetric way duality for l.c. groups. This class included besides usual l.c. groups and their duals also nontrivial (i.e., non-commutative and non-cocommutative) objects [19], [20]. The general Kac algebra theory was completed independently on the one hand by G. I. Kac and the second author [21] and on the other hand by M. Enock and J.-M. Schwartz (for a survey see [12]). However, this theory was not general enough to cover important new examples constructed starting from the eighties [3], [24], [25], [31], [41], [42], [52], [56], [60]–[66], which motivated essential efforts to get a generalization that would cover these examples and that would be as elegant and symmetric as the theory of Kac algebras. Important steps in this direction were made by S. Baaj and G. Skandalis [4], S. L. Woronowicz [58], [59], [60], T. Masuda and Y. Nakagami [34] and A. Van Daele [54]. The general theory of l.c. quantum groups was proposed by J. Kustermans and the first author [26], [27] (see [28] for an overview). Some motivations and applications of this theory can be found in the recent lecture notes [29].

The mentioned examples of l.c. quantum groups are, first of all, formulated algebraically, in terms of generators of Hopf*-algebras and commutation relations between them. Then one represents the generators as (typically, unbounded) operators on a Hilbert space and tries to give a meaning to the commutation relations as relations between these operators. There is no general approach to this nontrivial problem, and one elaborates specific methods in each specific case. Finally, it is necessary to associate an operator algebra with the above system of operators and commutation relations and to construct comultiplication, antipode and invariant weights as applications related to this algebra. This problem is even more difficult than the previous one and again one must consider separately each specific case (see the same papers). So, it would be desirable to have some general constructions of l.c. quantum groups which would allow to construct systematically concrete examples in a unified way.

One of such possibilities is offered by the *cocycle bicrossed product construction*. According to G. I. Kac [17], in the simplest case the needed data for this construction contains:

1. A pair of finite groups G_1 and G_2 equipped with their mutual actions on each other (as on sets) or, equivalently, G_1 and G_2 must be subgroups of a certain group G such that $G_1 \cap G_2 = \{e\}$ and any $g \in G$ can be written as $g = g_1 g_2$ ($g_1 \in G_1, g_2 \in G_2$) - we write briefly $G = G_1 G_2$. We then say, that G_1 and G_2 form *a matched pair of groups* [47].

2. A pair of compatible 2-cocycles for these actions, so G_1 and G_2 must form *a cocycle matched pair* (in what follows we often write simply "cocycle" rather then "2-cocycle").

Then, due to [17], one can construct a finite-dimensional Kac algebra from cocycle crossed products of the algebras of functions on each of the groups G_1 and G_2 with the cocycle action of the other group, and this construction gives exactly all extensions of the above groups in the category of finite-dimensional Kac algebras.

It is tempting to similarly treat Lie groups instead of finite groups, being supported by the theory of cocycle bicrossed products and extensions of l.c. groups developed in [51] (in fact, in [51], the general theory of cocycle bicrossed products and extensions of l.c. *quantum* groups was developed). But first of all it turns out that the above definition of a matched pair of groups in terms of the equality $G = G_1 G_2$ does not cover all interesting examples, see [5]. Following S. Baaj and G. Skandalis, one can just require $G_1 G_2$ to be an open subset of G with complement of measure zero. Then, the Lie algebras $\mathfrak{g}_1$ of G_1 and $\mathfrak{g}_2$ of G_2 are Lie subalgebras of the Lie algebra $\mathfrak{g}$ of G and $\mathfrak{g} = \mathfrak{g}_1 \oplus \mathfrak{g}_2$ as the direct sum of vector spaces, i.e., they form a matched pair of Lie algebras [33], 8.3. Thus, to get matched pairs of Lie groups one can start with matched pairs of Lie algebras (which are easier to find) and then try to exponentiate them.

To construct in this way cocycle matched pairs of Lie groups, one has to resolve two problems. First, given a matched pair of Lie algebras $(\mathfrak{g}_1, \mathfrak{g}_2)$ with $\mathfrak{g} = \mathfrak{g}_1 \oplus \mathfrak{g}_2$, one can always exponentiate $\mathfrak{g}$ to a connected and simply connected Lie group G and then find Lie subgroups G_1 and G_2 whose Lie algebras are $\mathfrak{g}_1$ and $\mathfrak{g}_2$, respectively. However, such a choice of G does not guarantee that G_1G_2 is dense in G, even if $\dim(G_1) = \dim(G_2) = 1$ [33], [44], [51], and it also may happen that $G_1 \cap G_2 \neq \{e\}$. So, it is necessary to pass to some non-connected Lie group G with the same Lie algebra $\mathfrak{g}$ in order to find a matched pair of its subgroups G_1 and G_2 [49], [51]. Secondly, given a matched pair of Lie groups, one has to find the corresponding cocycles. We give a solution of both these problems for *real* Lie groups G_1 and G_2 with $\dim(G_1) = 1$, $\dim(G_2) \leq 2$ and construct essentially all possible (up to obvious redundancies) matched pairs of such Lie groups having at most 2 connected components. Then, using the machinery of cocycle bicrossed products developed in [51], we construct l.c. quantum groups which are extensions of the mentioned Lie groups. Our discussion is motivated, apart from the above work by G. I. Kac, also by the works by S. Majid [30]–[33], S. Baaj and G. Skandalis [4], [5], [44], and by the works on extensions of Hopf algebras [1], [2], [43].

The material is organized as follows. In Section 2, we recall the necessary facts of the theory of l.c. quantum groups and, following [51], the main features of the cocycle bicrossed product construction for l.c. groups in connection with the theory of extensions. In the last subsection we introduce the notion of a closed normal quantum subgroup of a l.c. quantum group and explain its relation to the theory of extensions. As we explained above, the basic notion of this theory is that of a matched pair of l.c. groups. If the groups forming a matched pair are Lie groups, we naturally have a matched pair of their Lie algebras. But the converse problem, to construct a matched pair of Lie groups from a given matched pair $(\mathfrak{g}_1, \mathfrak{g}_2)$ of Lie algebras, is much more subtle. In particular, in Section 3 we show that any matched pair with $\mathfrak{g}_1 = \mathfrak{g}_2 = \mathbb{C}$ can be exponentiated to a matched pair of complex Lie groups, but there are simple examples of matched pairs of real and complex Lie algebras for which the exponentiation is impossible.

The study of matched pairs of Lie algebras with $\dim \mathfrak{g}_1 = n$, $\dim \mathfrak{g}_2 = 1$ in Section 4 splits in three cases. In case 1, when $\mathfrak{g}_1$ is an ideal in $\mathfrak{g}$, G can be constructed as semi-direct product of connected and simply connected Lie groups corresponding to $\mathfrak{g}_1$ and $\mathfrak{g}_2$ (this is possible also for $\dim \mathfrak{g}_2 > 1$). In case 2, when $\mathfrak{g}_1$ contains an ideal of codimension 1, the results of Section 3 show that for complex Lie algebras the exponentiation always exists when $n = 1$ and it does not exist in general if $n \geq 2$. For real Lie algebras we show that for $n \leq 4$ there always exists the exponentiation to a matched pair of Lie groups with at most two connected components, and for $n \geq 5$ the exponentiation does not exist in general. In the remaining case 3, for complex Lie algebras the exponentiation always exists when $n \leq 3$ and it does not exist in general if $n \geq 4$. For real Lie algebras we show that the exponentiation always exists when $n \leq 4$.

Section 5 is devoted to the complete classification of all matched pairs of real Lie algebras $\mathfrak{g}_1$ and $\mathfrak{g}_2$ when $\dim(\mathfrak{g}_1) = 1$, $\dim(\mathfrak{g}_2) \leq 2$ and to their explicit exponentiation to matched pairs of real Lie groups having at most 2 connected components. Here, we also describe the l.c. quantum groups obtained from these matched pairs by the bicrossed product construction. In Section 6, we calculate the cocycles for all the above mentioned matched pairs. Finally, Section 7 is devoted to the description of l.c. quantum groups with 2 and 3 generators and their infinitesimal objects – Hopf $*$-algebras and Lie bialgebras, having the structure of a cocycle bicrossed product – equivalently, those that can be obtained as extensions (we call them decomposable). At last, to complete the picture of low-dimensional l.c. quantum groups, we review the indecomposable ones and their infinitesimal objects.

Acknowledgements. The first author would like to thank the research group Analysis of the Department of Mathematics of the K.U.Leuven for the nice working atmosphere while this work was initiated. He also wants to thank the whole Operator Algebra team of the Institut de Mathématiques de Jussieu in Paris for their warm hospitality and the many useful discussions while this work was finalized. The second author is grateful to the research group Analysis of the Department of Mathematics of the K.U. Leuven, to l'Institut de Recherche Mathématique Avancée de Strasbourg and to Max-Planck-Institut für Mathematik in Bonn for the warm hospitality and financial support during his work on this paper.

2 Preliminaries

General notations. Let $B(H)$ denote the algebra of all bounded linear operators on a Hilbert space H, let $\otimes$ denote the tensor product of Hilbert spaces or von Neumann algebras and Σ (resp., σ) the flip map on it. If H, K and L are Hilbert spaces and $X \in B(H \otimes L)$ (resp., $X \in B(H \otimes K)$, $X \in B(K \otimes L)$), we denote by X_{13} (resp., X_{12}, X_{23}) the operator $(1 \otimes \Sigma^*)(X \otimes 1)(1 \otimes \Sigma)$ (resp., $X \otimes 1$, $1 \otimes X$) defined on $H \otimes K \otimes L$. Sometimes, when $H = H_1 \otimes H_2$ itself is a tensor product of two Hilbert spaces, we switch from the above leg-numbering notation with respect to $H \otimes K \otimes L$ to the one with respect to the finer tensor product $H_1 \otimes H_2 \otimes K \otimes L$, for example, from X_{13} to X_{124}. There is no confusion here, because the number of legs changes.

Given a comultiplication Δ, denote by Δ^{op} the opposite comultiplication $\sigma\Delta$. Our general reference to the modular theory of *normal semi-finite faithful* (n.s.f.) weights on von Neumann algebras is [45]. For any weight θ on a von Neumann algebra N, we use the notations

$$\mathcal{M}_\theta^+ = \{x \in N^+ \mid \theta(x) < +\infty\}, \quad \mathcal{N}_\theta = \{x \in N \mid x^*x \in \mathcal{M}_\theta^+\} \quad \text{and}$$
$$\mathcal{M}_\theta = \operatorname{span} \mathcal{M}_\theta^+.$$

L.c. quantum groups. A pair (M, Δ) is called a (von Neumann algebraic) l.c. quantum group [27] when

- M is a von Neumann algebra and $\Delta : M \to M \otimes M$ is a normal and unital $*$-homomorphism satisfying the coassociativity relation: $(\Delta \otimes \iota)\Delta = (\iota \otimes \Delta)\Delta$.
- There exist n.s.f. weights φ and ψ on M such that
  - φ is left invariant in the sense that $\varphi\big((\omega \otimes \iota)\Delta(x)\big) = \varphi(x)\omega(1)$ for all $x \in \mathcal{M}_\varphi^+$ and $\omega \in M_*^+$,
  - ψ is right invariant in the sense that $\psi\big((\iota \otimes \omega)\Delta(x)\big) = \psi(x)\omega(1)$ for all $x \in \mathcal{M}_\psi^+$ and $\omega \in M_*^+$.

Left and right invariant weights are unique up to a positive scalar [26], Theorem 7.14.

Represent M on the Hilbert space of a GNS-construction (H, ι, Λ) for the left invariant n.s.f. weight φ and define a unitary W on $H \otimes H$ by

$$W^*(\Lambda(a) \otimes \Lambda(b)) = (\Lambda \otimes \Lambda)(\Delta(b)(a \otimes 1)) \quad \text{for all } a, b \in N_\phi.$$

Here, $\Lambda \otimes \Lambda$ denotes the canonical GNS-map for the tensor product weight $\varphi \otimes \varphi$. One proves that W satisfies the pentagonal equation: $W_{12}W_{13}W_{23} = W_{23}W_{12}$, and we say that W is a multiplicative unitary. The von Neumann algebra M and the comultiplication on it can be given in terms of W respectively as

$$M = \{(\iota \otimes \omega)(W) \mid \omega \in B(H)_*\}^{-\,\sigma\text{-strong}^*}$$

and $\Delta(x) = W^*(1 \otimes x)W$, for all $x \in M$. Next, the l.c. quantum group (M, Δ) has an antipode S, which is the unique σ-strong* closed linear map from M to M satisfying $(\iota \otimes \omega)(W) \in \mathcal{D}(S)$ for all $\omega \in B(H)_*$ and $S(\iota \otimes \omega)(W) = (\iota \otimes \omega)(W^*)$ and such that the elements $(\iota \otimes \omega)(W)$ form a σ-strong* core for S. S has a polar decomposition $S = R\tau_{-i/2}$ where R is an anti-automorphism of M and (τ_t) is a strongly continuous one-parameter group of automorphisms of M. We call R the unitary antipode and (τ_t) the scaling group of (M, Δ). From [26], Proposition 5.26 we know that $\sigma(R \otimes R)\Delta = \Delta R$. So φR is a right invariant weight on (M, Δ) and we take $\psi := \varphi R$.

Let us denote by (σ_t) the modular automorphism group of φ. From [26], Proposition 6.8 we get the existence of a number $\nu > 0$, called the scaling constant, such that $\psi\,\sigma_t = \nu^{-t}\,\psi$ for all $t \in \mathbb{R}$. Hence, we get the existence of a unique positive, self-adjoint operator δ_M affiliated to M, such that $\sigma_t(\delta_M) = \nu^t\,\delta_M$ for all $t \in \mathbb{R}$ and $\psi = \varphi_{\delta_M}$, see [26], Definition 7.1. Formally this means that $\psi(x) = \varphi(\delta_M^{1/2} x \delta_M^{1/2})$, and for a precise definition of φ_{δ_M} we refer to [50]. The operator δ_M is called the modular element of (M, Δ). If $\delta_M = 1$ we call (M, Δ) unimodular. The scaling constant can be characterized as well by the relative invariance $\varphi\,\tau_t = \nu^{-t}\,\varphi$.

The dual l.c. quantum group $(\hat{M}, \hat{\Delta})$ is defined in [26], Section 8. Its von Neumann algebra $\hat{M}$ is

$$\hat{M} = \{(\omega \otimes \iota)(W) \mid \omega \in B(H)_*\}^{-\,\sigma\text{-strong}^*}$$

and the comultiplication $\hat{\Delta}(x) = \Sigma W(x \otimes 1)W^*\Sigma$ for all $x \in \hat{M}$. If we turn the predual M_* into a Banach algebra with product $\omega\,\mu = (\omega \otimes \mu)\Delta$ and define

$$\lambda : M_* \to \hat{M} : \lambda(\omega) = (\omega \otimes \iota)(W),$$

then λ is a homomorphism and $\lambda(M_*)$ is a σ-strong* dense subalgebra of $\hat{M}$. To construct explicitly a left invariant n.s.f. weight $\hat{\varphi}$ with GNS-construction $(H, \iota, \hat{\Lambda})$, first introduce the space

$$\mathcal{I} = \{\omega \in M_* \mid \text{ there exists } \xi(\omega) \in H \text{ s.t. } \omega(x^*) = \langle \xi(\omega), \Lambda(x)\rangle \text{ when } x \in \mathcal{N}_\varphi\}.$$

If $\omega \in \mathcal{I}$, then such a vector $\xi(\omega)$ clearly is uniquely determined. Next, one proves that there exists a unique n.s.f. weight $\hat{\varphi}$ on $\hat{M}$ with GNS-construction $(H, \iota, \hat{\Lambda})$ such that $\lambda(\mathcal{I})$ is a core for $\hat{\Lambda}$ (when we equip $\hat{M}$ with the σ-strong* topology and H with the norm topology) and such that

$$\hat{\Lambda}(\lambda(\omega)) = \xi(\omega) \quad \text{for all } \omega \in \mathcal{I}.$$

One proves that the weight $\hat{\varphi}$ is left invariant, and the associated multiplicative unitary is denoted by $\hat{W}$. From [26], Proposition 8.16 it follows that $\hat{W} = \Sigma W^*\Sigma$.

Since $(\hat{M}, \hat{\Delta})$ is again a l.c. quantum group, we can introduce the antipode $\hat{S}$, the unitary antipode $\hat{R}$ and the scaling group $(\hat{\tau}_t)$ exactly as we did it for (M, Δ). Also, we can again construct the dual of $(\hat{M}, \hat{\Delta})$, starting from the left invariant weight $\hat{\varphi}$ with GNS-construction $(H, \iota, \hat{\Lambda})$. From [26], Theorem 8.29 we have that the bidual l.c. quantum group $(\hat{\hat{M}}, \hat{\hat{\Delta}})$ is isomorphic to (M, Δ).

We denote by $(\hat{\sigma}_t)$ the modular automorphism groups of the weight $\hat{\varphi}$. The modular conjugations of the weights φ and $\hat{\varphi}$ will be denoted by J and $\hat{J}$ respectively. Then it is worthwhile to mention that

$$R(x) = \hat{J}x^*\hat{J} \quad \text{for all } x \in M \quad \text{and} \quad \hat{R}(y) = Jy^*J \quad \text{for all } y \in \hat{M}.$$

Let us mention important special cases of l.c. quantum groups.

a) *Kac algebras* [12]. From [12], we know that (M, Δ) is a Kac algebra if and only if (τ_t) is trivial and $\sigma_t\, R = R\,\sigma_{-t}$ for all $t \in \mathbb{R}$. Now, denote by (σ'_t) the modular automorphism group of ψ. Because $\psi = \varphi R$ we get that $\sigma'_t\, R = R\,\sigma_{-t}$ for all $t \in \mathbb{R}$. Hence (M, Δ) is a Kac algebra if and only if (τ_t) is trivial and $\sigma' = \sigma$. From [50], we know that $\sigma'_t(x) = \delta_M^{it}\sigma_t(x)\delta_M^{-it}$ for all $x \in M$ and $t \in \mathbb{R}$. Hence $\sigma' = \sigma$ if and only if δ_M is affiliated to the center of M.

In particular, (M, Δ) is a Kac algebra if M is commutative. Then (M, Δ) is generated by a usual l.c. group G : $M = L^\infty(G)$, $(\Delta f)(g, h) = f(gh)$ $(Sf)(g) =$

$f(g^{-1})$, $\varphi(f) = \int f(g)\,dg$, where $f \in L^\infty(G)$, $g, h \in G$ and we integrate with respect to the left Haar measure dg on G. The right invariant weight ψ is given by $\psi(f) = \int f(g^{-1})\,dg$. The modular element δ_M is given by the strictly positive function $g \mapsto \delta_G(g)^{-1}$.

The von Neumann algebra $M = L^\infty(G)$ acts on $H = L^2(G)$ by multiplication and

$$(W_G\xi)(g, h) = \xi(g, g^{-1}h)$$

for all $\xi \in H \otimes H = L^2(G \times G)$. Then $\hat{M} = \mathcal{L}(G)$ is the group von Neumann algebra generated by the operators $(\lambda_g)_{g \in G}$ of the left regular representation of G and $\hat{\Delta}(\lambda_g) = \lambda_g \otimes \lambda_g$. Clearly, $\hat{\Delta}^{\mathrm{op}} := \sigma\hat{\Delta} = \hat{\Delta}$, so $\hat{\Delta}$ is cocommutative.

b) A l.c. quantum group is called *compact* if its Haar measure is finite: $\varphi(1) < +\infty$, which is equivalent to the fact that the norm closure of $\{(\iota \otimes \omega)(W) | \omega \in B(H)_*\}$ is a unital C^*-algebra. A l.c. quantum group (M, Δ) is called *discrete* if $(\hat{M}, \hat{\Delta})$ is compact.

Crossed and bicrossed products. An *action* of a l.c. quantum group (M, Δ) on a von Neumann algebra N is a normal, injective and unital $*$-homomorphism $\alpha : N \to M \otimes N$ such that $(\iota \otimes \alpha)\alpha(x) = (\Delta \otimes \iota)\alpha(x)$ for all $x \in N$. This generalizes the definition of an action of a (separable) l.c. group G on a (σ-finite) von Neumann algebra N, as a continuous map $G \to \operatorname{Aut} N : s \mapsto \alpha_s$ such that $\alpha_{st} = \alpha_s\alpha_t$ for all $s, t \in G$. Indeed, putting $M = L^\infty(G)$, one can identify $M \otimes N$ with $L^\infty(G, N)$ and $M \otimes M \otimes N$ with $L^\infty(G \times G, N)$ and define the above homomorphism α by $(\alpha(x))(s) = \alpha_{s^{-1}}(x)$. The fixed point algebra of an action α is defined by $N^\alpha = \{x \in N \mid \alpha(x) = 1 \otimes x\}$.

A *cocycle* for an action of a l.c. group G on a commutative von Neumann algebra N is a Borel map $u : G \times G \to N$ such that $\alpha_r(u(s, t))\, u(r, st) = u(r, s)\, u(rs, t)$ nearly everywhere. Then, putting $M = L^\infty(G)$, one can define a unitary $\mathcal{U} \in M \otimes M \otimes N$ by $\mathcal{U}(s, t) = u(t^{-1}, s^{-1})$ satisfying

$$(\iota \otimes \iota \otimes \alpha)(\mathcal{U})(\Delta \otimes \iota \otimes \iota)(\mathcal{U}) = (1 \otimes \mathcal{U})(\iota \otimes \Delta \otimes \iota)(\mathcal{U}).$$

For the general definition of a cocycle action of a l.c. quantum group on an arbitrary von Neumann algebra, we refer to Definition 1.1 in [51].

The *cocycle crossed product* $G\,{}_{\alpha,\mathcal{U}}\!\ltimes N$ is the von Neumann subalgebra of $B(L^2(G)) \otimes N$ generated by

$$\alpha(N) \quad \text{and} \quad \{(\omega \otimes \iota \otimes \iota)(\tilde{W}) \mid \omega \in L^1(G)\},$$

where $\tilde{W} = (W_G \otimes 1)\mathcal{U}^*$. This is a von Neumann algebraic version of the twisted C^*-algebraic crossed product [40]. There exists a unique action $\hat{\alpha}$ of $(\mathcal{L}(G), \hat{\Delta})$ on $G\,{}_{\alpha,\mathcal{U}}\!\ltimes N$ such that

$$\hat{\alpha}(\alpha(x)) = 1 \otimes \alpha(x) \quad \text{for all } x \in N,$$
$$(\iota \otimes \hat{\alpha})(\tilde{W}) = W_{G,12}\tilde{W}_{134},$$

and for any n.s.f. weight θ on N, we can define the *dual* n.s.f. weight $\tilde{\theta}$ on $G\,{}_{\alpha,\mathcal{U}}\!\ltimes N$ by the formula

$$\tilde{\theta} = \theta\alpha^{-1}\ (\hat{\varphi} \otimes \iota \otimes \iota)\hat{\alpha}.$$

Definition 2.1. (see [6]) Let G, G_1 and G_2 be (separable) l.c. groups and let a homomorphism $i : G_1 \to G$ and an anti-homomorphism $j : G_2 \to G$ have closed images and be homeomorphisms onto these images. Suppose that $i(G_1) \cap j(G_2) = \{e\}$ and that the complement of $i(G_1)j(G_2)$ in G has measure zero. Then we call G_1 and G_2 a matched pair of l.c. groups.

Observe that this definition of a matched pair of l.c. groups, due to Baaj, Skandalis and the first author, is more general than the one studied in [5] and [51]. Indeed, in [6], there is given an example of a matched pair in the sense of the definition above, which does not fit in the definition of [5]. More specifically, consider the map

$$\theta : G_1 \times G_2 \to G : (g, s) \mapsto i(g)j(s),$$

which is clearly injective. In [5] and [51], the map θ is supposed to have a range Ω which is open in G, with complement of measure zero and such that θ is a homeomorphism of $G_1 \times G_2$ onto Ω. In the example of [6], the range of θ has an empty interior. However, the following proposition holds:

Proposition 2.2. *If, in Definition* 2.1, *G is a Lie group, then the map*

$$\theta : G_1 \times G_2 \to G : (g, s) \mapsto i(g)j(s)$$

has an open range Ω and is a diffeomorphism of $G_1 \times G_2$ onto Ω, where G_1 and G_2 are Lie groups under the identification with closed subgroups of G.

Proof. Denote by $\mathfrak{g}$, $\mathfrak{g}_1$, $\mathfrak{g}_2$ the Lie algebras of G, G_1, G_2, respectively. Then, we have an injective homomorphism and anti-homomorphism

$$\mathrm{d}i : \mathfrak{g}_1 \to \mathfrak{g} \quad \text{and} \quad \mathrm{d}j : \mathfrak{g}_2 \to \mathfrak{g}.$$

Because $i(G_1) \cap j(G_2) = \{e\}$, we get $\mathrm{d}i(\mathfrak{g}_1) \cap \mathrm{d}j(\mathfrak{g}_2) = \{0\}$ (otherwise, the exponential mapping produces elements in $i(G_1) \cap j(G_2)$). Hence, we can take a linear subspace $\mathfrak{k}$ of $\mathfrak{g}$ (not necessarily a Lie subalgebra) such that $\mathfrak{g} = \mathrm{d}i(\mathfrak{g}_1) \oplus \mathrm{d}j(\mathfrak{g}_2) \oplus \mathfrak{k}$ as vector spaces. We first prove that $\mathfrak{k} = \{0\}$.

Denote by $\exp_{\mathfrak{g}}$ the exponential mapping of G and analogously for $\exp_{\mathfrak{g}_{1,2}}$. Take open subsets $\mathcal{U}_i \subset \mathfrak{g}_i$, $\mathcal{V} \subset \mathfrak{k}$ containing 0 such that $\exp_{\mathfrak{g}}$ is a diffeomorphism of $\mathrm{d}i(\mathcal{U}_1) \times \mathrm{d}j(\mathcal{U}_2) \times \mathcal{V}$ onto an open subset of G. Define

$$\begin{aligned}\rho : \mathcal{U}_1 \times \mathcal{U}_2 \times \mathcal{V} \to G : \rho(v, w, z) &= i(\exp_{\mathfrak{g}_1}(v))\ j(\exp_{\mathfrak{g}_2}(w))\ \exp_{\mathfrak{g}}(z) \\ &= \exp_{\mathfrak{g}}(\mathrm{d}i(v))\ \exp_{\mathfrak{g}}(\mathrm{d}j(w))\ \exp_{\mathfrak{g}}(z).\end{aligned}$$

Because $\mathfrak{g} = \mathrm{d}i(\mathfrak{g}_1) \oplus \mathrm{d}j(\mathfrak{g}_2) \oplus \mathfrak{k}$, we find that $\mathrm{d}\rho(0, 0, 0)$ is bijective. So, for $\mathcal{U}_1, \mathcal{U}_2, \mathcal{V}$ small enough, ρ is a diffeomorphism onto an open subset of $\mathcal{W}$ of G containing e and $\exp_{\mathfrak{g}_i}$ will be a diffeomorphism of $\mathcal{U}_i$ onto an open subset $\mathcal{W}_i$ of G_i.

It is clear that $\theta(\mathcal{W}_1 \times \mathcal{W}_2) \subset \mathcal{W}$ and $\rho^{-1}(\theta(\mathcal{W}_1 \times \mathcal{W}_2)) = \mathcal{U}_1 \times \mathcal{U}_2 \times \{0\}$. As a diffeomorphism, ρ is a Borel isomorphism and so, if $\mathfrak{k} \neq \{0\}$, $\theta(\mathcal{W}_1 \times \mathcal{W}_2)$ has measure zero in G. This contradicts the result of [6], saying that θ is automatically a Borel isomorphism. Hence, $\mathfrak{k} = \{0\}$.

But then, $\theta : \mathcal{W}_1 \times \mathcal{W}_2 \to \mathcal{W}$ is a diffeomorphism. In particular, $\mathcal{W} \subset \Omega$. If now $i(g_0)j(s_0) \in \Omega$, it follows that $i(g_0)j(s_0) \in i(g_0)\mathcal{W}j(s_0) \subset \Omega$. Hence, Ω is open in G and θ is a diffeomorphism of $G_1 \times G_2$ onto Ω, because we know that θ is injective. □

In [6], it is proved that θ is automatically a Borel isomorphism, i.e. it induces an isomorphism between $L^\infty(G_1 \times G_2)$ and $L^\infty(G)$. Hence, this data allows to construct as follows two actions: α of G_1 on $M_2 = L^\infty(G_2)$ and β of G_2^{op} on $M_1 = L^\infty(G_1)$ verifying certain compatibility relations.

Define Ω to be the image of θ and define the Borel isomorphism

$$\rho : G_1 \times G_2 \to \Omega^{-1} : (g, s) \mapsto j(s)i(g).$$

So $\mathcal{O} = \theta^{-1}(\Omega \cap \Omega^{-1})$ and $\mathcal{O}' = \rho^{-1}(\Omega \cap \Omega^{-1})$ are Borel subsets of $G_1 \times G_2$, with complement of measure zero, and $\rho^{-1}\theta$ is a Borel isomorphism of $\mathcal{O}$ onto $\mathcal{O}'$. For all $(g, s) \in \mathcal{O}$ define $\beta_s(g) \in G_1$ and $\alpha_g(s) \in G_2$ such that

$$\rho^{-1}(\theta(g, s)) = (\beta_s(g), \alpha_g(s)).$$

Hence we get $j\big(\alpha_g(s)\big)\, i\big(\beta_s(g)\big) = i(g)j(s)$ for all $(g, s) \in \mathcal{O}$.

Lemma 2.3 ([51], Lemma 4.8). *Let $(g, s) \in \mathcal{O}$ and $h \in G_1$. Then $(hg, s) \in \mathcal{O}$ if and only if $(h, \alpha_g(s)) \in \mathcal{O}$, and in that case*

$$\alpha_{hg}(s) = \alpha_h\big(\alpha_g(s)\big) \quad \textit{and} \quad \beta_s(hg) = \beta_{\alpha_g(s)}(h)\, \beta_s(g).$$

Let $(g, s) \in \mathcal{O}$ and $t \in G_2$. Then $(g, ts) \in \mathcal{O}$ if and only if $(\beta_s(g), t) \in \mathcal{O}$ and in that case

$$\beta_{ts}(g) = \beta_t\big(\beta_s(g)\big) \quad \textit{and} \quad \alpha_g(ts) = \alpha_{\beta_s(g)}(t)\, \alpha_g(s).$$

Finally, for all $g \in G_1$ and $s \in G_2$ we have $(g, e) \in \mathcal{O}$, $(e, s) \in \mathcal{O}$, and

$$\alpha_g(e) = e, \quad \alpha_e(s) = s, \quad \beta_s(e) = e \quad \textit{and} \quad \beta_e(g) = g.$$

This can be viewed as a definition of a matched pair of l.c. groups in terms of mutual actions.

The cocycles for the above actions can be introduced as measurable maps $\mathcal{U} : G_1 \times G_1 \times G_2 \to U(1)$ and $\mathcal{V} : G_1 \times G_2 \times G_2 \to U(1)$, where $U(1)$ is the unit circle in $\mathbb{C}$, satisfying

$$\begin{aligned} \mathcal{U}(g,h,\alpha_k(s))\ \mathcal{U}(gh,k,s) &= \mathcal{U}(h,k,s)\ \mathcal{U}(g,hk,s), \\ \mathcal{V}(\beta_s(g),t,r)\ \mathcal{V}(g,s,rt) &= \mathcal{V}(g,s,t)\ \mathcal{V}(g,ts,r), \\ \mathcal{V}(gh,s,t)\ \bar{\mathcal{U}}(g,h,ts) &= \bar{\mathcal{U}}(g,h,s)\ \bar{\mathcal{U}}(\beta_{\alpha_h(s)}(g),\beta_s(h),t) \\ &\qquad \mathcal{V}(g,\alpha_h(s),\alpha_{\beta_s(h)}(t))\ \mathcal{V}(h,s,t) \end{aligned} \tag{2.1}$$

nearly everywhere. Then we have a definition of a cocycle matched pair of l.c. groups.

Fixing a cocycle matched pair of l.c. groups G_1 and G_2, denoting $H_i = L^2(G_i)$ $(i = 1, 2)$, $H = H_1 \otimes H_2$ and identifying $\mathcal{U}$ and $\mathcal{V}$ with unitaries in $M_1 \otimes M_1 \otimes M_2$ and in $M_1 \otimes M_2 \otimes M_2$ respectively, define unitaries W and $\hat{W}$ on $H \otimes H$ by

$$\hat{W} = (\beta \otimes \iota \otimes \iota)\big((W_{G_1} \otimes 1)\mathcal{U}^*\big)\ (\iota \otimes \iota \otimes \alpha)\big(\mathcal{V}(1 \otimes \hat{W}_{G_2})\big) \text{ and } W = \Sigma \hat{W}^* \Sigma.$$

On the von Neumann algebra $M = G_1\, {}_{\alpha,\mathcal{U}}\!\ltimes L^\infty(G_2)$, let us define a faithful $*$-homomorphism

$$\Delta : M \to B(H \otimes H) : \Delta(z) = W^*(1 \otimes z)W \ (\forall z \in M)$$

and denote by φ the dual weight of the canonical left invariant trace φ_2 on $L^\infty(G_2)$. Then, Theorem 2.13 of [51] shows that (M, Δ) is a l.c. quantum group with φ as a left invariant weight, which we call the *cocycle bicrossed product* of G_1 and G_2. One can also show that its scaling constant is 1. The dual l.c. quantum group is $(\hat{M}, \hat{\Delta})$, where $\hat{M} = L^\infty(G_1) \rtimes_{\beta,\mathcal{V}} G_2$ and $\hat{\Delta}(z) = \hat{W}^*(1 \otimes z)\hat{W}$ for all $z \in \hat{M}$.

One can get explicit formulas for the modular operators, modular conjugations of the left invariant weights, unitary antipodes, scaling groups and modular elements of both (M, Δ) and its dual in terms of the above mutual actions, the cocycles and the modular functions δ, δ_1 and δ_2 of the l.c. groups G, G_1 and G_2. In particular, one can characterize all cocycle bicrossed products of l.c. groups which are Kac algebras.

Proposition 2.4. *The l.c. quantum group (M, Δ) is a Kac algebra if and only if*

$$\delta\big(i(g\beta_s(g)^{-1})\big)\ \delta_1\big(g^{-1}\beta_s(g)\big)\ \delta_2\big(\alpha_g(s)s^{-1}\big) = 1 \quad \text{and} \quad \frac{\delta_1\big(\beta_s(g)\big)}{\delta_1(g)} = \frac{\delta_2\big(\alpha_g(s)\big)}{\delta_2(s)}.$$

This proposition implies three helpful corollaries.

Corollary 2.5. *If α or β is trivial, (M, Δ) and $(\hat{M}, \hat{\Delta})$ are Kac algebras.*

Corollary 2.6. *If both α and β preserve modular functions and Haar measures, then (M, Δ) and $(\hat{M}, \hat{\Delta})$ are Kac algebras.*

Remark that the conditions of this corollary are fulfilled if both groups are discrete. Indeed, any discrete group is unimodular and the Haar measure is constant at an arbitrary point of such a group.

Corollary 2.7. *If (G_1, G_2) is a fixed matched pair of l.c. groups and cocycles $\mathcal{U}$ and $\mathcal{V}$ satisfy* (2.1), *we get a cocycle bicrossed product (M, Δ). If one of these cocycle bicrossed products is a Kac algebra, then all of them are Kac algebras.*

Proof. The necessary and sufficient conditions for (M, Δ) to be a Kac algebra in Proposition 2.4 are independent of $\mathcal{U}$ and $\mathcal{V}$. □

It is easy to check that the above measurable mutual actions α_g and β_s of G_1 and G_2 are in fact the restrictions of the canonical continuous actions $\tilde{\alpha}_g$ of G_1 on G/G_1 and $\tilde{\beta}_s$ of G_2 on $G_2\backslash G$ (topologies on G_1 and $G_2\backslash G$ and, respectively, on G_2 and G/G_1, are in general different). This allows, in particular, to express the C^*-algebras of the C^*-algebraic versions of the split extension (i.e. with trivial cocycles) and its dual respectively as $G_1 {}_{\tilde{\alpha}}\ltimes C_0(G/G_1)$ and $C_0(G_2\backslash G) \rtimes_{\tilde{\beta}} G_2$, see [6].

Extensions of l.c. groups. To clarify the following definition, recall that any normal $*$-homomorphism $\beta : M_1 \to \hat{M}$ of l.c. quantum groups satisfying $\hat{\Delta}\beta = (\beta \otimes \beta)\Delta_1$ generates two canonical actions: μ of $(\hat{M}_1, \hat{\Delta}_1)$ on M and θ of $(\hat{M}_1, \hat{\Delta}_1^{\mathrm{op}})$ on M ([51], Proposition 3.1). On a formal level, this can be understood easily: the morphism β gives rise to a dual morphism $\tilde{\beta} : M \to \hat{M}_1$ and μ should be thought of as $\mu = (\tilde{\beta} \otimes \iota)\Delta$, while θ should be thought of as $\theta = (\tilde{\beta} \otimes \iota)\Delta^{\mathrm{op}}$.

Definition 2.8. Let G_i $(i = 1, 2)$ be l.c. groups and let (M, Δ) be a l.c. quantum group. We call

$$(L^\infty(G_2), \Delta_2) \xrightarrow{\alpha} (M, \Delta) \xrightarrow{\beta} (\mathcal{L}(G_1), \hat{\Delta}_1)$$

a short exact sequence, if

$$\alpha : L^\infty(G_2) \to M \quad \text{and} \quad \beta : L^\infty(G_1) \to \hat{M}$$

are normal, faithful $*$-homomorphisms satisfying

$$\Delta\alpha = (\alpha \otimes \alpha)\Delta_2 \quad \text{and} \quad \hat{\Delta}\beta = (\beta \otimes \beta)\Delta_1$$

and if $\alpha(L^\infty(G_2)) = M^\theta$, where θ is the canonical action of $(\mathcal{L}(G_1), \hat{\Delta}_1^{\mathrm{op}})$ on M generated by the morphism β. In this situation, we call (M, Δ) an extension of G_2 by $\hat{G}_1$.

The faithfulness of the morphisms α and β reflects the exactness of the sequence in the first and third place. The formula $\alpha(L^\infty(G_2)) = M^\theta$ reflects its exactness in the second place. Given a short exact sequence as above, one can check that the dual sequence

$$(L^\infty(G_1), \Delta_1) \xrightarrow{\beta} (\hat{M}, \hat{\Delta}) \xrightarrow{\alpha} (\mathcal{L}(G_2), \hat{\Delta}_2)$$

is exact as well.

Given a cocycle matched pair of l.c. groups, one can check that their cocycle bicrossed product is an extension in the sense of Definition 2.8. Moreover, it belongs to a special class of extensions, called cleft extensions ([51], Theorem 2.8). This theorem also shows that, conversely, all cleft extensions of l.c. groups (and of l.c. quantum groups) are given by the cocycle bicrossed products. This means that, whenever (M, Δ) is a cleft extension of G_2 by $\hat{G}_1$, the pair consisting of $(L^\infty(G_1), \Delta_1)$ and $(L^\infty(G_2), \Delta_2)$ is a cocycle matched pair in the sense of [51], Definition 2.1 and (M, Δ) is isomorphic to their cocycle bicrossed product. From the results of [6], it follows that this precisely means that (G_1, G_2) is a matched pair in the sense of Definition 2.1 with cocycles as in Equation (2.1).

By definition, two extensions

$$(L^\infty(G_2), \Delta_2) \xrightarrow{\alpha_a} (M_a, \Delta_a) \xrightarrow{\beta_a} (\mathcal{L}(G_1), \hat{\Delta}_1) \quad \text{and}$$
$$(L^\infty(G_2), \Delta_2) \xrightarrow{\alpha_b} (M_b, \Delta_b) \xrightarrow{\beta_b} (\mathcal{L}(G_1), \hat{\Delta}_1)$$

are called isomorphic, if there is an isomorphism $\pi : (M_a, \Delta_a) \to (M_b, \Delta_b)$ of l.c. quantum groups satisfying $\pi\alpha_a = \alpha_b$ and $\hat{\pi}\beta_a = \beta_b$, where $\hat{\pi}$ is the canonical isomorphism of $(\hat{M}_a, \hat{\Delta}_a)$ onto $(\hat{M}_b, \hat{\Delta}_b)$ associated with π.

Given a matched pair (G_1, G_2) of l.c. groups, any couple of cocycles $(\mathcal{U}, \mathcal{V})$ satisfying (2.1) generates as above a cleft extension

$$(L^\infty(G_2), \Delta_2) \xrightarrow{\alpha} (M, \Delta) \xrightarrow{\beta} (\mathcal{L}(G_1), \hat{\Delta}_1).$$

The extensions given by two pairs of cocycles $(\mathcal{U}_a, \mathcal{V}_a)$ and $(\mathcal{U}_b, \mathcal{V}_b)$, are isomorphic if and only if there exists a measurable map $\mathcal{R}$ from $G_1 \times G_2$ to $U(1)$, satisfying

$$\mathcal{U}_b(g, h, s) = \mathcal{U}_a(g, h, s)\ \mathcal{R}(h, s)\ \mathcal{R}(g, \alpha_h(s))\ \bar{\mathcal{R}}(gh, s)$$
$$\mathcal{V}_b(g, s, t) = \mathcal{V}_a(g, s, t)\ \mathcal{R}(g, s)\ \mathcal{R}(\beta_s(g), t)\ \bar{\mathcal{R}}(g, ts)$$

almost everywhere. If this is the case, the pairs $(\mathcal{U}_a, \mathcal{V}_a)$ and $(\mathcal{U}_b, \mathcal{V}_b)$ will be called cohomologous. Then the set of equivalence classes of cohomologous pairs of cocycles $(\mathcal{U}, \mathcal{V})$ satisfying (2.2), exactly corresponds to the set Γ of classes of isomorphic extensions associated with (G_1, G_2).

The set Γ can be given the structure of an abelian group by defining

$$\pi(\mathcal{U}_a, \mathcal{V}_a) \cdot \pi(\mathcal{U}_b, \mathcal{V}_b) = \pi(\mathcal{U}_a\mathcal{U}_b, \mathcal{V}_a\mathcal{V}_b)$$

where $\pi(\mathcal{U}, \mathcal{V})$ denotes the equivalence class containing the pair $(\mathcal{U}, \mathcal{V})$. The group Γ is called the group of extensions of $(L^\infty(G_2), \Delta_2)$ by $(\mathcal{L}(G_1), \hat{\Delta}_1)$ associated with the matched pair of l.c. groups (G_1, G_2). The unit of this group corresponds to the class of cocycles cohomologous to trivial. The corresponding extension is called *split extension*; all other extensions are called *non-trivial extensions*.

Closed normal quantum subgroups. Definition 2.8 is the partial case of the general definition of a short exact sequence

$$(M_2, \Delta_2) \xrightarrow{\alpha} (M, \Delta) \xrightarrow{\beta} (\hat{M}_1, \hat{\Delta}_1),$$

where (M_1, Δ_1), (M_2, Δ_2) and (M, Δ) are l.c. quantum groups, see Definition 3.2 in [51]. We explain the relation between this notion and the following notion of a closed normal quantum subgroup.

Definition 2.9. A l.c. quantum group (M_2, Δ_2) is called a closed quantum subgroup of (M, Δ) if there exists a normal, faithful *-homomorphism $\alpha : M_2 \to M$ such that $\Delta\alpha = (\alpha \otimes \alpha)\Delta_2$.

This definition might need some justification: in [23], J. Kustermans defines morphisms between l.c. quantum groups on the (natural) level of universal C*-algebraic quantum groups. So, it might seem strange to require the existence of a normal morphism on the von Neumann algebra level. We claim, however, that this precisely characterizes the closedness (or properness of the injective embedding). Let us illustrate this with an example. Consider the identity map from $\mathbb{R}_d$ with the discrete topology to $\mathbb{R}$ with its usual topology. Dualizing, we get a morphism $\alpha : C_0(\mathbb{R}) \to M(C_0(\mathbb{R}_d)) = C_b(\mathbb{R}_d)$ which is injective. It is clear that we want to exclude this type of morphisms. This is precisely achieved by requiring the normality (weak continuity) of the morphism. To conclude, we mention that in the case where $M_2 = \mathcal{L}(G_2)$ and $M = \mathcal{L}(G)$, we precisely are in the situation of an identification $\pi : G_2 \to G$ of G_2 with a closed subgroup of G and $\alpha(\lambda_g) = \lambda_{\pi(g)}$, see Theorem 6 in [46].

Next, we define normality of a closed quantum subgroup. Recall that when $\mathcal{A}_1$ is a Hopf subalgebra of a Hopf algebra $\mathcal{A}$, $\mathcal{A}_1$ is called normal if $\mathcal{A}_1$ is invariant under the adjoint action. Using Sweedler notation, this means

$$\sum a_{(1)} x S(a_{(2)}) \in \mathcal{A}_1 \quad \text{for all } \; x \in \mathcal{A}_1, a \in \mathcal{A}.$$

Recalling that $S((\iota \otimes \omega)(W)) = (\iota \otimes \omega)(W^*)$ and that

$$\Delta((\iota \otimes \omega)(W)) = (\iota \otimes \iota \otimes \omega)(W_{13} W_{23})$$

because of the pentagon equation, it is easy to verify that the operator algebraic version of normality is given as follows.

Definition 2.10. If $\alpha : M_2 \to M$ turns (M_2, Δ_2) into a closed quantum subgroup of the l.c. quantum group (M, Δ), we say that (M_2, Δ_2) is normal if

$$W(\alpha(M_2) \otimes 1)W^* \subset \alpha(M_2) \otimes B(H).$$

As could be expected, we now prove the bijective correspondence between closed normal quantum subgroups and short exact sequences.

Theorem 2.11. *Suppose that $\alpha : M_2 \to M$ turns (M_2, Δ_2) into a closed normal quantum subgroup of (M, Δ). Then, there exists a unique (up to isomorphism) l.c. quantum group (M_1, Δ_1) and a unique $\beta : M_1 \to \hat{M}$ such that*

$$(M_2, \Delta_2) \xrightarrow{\alpha} (M, \Delta) \xrightarrow{\beta} (\hat{M}_1, \hat{\Delta}_1)$$

is a short exact sequence.

If, conversely, we have a short exact sequence, then $\alpha : M_2 \to M$ *turns* (M_2, Δ_2) *into a closed normal quantum subgroup of the l.c. quantum group* (M, Δ).

Proof. Suppose first that we have a short exact sequence. Consider the coaction θ of $(\hat{M}_1, \hat{\Delta}_1^{\mathrm{op}})$ on M associated with β. By definition of exactness, we have $\alpha(M_2) = M^\theta$. Let $x \in M_2$. It suffices to prove that $(\theta \otimes \iota)(W(\alpha(x) \otimes 1)W^*) = W_{23}(1 \otimes \alpha(x) \otimes 1)W_{23}^*$. From Proposition 3.1 of [51] and with the notations introduced over there, it follows that it is sufficient to prove that $1 \otimes \alpha(x)$ commutes with Z_1, or equivalently, $\mu(\alpha(x)) = 1 \otimes \alpha(x)$. But,

$$\mu(\alpha(x)) = (\hat{R}_1 \otimes R)\theta(R(\alpha(x))) = (\hat{R}_1 \otimes R)\theta(\alpha(R_2(x))) = 1 \otimes \alpha(x).$$

This proves the most easy, second part of the theorem.

Next, suppose that we have a closed normal quantum subgroup (M_2, Δ_2) of (M, Δ). Using Proposition 3.1 from [51], the morphism α generates two actions: $\hat{\mu}$ is an action of $(\hat{M}_2, \hat{\Delta}_2)$ on $\hat{M}$ and $\hat{\theta}$ is an action of $(\hat{M}_2, \hat{\Delta}_2^{\mathrm{op}})$ on $\hat{M}$ and they are determined by

$$\hat{\mu}(x) = \hat{Z}_1(1 \otimes x)\hat{Z}_1^* \quad \text{and} \quad \hat{\theta}(x) = \hat{Z}_2(1 \otimes x)\hat{Z}_2^* \quad \text{for all } x \in \hat{M},$$

where

$$\hat{Z}_1 = (\iota \otimes \alpha)(\hat{W}_2^*) \quad \text{and} \quad \hat{Z}_2 = (J_2 \otimes J)\hat{Z}_1(J_2 \otimes J).$$

The actions $\hat{\mu}$ and $\hat{\theta}$ are related by the formula $\hat{\theta}(x) = (\hat{R}_2 \otimes \hat{R})\hat{\mu}(\hat{R}(x))$ and satisfy

$$(\hat{\mu} \otimes \iota)(\hat{W}) = (\iota \otimes \alpha)(\hat{W}_2)_{13}\hat{W}_{23} \quad \text{and} \quad (\hat{\theta} \otimes \iota)(\hat{W}) = \hat{W}_{23}(\iota \otimes \alpha)(\hat{W}_2)_{13}.$$

Using the definition of the left invariant weight $\hat{\varphi}$ on the dual $(\hat{M}, \hat{\Delta})$ of (M, Δ), we easily conclude that $\hat{\varphi}$ is invariant under the action $\hat{\mu}$ and moreover, for all $x \in \mathcal{N}_{\hat{\varphi}}$ and $\omega \in \hat{M}_{2,*}$, we have $(\omega \otimes \iota)\hat{\mu}(x) \in \mathcal{N}_{\hat{\varphi}}$ and

$$\hat{\Lambda}((\omega \otimes \iota)\hat{\mu}(x)) = (\omega \otimes \iota)(\hat{Z}_1)\hat{\Lambda}(x).$$

From Proposition 4.3 of [48], it then follows that $\hat{Z}_1$ is the canonical implementation of the action $\hat{\mu}$ (in the sense of Definition 3.6 of [48]). We want to prove that $\hat{\mu}$ is integrable (see Definition 1.4 in [48]) and we will use Theorem 5.3 of [48] to do this. So, we have to construct a normal *-homomorphism $\rho : \hat{M}_2 \,{}_{\hat{\mu}}\!\ltimes \hat{M} \to B(H)$ such that

$$\rho(\hat{\mu}(x)) = x \quad \text{for all } x \in \hat{M} \quad \text{and} \quad (\iota \otimes \rho)(\hat{W}_2 \otimes 1) = \hat{Z}_1^*.$$

We first define

$$\tilde{\rho} : \hat{M}_2 \,{}_{\hat{\mu}}\!\ltimes \hat{M} \to B(H \otimes H) : \tilde{\rho}(z) = \mathcal{V}(\alpha \otimes \iota)(\hat{Z}_1^* z \hat{Z})\mathcal{V}^*,$$

where $\mathcal{V} = (\hat{J} \otimes \hat{J})W(\hat{J} \otimes \hat{J})$ has the properties $\mathcal{V} \in M \otimes \hat{M}'$ and $\Delta^{\mathrm{op}}(y) = \mathcal{V}^*(1 \otimes y)\mathcal{V}$ for all $y \in M$. This map $\tilde{\rho}$ is well-defined for the following reasons. For $x \in \hat{M}$, we have $\hat{Z}_1^*\hat{\mu}(x)\hat{Z}_1 = 1 \otimes x$. We can apply $\alpha \otimes \iota$ and because $\mathcal{V} \in M \otimes \hat{M}'$, we find that $\tilde{\rho}(\hat{\mu}(x)) = 1 \otimes x$. Next, for $\omega \in \hat{M}_{2,*}$, we find

$$\begin{aligned}\hat{Z}_1^*((\omega \otimes \iota)(\hat{W}_2) \otimes 1)\hat{Z}_1 &= (\omega \otimes \iota \otimes \iota)(\iota \otimes \iota \otimes \alpha)(\hat{W}_{2,23}\hat{W}_{2,12}\hat{W}_{2,23}^*)\\ &= (\omega \otimes \iota \otimes \iota)(\hat{W}_{2,12}(\iota \otimes \alpha)(\hat{W}_2)_{13}).\end{aligned}$$

Again, it is possible to apply $\alpha \otimes \iota$ and we find

$$(\iota \otimes \tilde{\rho})(\hat{W}_2 \otimes 1) = \mathcal{V}_{23}(\iota \otimes \alpha \otimes \alpha)(\iota \otimes \Delta_2^{\mathrm{op}})(\hat{W}_2)\mathcal{V}_{23}^* = \hat{Z}_{1,13}^*,$$

because α is a morphism and $\mathcal{V}^*$ implements Δ^{op}. The ρ that we were looking for, is then obtained as $\tilde{\rho}(z) = 1 \otimes \rho(z)$ for all $z \in \hat{M}_2 \,{}_{\hat{\mu}}\!\ltimes \hat{M}$. Hence, $\hat{\mu}$ is integrable.

Define the von Neumann algebra $M_1 := \hat{M}^{\hat{\mu}}$, the fixed point algebra of $\hat{\mu}$. Because $(\iota \otimes \hat{\Delta})\hat{\mu} = (\hat{\mu} \otimes \iota)\hat{\Delta}$, it is clear that $\hat{\Delta}(M_1) \subset M_1 \otimes \hat{M}$. We claim that also $\hat{\Delta}(M_1) \subset \hat{M} \otimes M_1$. For this, we will need the normality. Observe that the right leg of $\hat{Z}_1$ generates $\alpha(M_2)$. Hence, by definition, $M_1 = \hat{M} \cap \alpha(M_2)'$. Because $R\alpha = \alpha R_2$ and $R(x) = \hat{J}x^*\hat{J}$, we conclude that $\hat{J}M_1'\hat{J} = (\hat{M} \cup \alpha(M_2))''$. By normality, we know that $W(\alpha(M_2) \otimes 1)W^* \subset \alpha(M_2) \otimes \hat{M}$. Because $W(\hat{M} \otimes 1)W^* = \hat{\Delta}^{\mathrm{op}}(\hat{M}) \subset \hat{M} \otimes \hat{M}$, we get $W(\hat{J}M_1'\hat{J} \otimes 1)W^* \subset \hat{J}M_1'\hat{J} \otimes \hat{M}$. Writing $\hat{J} \otimes J$ around this equation, we find $W^*(M_1' \otimes 1)W \subset M_1' \otimes B(H)$. Because $W \in M \otimes \hat{M}$, it follows that $W^*(M_1' \otimes \hat{M}')W \subset M_1' \otimes B(H)$. Taking commutants, we conclude that $M_1 \otimes 1 \subset W^*(M_1 \otimes \hat{M})W$. Bringing the W to the other side, we have proven our claim that $\hat{\Delta}^{\mathrm{op}}(M_1) \subset M_1 \otimes \hat{M}$.

Defining Δ_1 to be the restriction of $\hat{\Delta}$ to M_1, we have found a von Neumann algebra with comultiplication (M_1, Δ_1). In order to produce invariant weights, we first prove that $\hat{R}(M_1) \subset M_1$. Verifying the following equality on a slice of $\hat{W}$, we easily arrive at the formula

$$(\iota \otimes \hat{\mu})\hat{\Delta}(x) = ((\hat{\theta} \otimes \iota)\hat{\Delta}(x))_{213} \quad \text{for all } x \in \hat{M}.$$

Let now $x \in M_1$. Then $\hat{\Delta}(x) \in M_1 \otimes M_1$, so that

$$\hat{\Delta}(x)_{13} = (\iota \otimes \hat{\mu})\hat{\Delta}(x) = ((\hat{\theta} \otimes \iota)\hat{\Delta}(x))_{213}.$$

So,

$$\hat{\Delta}(x) \in \hat{M}^{\hat{\theta}} \otimes M_1 = \hat{R}(M_1) \otimes M_1, \tag{2.2}$$

because of the relation between $\hat{\mu}$ and $\hat{\theta}$. If we regard the restriction of $\hat{\Delta}^{\mathrm{op}}$ as a map from M_1 to $\hat{M} \otimes M_1$, then it will be an action of $(\hat{M}, \hat{\Delta}^{\mathrm{op}})$ on M_1. But then we know that the σ-strong* closure of

$$\{(\omega \otimes \iota)\hat{\Delta}^{\mathrm{op}}(x) \mid x \in M_1, \omega \in \hat{M}_*\}$$

equals M_1. Combining this with Equation (2.2), we find that $M_1 \subset \hat{R}(M_1)$. Applying $\hat{R}$, we get the equality $M_1 = \hat{R}(M_1)$. In particular, we also have $M_1 = \hat{M}^{\hat{\theta}}$.

Because the restriction of $\hat{R}$ to M_1 will be an anti-automorphism of M_1 anti-commuting with the comultiplication Δ_1, it now suffices to produce a left invariant weight on (M_1, Δ_1), in order to get that (M_1, Δ_1) is a l.c. quantum group. Choose an arbitrary n.s.f. weight η on M_1. Because $\hat{\mu}$ is integrable, also $\hat{\theta}$ is integrable and we can define an n.s.f. operator valued weight T from $\hat{M}$ to $M_1 = \hat{M}^{\hat{\theta}}$ by the formula $T(z) = (\hat{\varphi}_2 \otimes \iota)\hat{\theta}(z)$ for all $z \in \hat{M}^+$. Defining $\tilde{\eta} = \eta T$, we get an n.s.f. weight $\tilde{\eta}$ on $\hat{M}$. We claim that the weight $\tilde{\eta}$ is invariant under the action $\hat{\mu}$. In fact, by verifying the next formula on slices of $\hat{W}$, we easily get that

$$(\iota \otimes \hat{\theta})\hat{\mu}(x) = ((\iota \otimes \hat{\mu})\hat{\theta}(x))_{213} \quad \text{for all } x \in \hat{M}.$$

So, for all $z \in \hat{M}^+$, $\omega \in \hat{M}^+_{2,*}$ and $\omega' \in \hat{M}^+_*$, we get

$$\omega'(T((\omega \otimes \iota)\hat{\mu}(z))) = (\omega \otimes \omega')\hat{\mu}(T(z)) = \omega(1)\, \omega'(T(z)),$$

because $T(z)$ belongs to the extended positive part of M_1. Hence, $\tilde{\eta}((\omega \otimes \iota)\hat{\mu}(z)) = \omega(1)\, \tilde{\eta}(z)$, proving our claim.

Above, we already observed that $\hat{\varphi}$ is invariant under $\hat{\mu}$. It then follows from Lemma 3.9 in [48] that the Connes cocycle $u_t = [D\hat{\varphi} : D\tilde{\eta}]_t$ belongs to M_1 for all $t \in \mathbb{R}$. From the theory of operator valued weights, we know that $\sigma_t^{\tilde{\eta}}(x) = \sigma_t^{\eta}(x)$ for all $x \in M_1$. Hence, (u_t) is a cocycle with respect to the modular group (σ_t^{η}) on M_1. So, there exists a unique n.s.f. weight φ_1 on M_1 such that $[D\varphi_1 : D\eta]_t = u_t$. Define $\tilde{\varphi}_1 = \varphi_1 T$. From operator valued weight theory, we get

$$[D\tilde{\varphi}_1 : D\tilde{\eta}]_t = [D\varphi_1 : D\eta]_t = u_t = [D\hat{\varphi} : D\tilde{\eta}]_t,$$

which yields $\tilde{\varphi}_1 = \hat{\varphi}$. Let now $x \in \hat{M}^+$. Because

$$(\iota \otimes \hat{\theta})\hat{\Delta}(x) = ((\iota \otimes \hat{\Delta})\hat{\theta}(x))_{213},$$

we find, for all $\omega, \omega' \in \hat{M}^+_*$,

$$\omega'(T((\omega \otimes \iota)\hat{\Delta}(x))) = (\omega \otimes \omega')\Delta_1(T(x)).$$

When $x \in \hat{M}^+$ is such that $T(x)$ is bounded, we conclude that

$$\tilde{\varphi}_1((\omega \otimes \iota)\hat{\Delta}(x)) = \varphi_1((\omega \otimes \iota)\Delta_1(T(x))).$$

Because $\tilde{\varphi}_1 = \hat{\varphi}$, the left hand side equals $\omega(1)\, \tilde{\varphi}_1(x) = \omega(1)\, \varphi_1(T(x))$. Hence, for all $x \in \hat{M}^+$ such that $T(x)$ is bounded and for all $\omega \in M^+_{1,*}$, we find

$$\omega(1)\, \varphi_1(T(x)) = \varphi_1((\omega \otimes \iota)\Delta_1(T(x))).$$

Take an increasing net (u_i) in $\hat{M}^+$ such that $T(u_i)$ converges increasingly to 1. Take $y \in M_1$. By lower semi-continuity, we get

$$\begin{aligned}\omega(1)\, \varphi_1(y^*y) = \sup_i \omega(1)\, \varphi_1(T(y^*u_i y)) &= \sup_i \varphi_1((\omega \otimes \iota)\Delta_1(T(y^*u_i y))) \\ &= \varphi_1((\omega \otimes \iota)\Delta_1(y^*y)).\end{aligned}$$

Hence, φ_1 is an n.s.f. left invariant weight on (M_1, Δ_1) and the latter is a l.c. quantum group.

Define β to be the identity map, embedding M_1 into $\hat{M}$. In order to obtain that

$$(M_2, \Delta_2) \xrightarrow{\alpha} (M, \Delta) \xrightarrow{\beta} (\hat{M}_1, \hat{\Delta}_1)$$

is a short exact sequence, it remains to show that $\alpha(M_2) = M^\theta$, where θ is the canonical action of $(\hat{M}_1, \hat{\Delta}_1^{\mathrm{op}})$ on M, associated to β (see Proposition 3.1 in [51]). Because $\theta = (\hat{R}_1 \otimes R)\mu R$ and $R(\alpha(M_2)) = \alpha(M_2)$, it suffices to show that $\alpha(M_2) = M^\mu$. From Proposition 3.1 in [51], we immediately deduce that $M^\mu = M \cap \beta(M_1)'$. Above we already saw that $M_1 = \hat{M} \cap \alpha(M_2)'$. Because $\alpha(M_2)$ is a two-sided coideal of (M, Δ), it follows from Théorème 3.3 in [11] that

$$M \cap (\hat{M} \cap \alpha(M_2)')' = \alpha(M_2).$$

So, we have a short exact sequence of l.c. quantum groups.

Finally, we should prove the uniqueness of this short exact sequence up to isomorphism. Suppose that we have another short exact sequence

$$(M_2, \Delta_2) \xrightarrow{\alpha} (M, \Delta) \xrightarrow{\gamma} (\hat{M}_3, \hat{\Delta}_3).$$

We still have the same action $\hat{\mu}$ of $(\hat{M}_2, \hat{\Delta}_2)$ on $\hat{M}$ and a reasoning as in the previous paragraph yields that $\gamma(M_3) = \hat{M}^{\hat{\mu}}$. Hence, it follows that γ gives an isomorphism of l.c. quantum groups between (M_3, Δ_3) and the l.c. quantum group (M_1, Δ_1) constructed above. □

3 Matched pairs of Lie groups and Lie algebras

In what follows we consider Lie groups and Lie algebras over the field $k = \mathbb{C}$ or $\mathbb{R}$.

Definition 3.1. We call (G_1, G_2) a matched pair of Lie groups if, in Definition 2.1, G is a Lie group.

Observe that it follows from Proposition 2.2 that θ is a diffeomorphism of $G_1 \times G_2$ onto the open subset Ω of G.

The infinitesimal form of this definition is as follows (see [33]).

Definition 3.2. We call $(\mathfrak{g}_1, \mathfrak{g}_2)$ a matched pair of Lie algebras, if there exists a Lie algebra $\mathfrak{g}$ with Lie subalgebras $\mathfrak{g}_1$ and $\mathfrak{g}_2$ such that $\mathfrak{g} = \mathfrak{g}_1 \oplus \mathfrak{g}_2$ as vector spaces.

These conditions are equivalent to the existence of a left action $\rhd : \mathfrak{g}_2 \otimes \mathfrak{g}_1 \to \mathfrak{g}_1$ and a right action $\lhd : \mathfrak{g}_2 \otimes \mathfrak{g}_1 \to \mathfrak{g}_2$, so that $\mathfrak{g}_1$ is a left $\mathfrak{g}_2$-module and $\mathfrak{g}_2$ is a right $\mathfrak{g}_1$-module and

1. $x \rhd [a, b] = [x \rhd a, b] + [a, x \rhd b] + (x \lhd a) \rhd b - (x \lhd b) \rhd a,$

2. $[x, y] \lhd a = [x, y \lhd a] + [x \lhd a, y] + x \lhd (y \rhd a) - y \lhd (x \rhd a),$

for all $a, b \in \mathfrak{g}_1,\ x, y \in \mathfrak{g}_2$. Then, for the decomposition of vector spaces above we have

$$[a \oplus x, b \oplus y] = ([a, b] + x \triangleright b - y \triangleright a) \oplus ([x, y] + x \triangleleft b - y \triangleleft a)$$

(see [33], Proposition 8.3.2).

Two matched pairs of Lie algebras, $(\mathfrak{g}_1, \mathfrak{g}_2)$ and $(\mathfrak{g}'_1, \mathfrak{g}'_2)$, are called isomorphic if there is an isomorphism of the corresponding Lie algebras $\mathfrak{g}$ and $\mathfrak{g}'$ sending $\mathfrak{g}_i$ onto $\mathfrak{g}'_i$ $(i = 1, 2)$.

Let us explain the relation between the two notions of a matched pair.

Proposition 3.3. *Let (G_1, G_2) be a matched pair of Lie groups in the sense of Definition* 3.1. *If $\mathfrak{g}$ denotes the Lie algebra of G, and if $\mathfrak{g}_1$, resp. $\mathfrak{g}_2$, are the Lie subalgebras corresponding to the closed subgroups $i(G_1)$, resp. $j(G_2)$, then $(\mathfrak{g}_1, \mathfrak{g}_2)$ is a matched pair of Lie algebras.*

Proof. The fact that $\mathfrak{g} = \mathfrak{g}_1 \oplus \mathfrak{g}_2$ as vector spaces follows from the fact that θ is a diffeomorphism in the neighbourhood of the unit element. □

The converse problem, to construct a matched pair of Lie groups from a given matched pair $(\mathfrak{g}_1, \mathfrak{g}_2)$ of Lie algebras, is much more subtle. Indeed, one can take, of course, the connected, simply connected Lie group G of the corresponding $\mathfrak{g}$ and find unique connected, closed subgroups G_1 and G_2 of G whose tangent Lie algebras are $\mathfrak{g}_1$ and $\mathfrak{g}_2$, respectively. However, in the proof of the following proposition, we see that (G_1, G_2) is not necessarily a matched pair of Lie groups even if $\dim \mathfrak{g}_1 = \dim \mathfrak{g}_2 = 1$.

Proposition 3.4. *Every matched pair of complex Lie algebras $\mathfrak{g}_1 = \mathfrak{g}_2 = \mathbb{C}$ can be exponentiated to a matched pair of Lie groups (G_1, G_2) where G_1, G_2 are either $(\mathbb{C}, +)$ or $(\mathbb{C} \setminus \{0\}, \cdot)$.*

Every matched pair of real Lie algebras $\mathfrak{g}_1 = \mathfrak{g}_2 = \mathbb{R}$ can be exponentiated to a matched pair of Lie groups (G_1, G_2) where G_1, G_2 are either $(\mathbb{R}, +)$ or $(\mathbb{R} \setminus \{0\}, \cdot)$.

Proof. Consider first the complex case. The only two-dimensional complex Lie algebras are the abelian one and the one with generators X, Y and relation $[X, Y] = Y$. If $\mathfrak{g}$ is abelian, the mutual actions of $\mathfrak{g}_1$ and $\mathfrak{g}_2$ on each other are trivial and the exponentiation is obviously a direct sum.

If $\mathfrak{g}$ is generated by $[X, Y] = Y$, we either have that $\mathfrak{g}_1$ or $\mathfrak{g}_2$ is equal to $\mathbb{C}Y$, in which case one of the actions is trivial and G can be constructed as semi-direct product of the connected, simply connected Lie groups of $\mathfrak{g}_1$ and $\mathfrak{g}_2$, or we have that both $\mathfrak{g}_1$ and $\mathfrak{g}_2$ differ from $\mathbb{C}Y$. In the latter case, there is, up to isomorphism, only one possibility, namely $\mathfrak{g}_1 = \mathbb{C}X$, $\mathfrak{g}_2 = \mathbb{C}(X + Y)$. Define on $\mathbb{C} \setminus \{0\} \times \mathbb{C}$ the Lie group with product

$$(t, s)(t', s') = (tt', s + ts').$$

Define $G_1 = G_2 = \mathbb{C} \setminus \{0\}$ with embeddings $i(g) = (g, 0)$ and $j(s) = (s, s-1)$, we indeed get a matched pair of complex Lie groups with mutual actions

$$\alpha_g(s) = g(s-1) + 1, \quad \beta_s(g) = \frac{sg}{g(s-1)+1}. \tag{3.1}$$

The real case is completely analogous. □

Remark 3.5. The connected simply connected complex Lie group G of $\mathfrak{g}$ consists of all pairs (t, s) with $t, s \in \mathbb{C}$ and the product

$$(t, s)(t', s') = (t + t', s + \exp(t)s')$$

(see, for example, [14], §10.1), and its closed subgroups G_1 and G_2 corresponding to the decomposition $\mathfrak{g} = \mathbb{C}X \oplus \mathbb{C}(X + Y)$ above consist respectively of all pairs of the form $(g, 0)$ and $(s, \exp(s) - 1)$ with $g, s \in \mathbb{C}$. These groups do not form a matched pair because $G_1 \cap G_2 = \{(2\pi i n, 0) | n \in \mathbb{Z}\} = Z(G)$. So, it is crucial not to take G simply connected above.

Taking g, t, s above real, we come to the example of a matched pair of real Lie groups from [51], Section 5.3. Here $\mathfrak{g}$ is a real Lie algebra generated by X and Y subject to the relation $[X, Y] = Y$ and one considers the decomposition $\mathfrak{g} = \mathbb{R}X \oplus \mathbb{R}(X+Y)$. Then, to get a matched pair of Lie groups, we consider G as the variety $\mathbb{R} \setminus \{0\} \times \mathbb{R}$ with the product

$$(s, x)(t, y) = (st, x + sy)$$

and embed $G_1 = G_2 = \mathbb{R} \setminus \{0\}$ by the formulas $i(g) = (g, 0)$ and $j(s) = (s, s-1)$. Remark that here, it is impossible to take the connected component of the unity of the group of affine transformations of the real line as G, because it is easy to see that for its closed subgroups G_1 and G_2 corresponding to the above mentioned subalgebras, the set $G_1 G_2$ is not dense in G.

The next example shows that in general, for a given matched pair of Lie algebras, it is even possible that $G_1 \cap G_2 \neq \{e\}$ for *any* corresponding pair of Lie groups, which means that such a matched pair of Lie algebras cannot be exponentiated to a matched pair of Lie groups in the sense of Definition 3.1.

Example 3.6. Consider a family of complex Lie algebras $\mathfrak{g} = \operatorname{span}\{X, Y, Z\}$ with $[X, Y] = Y$, $[X, Z] = \alpha Z$, $[Y, Z] = 0$, where $\alpha \in \mathbb{C} \setminus \{0\}$, and the decomposition $\mathfrak{g} = \operatorname{span}\{X, Y\} \oplus \mathbb{C}(X + \alpha Z)$. The corresponding connected simply connected complex Lie group H consists of all triples (t, u, v) with $t, u, v \in \mathbb{C}$ and the product

$$(t, u, v)(t', u', v') = (t + t', u + \exp(t)u', v + \exp(\alpha t)v')$$

(see, for example, [14], §10.3), and its closed subgroups H_1 and H_2 corresponding to the decomposition above consist respectively of all triples of the form $(t, u, 0)$ and $(s, 0, \exp(\alpha s) - 1)$ with $t, u, s \in \mathbb{C}$. These groups do not form a matched pair because $H_1 \cap H_2 = \{(\frac{2\pi i n}{\alpha}, 0, 0) | n \in \mathbb{Z}\}$.

We claim that, if $1/\alpha \notin \mathbb{Z}$ and if G is any complex Lie group with Lie algebra $\mathfrak{g}$, such that G_1, G_2 are closed subgroups of G with tangent Lie algebras $\mathfrak{g}_1$, resp. $\mathfrak{g}_2$, then $G_1 \cap G_2 \neq \{e\}$. Indeed, since the Lie group H is connected and simply connected, the connected component $G^{(e)}$ of e in G can be identified with the quotient of H by a discrete central subgroup. If $\alpha \notin \mathbb{Q}$, the center of H is trivial, so that we can identify $G^{(e)}$ and H. Under this identification, the connected components of e in G_1, G_2 agree with H_1, H_2. Because $H_1 \cap H_2 \neq \{e\}$, our claim follows. If $\alpha = \frac{m}{n}$ for $m, n \in \mathbb{Z} \setminus \{0\}$ mutually prime, the center of H consists of the elements $\{(2\pi n N, 0, 0) \mid N \in \mathbb{Z}\}$. Hence, the different possible quotients of H are labeled by $N \in \mathbb{Z}$ and are given by the triples $(a, u, v) \in \mathbb{C}^3$, $a \neq 0$ and the product

$$(a, u, v)(a', u', v') = (aa', u + a^{nN} u', v + a^{mN} v'). \tag{3.2}$$

The closed subgroups corresponding to $\mathfrak{g}_1$ and $\mathfrak{g}_2$ are given by $(a, u, 0)$ and $(b, 0, b^{mN} - 1)$ with $a, b, u \in \mathbb{C}$ and $a, b \neq 0$. The intersection of both subgroups is non-trivial whenever $mN \neq \pm 1$. This proves our claim.

Considering now the complex Lie algebras above as real Lie algebras with generators X, iX, Y, iY, Z, iZ and the decomposition above as a decomposition of real Lie algebras, we get a matched pair of real Lie algebras which cannot be exponentiated to a matched pair of real Lie groups.

In the remaining case $\alpha = 1/n$ with $n \in \mathbb{Z} \setminus \{0\}$, we can consider the Lie group G defined by Equation (3.2) with $m = N = 1$. Consider $G_1 = \mathbb{C} \setminus \{0\} \times \mathbb{C}$ with $(a, u)(a', u') = (aa', u + a^n u')$ and $G_2 = \mathbb{C} \setminus \{0\}$. Writing $i(a, u) = (a, u, 0)$ and $j(v) = (v, 0, v - 1)$, we get a matched pair of Lie groups with mutual actions

$$\alpha_{(a,u)}(v) = a(v-1) + 1 \quad \text{and} \quad \beta_v(a, u) = \left(\frac{va}{a(v-1)+1}, \frac{u}{(a(v-1)+1)^n} \right).$$

In the next section, we study more closely the exponentiation of a matched pair of Lie algebras when one of the Lie algebras has dimension 1.

4 Matched pairs of Lie groups and Lie algebras in dimension $n + 1$

We use systematically the following terminology.

Terminology 4.1. A matched pair $(\mathfrak{g}_1, \mathfrak{g}_2)$, resp. (G_1, G_2), is said to be of dimension $n_1 + n_2$ if the dimension of $\mathfrak{g}_i$, resp. G_i, is n_i.

Suppose that $\mathfrak{g} = \mathfrak{g}_1 \oplus \mathfrak{g}_2$ is a matched pair of Lie algebras with $\dim \mathfrak{g}_2 = 1$. Put $\mathfrak{g}_2 = kA$. For all $X \in \mathfrak{g}_1$, we define $\beta(X) \in \mathfrak{g}_1$ and $\chi(X) \in k$ such that

$$[X, A] = \beta(X) + \chi(X) A.$$

Then, β and χ are linear, and, for all $X, Y \in \mathfrak{g}_1$, the Jacobi identity for $\mathfrak{g}$ gives:

$$\begin{aligned}&\chi([X,Y]) = 0,\\ &\beta([X,Y]) = [X,\beta(Y)] + [\beta(X),Y] + \beta(X)\chi(Y) - \beta(Y)\chi(X)\,.\end{aligned} \tag{4.1}$$

By induction, one verifies that

$$\begin{aligned}\beta^n([X,Y]) &= \sum_{k=0}^{n}\binom{n}{k}[\beta^k(X),\beta^{n-k}(Y)]\\ &+\sum_{k=1}^{n}\binom{n}{k-1}\big(\beta^k(X)\chi(\beta^{n-k}(Y)) - \beta^k(Y)\chi(\beta^{n-k}(X))\big).\end{aligned}$$

Hence,

$$\chi(\beta^n([X,Y])) = n\big(\chi(\beta^n(X))\chi(Y) - \chi(\beta^n(Y))\chi(X)\big). \tag{4.2}$$

Then, we claim that the linear forms χ, $\chi\beta$ and $\chi\beta^2$ are linearly dependent. If not, we find $X_0, X_1, X_2 \in \mathfrak{g}_1$, such that $\chi(\beta^i(X_j)) = \delta_{ij}$ for $i, j \in \{0,1,2\}$, where δ_{ij} is the Kronecker symbol. Because $\chi(X_1) = \chi(X_2) = 0$, we get $\chi(\beta^n([X_1,X_2])) = 0$ for all n. Define

$$\mathfrak{g}_0 = \bigcap_{i=0}^{2} \operatorname{Ker} \chi\beta^i .$$

Then, $\beta([X_1,X_2]) \in \mathfrak{g}_0$. Using Equation (4.1), we get

$$\beta([X_1,X_2]) = [X_1,\beta(X_2)] + [\beta(X_1),X_2]. \tag{4.3}$$

On the other hand, using Equation (4.2), it follows that $[X_1,\beta(X_2)] \in \mathfrak{g}_0$, because $\chi(\beta(X_2)) = \chi(X_1) = 0$. Combining this with Equation (4.3), we get that $[\beta(X_1),X_2] \in \mathfrak{g}_0$. Nevertheless, using once again Equation (4.2), we get that $\chi(\beta^2([\beta(X_1),X_2])) = -2$, contradicting the fact that $[\beta(X_1),X_2] \in \mathfrak{g}_0$. So, we have proved that χ, $\chi\beta$ and $\chi\beta^2$ are linearly dependent.

Hence, we can separate three different possibilities.

Case 1. $\chi = 0$.

Case 2. $\chi \neq 0$ and $\beta(\operatorname{Ker}\chi) \subset \operatorname{Ker}\chi$.

Case 3. χ and $\chi\beta$ linearly independent, and $\beta(\operatorname{Ker}\chi \cap \operatorname{Ker}\chi\beta) \subset \operatorname{Ker}\chi \cap \operatorname{Ker}\chi\beta$.

Case 1. The action of $\mathfrak{g}_1$ on $\mathfrak{g}_2$ is trivial and $\mathfrak{g}_2$ acts on $\mathfrak{g}_1$ by automorphisms. To exponentiate such a matched pair it suffices to use a semi-direct product of k and the connected, simply connected Lie group G_1 of $\mathfrak{g}_1$.

Case 2. In this case $\mathfrak{g}_0 := \operatorname{Ker}(\chi)$ is an ideal of $\mathfrak{g}$, on which $\mathfrak{g}_2$ acts as an automorphism group. There exists an $a \in k$ such that $\chi\beta = a\chi$, by assumption. Take $X_0 \in \mathfrak{g}_1$ such that $\chi(X_0) = 1$. Then, we get

$$[X_0, A] = A + aX_0 + Y_0 \quad (Y_0 \in \mathfrak{g}_0).$$

Putting $\tilde{A} = A + aX_0$, we get $[X_0, \tilde{A}] = \tilde{A} + Y_0$. This suggests how to exponentiate $\mathfrak{g}$. We start with the connected, simply connected Lie group G_0 of $\mathfrak{g}_0$ and observe that, due to the action of $\tilde{A}$, k acts by automorphisms $(\rho_x)_{x \in k}$ on G_0. We exponentiate $k\tilde{A} + \mathfrak{g}_0$ on the space $k \times G_0$ with product

$$(x, g)(y, h) = (x + y, g\rho_x(h)).$$

Next, we observe that, due to the action of X_0, k is acting by automorphisms on $k \ltimes G_0$, and one would suggest to make another semi-direct product. But in this case the subgroups corresponding to $\mathfrak{g}_1$ and kA do not form a matched pair of Lie groups if $a \neq 0$. The subgroups corresponding to $\mathfrak{g}_1$ and kA do form a matched pair of Lie groups when $a = 0$. So, in what follows, we treat the case $a \neq 0$. Example 3.6 shows that if $k = \mathbb{C}$, the exponentiation is in general impossible if $\dim \mathfrak{g}_1 \geq 2$. So, we restrict to the case $k = \mathbb{R}$.

We first prove some general results which allow to obtain the exponentiation whenever $n \leq 4$. Afterwards, we will give an example showing that if dimension $n \geq 5$, the exponentiation is in general impossible.

Writing $\mu = \operatorname{Ad} X_0$ and $\rho = \operatorname{Ad} \tilde{A}$ on $\mathfrak{g}_0$, we are given μ and ρ, derivations of $\mathfrak{g}_0$ and an element $Y_0 \in \mathfrak{g}_0$ such that $[\mu, \rho] = \rho + \operatorname{Ad} Y_0$. Further, we have $[X_0, \tilde{A}] = \tilde{A} + Y_0$, $\mathfrak{g}_1 = \mathbb{R}X_0 + \mathfrak{g}_0$ and $\mathfrak{g}_2 = \mathbb{R}A$, where $A = \tilde{A} - aX_0$.

We introduce the notation $\mathbb{R}^*$ for the Lie group $\mathbb{R} \setminus \{0\}$ with multiplication and $\mathbb{R}^*_+$ for the subgroup of elements $s > 0$.

Proposition 4.2. *The matched pair $(\mathfrak{g}_1, \mathfrak{g}_2)$ has an exponentiation with at most two connected components in the following cases:*

1. *ρ is inner and the center of $\mathfrak{g}_0$ is trivial.*
2. *$\mathfrak{g}_0$ is abelian.*
3. *$\mathfrak{g}_0 = \langle X, Y \rangle \oplus \mathfrak{z}_0$, with $[X, Y] = Y$ and $\mathfrak{z}_0$ central in $\mathfrak{g}_0$.*
4. *$\mathfrak{g}_0 = \langle X, Y, Z \rangle$ with $[X, Y] = Z$ and Z central.*

In the proof of this proposition we use systematically the following lemma.

Lemma 4.3. *Let G_0 be a Lie group with Lie algebra $\mathfrak{g}_0$, with center $\mathfrak{z}_0$. Suppose that $[\mathfrak{g}_0, \mathfrak{g}_0] \subset \mathfrak{z}_0$. Let (μ_s) be an action of $\mathbb{R}^*$ by automorphisms of G_0. Denote by μ the derivation of $\mathfrak{g}_0$ corresponding to $(\mu_s)_{s>0}$ and denote by θ the involutive automorphism $\theta := d\mu_{-1}$, giving rise to a decomposition $\mathfrak{g}_0 = \mathfrak{g}_0^+ \oplus \mathfrak{g}_0^-$.*

So, μ leaves $\mathfrak{g}_0^-$ invariant and we suppose that μ is invertible on $\mathfrak{g}_0^-$. Further, μ leaves $[\mathfrak{g}_0, \mathfrak{g}_0]$ invariant and we suppose that μ is invertible on $[\mathfrak{g}_0, \mathfrak{g}_0]$ as well.

Define $G := \mathbb{R}^* \ltimes G_0$ with product $(s, g)(t, h) = (st, g\mu_s(h))$ and Lie algebra $\mathbb{R}X_0 + \mathfrak{g}_0$, where X_0 denotes the canonical generator of $\mathfrak{g}$ corresponding to $\mathbb{R}^*$. Let $C \in \mathfrak{g}_0$. Then, there exists a closed subgroup

$$K = \{(s, v(s)) \mid s \in \mathbb{R}^*\}$$

of G, with tangent Lie algebra $\mathbb{R}(X_0 + C)$, where $v : \mathbb{R}^* \to G_0$ is a smooth function.

Proof. Denote by $\mathrm{Exp}_\mathfrak{g}$ the exponential mapping $\mathfrak{g} \to G$. Then, we get a smooth function $v : \mathbb{R}_+^* \to G_0$ such that $\mathrm{Exp}_\mathfrak{g}((\log s)(X_0 + C)) = (s, v(s))$ for $s > 0$ Then, $v(1) = e$ and $v'(1) = C$. Further, $v(st) = v(s)\mu_s(v(t))$ for all $s, t > 0$. We say that v is a (μ_s)-cocycle. So, we are looking for an element $g_0 \in G_0$, so that we can define $v(-1) = g_0$. Then, $(-1, g_0) \in K$ and, for $s > 0$,

$$(-1, g_0)(s, v(s)) = (-s, g_0\mu_{-1}(v(s))), \quad (s, v(s))(-1, g_0) = (-s, v(s)\mu_s(g_0)).$$

Hence, we are done if we can find an element $g_0 \in G_0$ such that $g_0\mu_{-1}(v(s)) = v(s)\mu_s(g_0)$, because than we can define $v(-s)$ to be this expression.

Denote by $\mathrm{Exp}_{\mathfrak{g}_0}$ the exponential mapping $\mathfrak{g}_0 \to G_0$. Then, we look for $D \in \mathfrak{g}_0$ such that

$$\mu_{-1}(v(s)) = \mathrm{Exp}_{\mathfrak{g}_0}(-D)v(s)\,\mathrm{Exp}_{\mathfrak{g}_0}(\exp((\log s)\mu)(D)). \tag{4.4}$$

We want to derive at $s = 1$. For this, observe that

$$\begin{aligned}&\mathrm{Exp}_{\mathfrak{g}_0}(-D)\,\mathrm{Exp}_{\mathfrak{g}_0}(\exp((\log s)\mu)(D))\\ &\qquad = \mathrm{Exp}_{\mathfrak{g}_0}\Big(-D + \exp((\log s)\mu)(D) - \frac{1}{2}[D, \exp((\log s)\mu)(D)]\Big),\end{aligned}$$

where we have used that $[\mathfrak{g}_0, \mathfrak{g}_0] \subset \mathfrak{z}_0$. Taking the derivative at $s = 1$ of Equation (4.4), we look for $D \in \mathfrak{g}_0$ such that

$$\theta(C) = (\mathrm{Ad}\,\mathrm{Exp}_{\mathfrak{g}_0}(-D))(C) + \mu(D) - \frac{1}{2}[D, \mu(D)].$$

The equation becomes

$$\theta(C) = \exp(-\mathrm{Ad}\,D)(C) + \mu(D) - \frac{1}{2}[D, \mu(D)].$$

Using once again that $[\mathfrak{g}_0, \mathfrak{g}_0] \subset \mathfrak{z}_0$, we get the equation

$$\theta(C) - C = \mu(D) - [D, C] - \frac{1}{2}[D, \mu(D)].$$

Define $Y_2 = \mu^{-1}(\theta(C) - C)$, which is possible because μ is invertible on $\mathfrak{g}_0^-$. Next, define $Z_2 := \frac{1}{2}\mu^{-1}([Y_2, \theta(C) + C])$, which is possible because μ is invertible on $[\mathfrak{g}_0, \mathfrak{g}_0]$. Then, $Z_2 \in \mathfrak{z}_0$ and we define $D = Y_2 + Z_2$. Then,

$$\begin{aligned}&\mu(D) - [D, C] - \frac{1}{2}[D, \mu(D)]\\ &\quad = \theta(C) - C + \frac{1}{2}[Y_2, \theta(C) + C] - [Y_2, C] - \frac{1}{2}[Y_2, \theta(C) - C]\\ &\quad = \theta(C) - C.\end{aligned}$$

So, writing $g_0 = \mathrm{Exp}_{\mathfrak{g}_0}(D)$, we may conclude that the left and right hand side of the to be proven equation

$$\mu_{-1}(v(s)) = g_0^{-1}v(s)\mu_s(g_0)$$

have the same derivative at $s = 1$. But, both sides of the equations are $(\mu_s)_{s>0}$-cocycles and hence, both sides are equal for all $s > 0$. □

Proof of Proposition 4.2. *Part* 1: *ρ is inner and the center of $\mathfrak{g}_0$ is trivial.*

Take the unique $B \in \mathfrak{g}_0$ such that $\rho = \operatorname{Ad} B$ on $\mathfrak{g}_0$. Take a new generator $\hat{A} = \tilde{A} - B$ in $\mathfrak{g}$. Because the center of $\mathfrak{g}_0$ is trivial and

$$0 = [\mu, \rho - \operatorname{Ad} B] = \rho + \operatorname{Ad}(Y_0 - \mu(B)) = \operatorname{Ad}(Y_0 + B - \mu(B)),$$

we get $Y_0 + B - \mu(B) = 0$ and hence $[X_0, \hat{A}] = \hat{A}$. Because $[\hat{A}, \mathfrak{g}_0] = \{0\}$, the connected, simply connected Lie group of $\mathfrak{h} := \mathbb{R}\hat{A} + \mathfrak{g}_0$ is given by $H := \mathbb{R} \oplus G_0$, where G_0 is the connected, simply connected Lie group of $\mathfrak{g}_0$. Using the derivation $\operatorname{Ad} X_0$ on $\mathfrak{h}$, we get an action (μ_s) of $\mathbb{R}^*_+$ on H of the form

$$\mu_s(x, g) = (sx, \eta_s(g)), \quad \text{where } s > 0, x \in \mathbb{R}, g \in G_0,$$

and where (η_s) is an action of $\mathbb{R}^*_+$ on G_0. We easily extend (μ_s) to an action of $\mathbb{R}^*$, defining $\mu_s(x, g) = (sx, \eta_{|s|}(g))$. So, we can define a Lie group $G := \mathbb{R}^* \ltimes H$, with Lie algebra $\mathfrak{g}$, on the space $\mathbb{R}^* \times \mathbb{R} \times G_0$ with product

$$(s, x, g)(t, y, h) = (st, x + sy, g\eta_{|s|}(h)).$$

Define the closed subgroup G_1 consisting of the elements $(s, 0, g)$, where $s \in \mathbb{R}^*$ and $g \in G_0$. The tangent Lie algebra of G_1 is precisely $\mathfrak{g}_1$. To define the closed subgroup G_2, recall that $A = \tilde{A} - aX_0 = \hat{A} - aX_0 + B$. Using the exponential mapping of G, we find a smooth function $v : \mathbb{R}^*_+ \to G_0$ such that

$$\left\{ \left(s, \frac{1}{a}(1 - s), v(s) \right) \mid s \in \mathbb{R}^*_+ \right\}$$

is a closed subgroup of G with tangent Lie algebra $\mathbb{R}A$. We define

$$G_2 := \left\{ \left(s, \frac{1}{a}(1 - s), v(|s|) \right) \mid s \in \mathbb{R}^* \right\}$$

and then, one verifies that G_2 is a closed subgroup of G with tangent Lie algebra $\mathbb{R}A$. It is easy to see that (G_1, G_2) is a matched pair of Lie groups.

Part 2: *$\mathfrak{g}_0$ is abelian.*

In this case, μ and ρ are linear transformations of $\mathfrak{g}_0$ satisfying $\mu\rho - \rho\mu = \rho$. Then, for any polynomial P, we get

$$P(\mu)\rho = \rho P(\mu + \iota). \tag{4.5}$$

Let V be the complexified vector space of the real vector space $\mathfrak{g}_0$ and consider μ as a linear operator on V, which we still denote by μ. Then, we have a direct sum decomposition

$$V = \bigoplus_{\lambda \in \mathbb{C}} E_\lambda,$$

where E_λ is the generalized eigenspace corresponding to $\lambda \in \mathbb{C}$. A vector $X \in V$ belongs to E_λ if and only if $(\mu - \lambda\iota)^n X = 0$ for n big enough. We also get a direct sum decomposition

$$\mathfrak{g}_0 = \bigoplus_{r\in\mathbb{R}} F_r,$$

where F_r is such that

$$\mathbb{C}F_r = \bigoplus_{\lambda,\mathrm{Re}(\lambda)=r} E_\lambda.$$

The subspaces F_r are invariant under μ. Also ρ extends to V and using Equation (4.5), it is clear that $\rho(E_\lambda) \subset E_{\lambda+1}$. Hence, $\rho(F_r) = F_{r+1}$. Denote by $\varepsilon(r)$ the entire part of $r \in \mathbb{R}$, such that $\varepsilon(r) \in \mathbb{Z}$ and $\varepsilon(r) \leq r < \varepsilon(r) + 1$. Then, we define

$$\mathfrak{g}_0^+ = \bigoplus_{\varepsilon(r) \text{ is even}} F_r \quad \text{and} \quad \mathfrak{g}_0^- = \bigoplus_{\varepsilon(r) \text{ is odd}} F_r.$$

Defining $\theta(X) = X$ for $X \in \mathfrak{g}_0^+$ and $\theta(X) = -X$ for $X \in \mathfrak{g}_0^-$, we obtain an involution θ of $\mathfrak{g}_0$ satisfying $\theta\mu = \mu\theta$ and $\theta\rho = -\rho\theta$. Observe that μ leaves the subspaces $\mathfrak{g}_0^+$ and $\mathfrak{g}_0^-$ globally invariant and that $\mu - \lambda\iota$ is invertible on $\mathfrak{g}_0^+$ when $\varepsilon(\lambda)$ is odd, while $\mu - \lambda\iota$ is invertible on $\mathfrak{g}_0^-$ when $\varepsilon(\lambda)$ is even.

Define the Lie algebra $\mathfrak{h} := \mathbb{R}\tilde{A} + \mathfrak{g}_0$. Its connected, simply connected Lie group H lives on the space $\mathbb{R} \times \mathfrak{g}_0$, with product

$$(x, X)(y, Y) = (x + y, X + \exp(x\rho)(Y)).$$

We extend the derivation $\mu = \mathrm{Ad}\, X_0$ to $\mathfrak{h}$ and observe that $\mu(\tilde{A}) = \tilde{A} + Y_0$. Defining $\theta(\tilde{A}) = -\tilde{A} + Z_0$, where

$$Z_0 = (\mu - \iota)^{-1}(\theta(Y_0) + Y_0) \in \mathfrak{g}_0^+,$$

one verifies that θ is an involutive automorphism of $\mathfrak{h}$ commuting with the derivation μ. Putting together μ and θ, we obtain an action (μ_s) of $\mathbb{R}^*$ on H such that $\mu_s(x, X) = (sx, \eta_s(X)u(s, x))$, where (η_s) is an action of $\mathbb{R}^*$ on $\mathfrak{g}_0$. We define the Lie group $G := \mathbb{R}^* \ltimes H$, which lives on the space $\mathbb{R}^* \times \mathbb{R} \times \mathfrak{g}_0$, with product

$$(s, x, X)(t, y, Y) = (st, x + sy, X + \exp(x\rho)(\eta_s(Y)u(s, y))).$$

Define the closed subgroup G_1 consisting of the elements $(s, 0, X)$, where $s \in \mathbb{R}^*$ and $X \in \mathfrak{g}_0$. The tangent Lie algebra of G_1 is precisely $\mathfrak{g}_1$. Finally, we have to find a closed subgroup G_2 with tangent Lie algebra $\mathbb{R}A = \mathbb{R}(\tilde{A} - aX_0)$, which consists of the elements $(s, \frac{1}{a}(1 - s), v(s))$, $s \in \mathbb{R}^*$ and $v : \mathbb{R}^* \to \mathfrak{g}_0$ a smooth function. Conjugating with the element $(1, -\frac{1}{a}, 0)$, an equivalent question is to find a closed subgroup with tangent Lie algebra $\mathbb{R}(X_0 + Z_0)$ (for a certain $Z_0 \in \mathfrak{g}_0$), consisting of the elements $(s, 0, w(s))$, $s \in \mathbb{R}^*$ and $w : \mathbb{R}^* \to G_0$ a smooth function. This is possible applying Lemma 4.3 to the action (η_s) of $\mathbb{R}^*$ on $\mathfrak{g}_0$. Then, it is clear that (G_1, G_2) form a matched pair of Lie groups.

Part 3: $\mathfrak{g}_0 = \langle X, Y\rangle \oplus \mathfrak{z}_0$, *with* $[X,Y] = Y$ *and* $\mathfrak{z}_0$ *central.*

Every derivation of $\mathfrak{g}_0$ leaves the center $\mathfrak{z}_0$ invariant. On the quotient Lie algebra $\mathfrak{g}_0/\mathfrak{z}_0$ every derivation is inner. So, changing the generators X_0 and $\tilde{A}$ of $\mathfrak{g}$, we may suppose that we are in the following situation:

$$[X_0, \tilde{A}] = \tilde{A} + Y_0, \quad [X_0, \mathfrak{g}_0] \subset \mathfrak{z}_0, \quad [\tilde{A}, \mathfrak{g}_0] \subset \mathfrak{z}_0,$$

and $\mathfrak{g}_1 = \mathbb{R}X_0 + \mathfrak{g}_0$, $\mathfrak{g}_2 = \mathbb{R}(\tilde{A} - aX_0 + B_1)$ for a certain element $B_1 \in \mathfrak{g}_0$.

Write $\mu = \operatorname{Ad} X_0$ and $\rho = \operatorname{Ad} \tilde{A}$ as derivations on $\mathfrak{g}_0$. Because μ and ρ preserve $[\mathfrak{g}_0, \mathfrak{g}_0]$, we get $\mu(Y_1) = \rho(Y_1) = 0$. Further, $[Y_0, \mathfrak{g}_0] = ([\mu,\rho] - \rho)(\mathfrak{g}_0) \subset \mathfrak{z}_0$. Because $[\mathfrak{g}_0, \mathfrak{g}_0] = \mathbb{R}Y_1$, we get $[Y_0, \mathfrak{g}_0] = \{0\}$, which gives $Y_0 \in \mathfrak{z}_0$ and so, $[\mu, \rho] = \rho$. Suppose $\mu(X_1) = Z_1$ and $\rho(X_1) = Z_2$. Because $[\mu,\rho] = \rho$, we get

$$\mu(Z_2) - Z_2 = \rho(Z_1). \tag{4.6}$$

As in part 2, we can find an involutive automorphism θ of $\mathfrak{z}_0$, such that θ commutes with μ and anti-commutes with ρ and such that $\mu - \lambda\iota$ is invertible on $\mathfrak{z}_0^+$ when $\varepsilon(\lambda)$ is odd and invertible on $\mathfrak{z}_0^-$ when $\varepsilon(\lambda)$ is even. Defining $Z_3 = \mu^{-1}(\theta(Z_1) - Z_1) \in \mathfrak{z}_0^-$, we can extend θ to an involutive automorphism of $\mathfrak{g}_0$, by putting $\theta(X_1) = X_1 + Z_3$ and $\theta(Y_1) = Y_1$. Then, θ commutes with μ, by definition of Z_3, and anti-commutes with ρ, because $\rho(Z_3) = -Z_2 - \theta(Z_2)$. This last equality can be deduced as follows: because $(\mu - \iota)\rho = \rho\mu$, we get $\rho\mu^{-1} = (\mu - \iota)^{-1}\rho$ on $\mathfrak{z}_0^-$; in particular,

$$\rho(Z_3) = (\mu - \iota)^{-1}(-\theta(\rho(Z_1)) - \rho(Z_1)) = -Z_2 - \theta(Z_2),$$

where we used Equation (4.6).

Next, we define the Lie algebra $\mathfrak{h} := \mathbb{R}\tilde{A} + \mathfrak{g}_0$. We extend θ to an involutive automorphism of $\mathfrak{h}$ by defining $\theta(\tilde{A}) = -\tilde{A} + Z_4$, where $Z_4 := (\mu - \iota)^{-1}(\theta(Y_0) + Y_0) \in \mathfrak{z}_0^+$. Extending also $\mu = \operatorname{Ad} X_0$ to $\mathfrak{h}$ (recall that $\mu(\tilde{A}) = \tilde{A} + Y_0$), we observe that θ and μ commute.

As above, the derivation ρ gives rise to an action of $\mathbb{R}$ on G_0, where G_0 lives on the space $\mathbb{R}^2 \times \mathfrak{z}_0$ with product $(x,y,Z)(x',y',Z') = (x + x', y + \exp(x)y', Z + Z')$. Then, $H := \mathbb{R}\,{}_\rho\!\ltimes G_0$ gives an exponentiation of $\mathfrak{h}$, on which the derivation μ and the involutive automorphism θ are combined to produce an action (μ_s) of $\mathbb{R}^*$ by automorphisms of H. We define $G = \mathbb{R}^*\,{}_\mu\!\ltimes H$ and the product is given by $(s,x,g)(t,y,h) = (st, x + sy, \ldots)$. The precise form of the product can be written as above. The closed subgroup G_1 consists again of the elements $(s, 0, g)$ for $s \in \mathbb{R}^*$ and $g \in G_0$. To find G_2, we have to construct, as above, the closed subgroup G_2 of G with tangent Lie algebra $\mathbb{R}(\tilde{A} - aX_0 + B_1)$, where $B_1 \in \mathfrak{g}_0$, and such that G_2 consists of the elements $(s, \frac{1}{a}(1-s), v(s))$, where $s \in \mathbb{R}^*$. Conjugating, the problem is reduced to finding a closed subgroup of G with tangent Lie algebra $\mathbb{R}(X_0 + B_2)$, for some arbitrary $B_2 \in \mathfrak{g}_0$, consisting of the elements $(s, 0, w(s))$, $s \in \mathbb{R}^*$. We solve this problem in the closed subgroup $K := \mathbb{R}^* \ltimes G_0$ of G, whose Lie algebra is $\mathfrak{k} := \mathbb{R}X_0 + \mathfrak{g}_0$. Suppose that $B_2 = x_1X_1 + y_1Y_1 \mod \mathfrak{z}_0$. If $b \in \mathbb{R}$,

$$\exp(\operatorname{Ad}(bY_1))(X_0 + B_2) = x_1X_1 + (y_1 - bx_1)Y_1 \mod \mathfrak{z}_0.$$

If $\mathrm{Exp}_{\mathfrak{k}}$ denotes the exponential mapping from $\mathfrak{k}$ to K, we observe that conjugation by $\mathrm{Exp}_{\mathfrak{k}}(bY_1)$ for a well chosen $b \in \mathbb{R}$, reduces the problem to either $x_1 = 0$ or $y_1 = 0$. Both cases are solved by Lemma 4.3, because $\mathbb{R}X_1 + \mathfrak{g}_0$ and $\mathbb{R}Y_1 + \mathfrak{g}_0$ are abelian. So, the proof of part 3 is done.

Part 4: $\mathfrak{g}_0 = \langle X, Y, Z\rangle$ *with* $[X, Y] = Z$ *and* Z *central*.

A general derivation μ of $\mathfrak{g}_0$ has the form

$$\mu(X) = x_1X + y_1Y + z_1Z, \quad \mu(Y) = x_2X + y_2Y + z_2Z, \quad \mu(Z) = (x_1 + y_2)Z.$$

Perturbing μ with $\mathrm{Ad}(-z_2X + z_1Y)$, μ is of the form

$$\mu\begin{pmatrix} X \\ Y \end{pmatrix} = P\begin{pmatrix} X \\ Y \end{pmatrix}, \quad \mu(Z) = \mathrm{Tr}(P)\, Z.$$

Because $[\mu, \rho] = \rho + \mathrm{Ad}\, Y_0$, we only have two possibilities. Either $\rho = 0$, or μ and ρ have after inner perturbation and a change of basis in $\mathfrak{g}_0$ (respecting the relations of $\mathfrak{g}_0$), the form

$$\begin{aligned} &\mu(X) = \alpha X, \quad &&\mu(Y) = (\alpha - 1)Y, \quad &&\mu(Z) = (2\alpha - 1)Z, \\ &\rho(X) = 0, &&\rho(Y) = X, &&\rho(Z) = 0. \end{aligned}$$

Then, necessarily, $Y_0 = \lambda Z$ for some $\lambda \in \mathbb{R}$. We have $[X_0, \tilde{A}] = \tilde{A} + \lambda Z$, $\mathfrak{g}_1 = \mathbb{R}X_0 + \mathfrak{g}_0$ and $\mathfrak{g}_2 = \mathbb{R}(\tilde{A} - aX_0 + B)$ for some $B \in \mathfrak{g}_0$. To prove the existence of an exponentiation, we apply $\exp(-\frac{1}{a}\,\mathrm{Ad}\,\tilde{A})$ and observe that it is an equivalent question to exponentiate $\mathfrak{g}_1 = \mathbb{R}(X_0 + \frac{1}{a}\tilde{A}) + \mathfrak{g}_0$ and $\mathfrak{g}_2 = \mathbb{R}(X_0 + C)$ for some $C \in \mathfrak{g}_1$. Write $C = eX + fY + gZ$. If we now replace X_0 by $X_0 + (g + 2ef)Z$, we see that none of the relations above change, because $\rho(Z) = 0$, but only C changes to $eX + fY - 2efZ$. So, we may suppose that C has this last form.

Define $\mathfrak{h} := \mathbb{R}\tilde{A} + \mathfrak{g}_0$ and exponentiate as $H := \mathbb{R}\,{}_{\rho}\!\ltimes G_0$. Define $\theta(\tilde{A}) = -\tilde{A}$, $\theta(X) = pX, \theta(Y) = -pY$ and $\theta(Z) = -Z$, where $p = \pm 1$ will be determined later. Then, θ is an involutive automorphism of $\mathfrak{h}$ that commutes with the derivation $\mu = \mathrm{Ad}\, X_0$ of $\mathfrak{h}$. Both combine to an action μ of $\mathbb{R}^*$ on H and we define $G = \mathbb{R}^*\,{}_{\mu}\!\ltimes H$. The only problem left is the definition of the good closed subgroup of G with tangent Lie algebra $\mathbb{R}(X_0+C)$. Suppose first that $\alpha \neq 1$ and $\alpha \neq \frac{1}{2}$. Then, we take $p = 1$ and observe that μ is invertible on $\mathfrak{g}_0^- = \langle Y, Z\rangle$ and on $[\mathfrak{g}_0, \mathfrak{g}_0] = \mathbb{R}Z$. So, Lemma 4.3 provides us with the needed subgroup. If $\alpha = 1$, we take $p = -1$ and we are done as well. Finally, take $\alpha = \frac{1}{2}$ and $p = 1$. Define $D = 4fY$ and observe that

$$\mu(D) - [D, C] - \frac{1}{2}[D, \mu(D)] = -2fY + 4efZ = \theta(C) - C.$$

Hence, it follows from the proof of Lemma 4.3 that we can define the right closed subgroup of G.

Finally, we have to consider the case where ρ is trivial. This is very much similar to part 1 of this proof, but simpler. □

Now we prove that at least one of the conditions of Proposition 4.2 is fulfilled when $\dim \mathfrak{g}_0 \leq 3$, up to one exceptional case, that we exponentiate explicitly 'by hands'.

Corollary 4.4. *In case 2, every real matched pair of dimension $n+1$ with $n \leq 4$ can be exponentiated to a matched pair of real Lie groups with at most two connected components.*

Proof. If $\dim \mathfrak{g}_0 = 1$ or 2, then either $\mathfrak{g}_0$ is abelian, or $\mathfrak{g}_0$ has trivial center and all derivations are inner. If $\dim \mathfrak{g}_0 = 3$ and $\mathfrak{g}_0$ is of rank 3, then $\mathfrak{g}_0 = \mathfrak{sl}_2$ or $\mathfrak{su}_2$. In both cases, every derivation is inner and the center is trivial.

When $\mathfrak{g}_0$ has rank 1 or rank 0, one can always apply Proposition 4.2.

Finally, there are only three real non-isomorphic 3-dimensional $\mathfrak{g}_0$ of rank 2 defined respectively by : a) $[H, X] = X,\ [H, Y] = \alpha Y,\ [X, Y] = 0$ ($\alpha \in \mathbb{R}$); b) $[H, X] = X+Y,\ [H, Y] = Y,\ [X, Y] = 0$; c) $[H, X] = rX+Y,\ [H, Y] = -X+rY,\ [X, Y] = 0$ ($r \in \mathbb{R}$) (see [14]).

All these cases (except a), $\alpha = 1$, which we will study separately), can be treated in a similar way. Namely, a general derivation μ of $\mathfrak{g}_0$ has the following form:

a) $\alpha \neq 1$: $\mu(H) = aX + bY,\quad \mu(X) = cX,\quad \mu(Y) = dY$, and it is inner if $d = \alpha c$.

b) $\mu(H) = xX + yY$, $\mu(X) = aX + bY$, $\mu(Y) = aY$, and it is inner if $a = b$.

c) $\mu(H) = xX + yY$, $\mu(X) = aX + bY$, $\mu(Y) = -bX + aY$, and it is inner if $a = rb$

Then, since μ and ρ are derivations of $\mathfrak{g}_0$, we observe that in all cases above $[\mu, \rho]$ is inner. Hence, $\rho = [\mu, \rho] - \operatorname{Ad} Y_0$ is inner. Also, the center of $\mathfrak{g}_0$ is always trivial.

At last, we study separately $\mathfrak{g}_0$ defined by $[H, X] = X$, $[H, Y] = Y$ and $[X, Y] = 0$. A general derivation μ of $\mathfrak{g}_0$ has the form

$$\mu(H) = xX + yY, \quad \mu(X) = aX + bY, \quad \mu(Y) = cX + dY,$$

which is inner if $b = c = 0$ and $a = d$. Because $[\mu, \rho] = \rho + \operatorname{Ad} Y_0$, we conclude that there exists a $\lambda \in \mathbb{R}$ such that, on $\langle X, Y \rangle$, $\mu\rho - \rho\mu = \rho + \lambda\iota$. It is easy to see that either $\rho = -\lambda\iota$, in which case ρ is inner and the exponentiation exists, or that we can change the basis X, Y and perturb μ and ρ innerly such that

$$\begin{aligned} \mu(H) &= 0, \quad \mu(X) = X, \quad \mu(Y) = 0, \\ \rho(H) &= 0, \quad \rho(X) = 0, \quad \rho(Y) = X. \end{aligned}$$

Our matched pair lives now in the Lie algebra $\mathfrak{g}$ with generators $X_0, \tilde{A}, X, Y, Z$ with relations $\operatorname{Ad} X_0 = \mu$, $\operatorname{Ad} \tilde{A} = \rho$ on $\mathfrak{g}_0$, $[X_0, \tilde{A}] = \tilde{A}$. Observe that Y_0 disappeared because, after the necessary inner perturbations of μ and ρ, we arrived at $[\mu, \rho] = \rho$. The Lie subalgebras $\mathfrak{g}_1$ and $\mathfrak{g}_2$ are $\langle X_0, H, X, Y \rangle$ and $\mathbb{R}(\tilde{A} - aX_0 + B)$, respectively, where $B \in \mathfrak{g}_0$ is arbitrary. We exponentiate $\mathbb{R}\tilde{A} + \mathfrak{g}_0$ on the space $\mathbb{R}^4$ with product

$$(a, h, x, y)(a', h', x', y') = (a+a', h+h', x+\exp(h)x' + a\exp(h)y', y+\exp(h)y')$$

and we denote this Lie group by H. Write $B = \alpha H + \beta X + \gamma Y$. Suppose first that $\alpha \neq 0$. Then, we make $\mathbb{R}^*$ act on H by the automorphisms $\mu_s(a, h, x, y) =$

$(sa, h, |s|x, \mathrm{Sgn}(s)y)$. Take $G = \mathbb{R}^* \ltimes H$ in which we consider the closed subgroup G_1 of elements $(s, 0, h, x, y)$, $s \in \mathbb{R}^*, h, x, y \in \mathbb{R}$. Next, we have to find a good (as in the proof of Proposition 4.2) closed subgroup G_2 with tangent Lie algebra $\mathbb{R}(\tilde{A} - aX_0 + B)$. Conjugating with the element $(1, -\frac{1}{a}, 0, 0, 0)$, we have to find a good closed subgroup of $\mathbb{R}^* \ltimes G_0$ with tangent Lie algebra $\mathbb{R}(X_0 + \alpha' H + \beta' X + \gamma' Y)$, where $\alpha' = -\frac{\alpha}{a} \neq 0$. If $\alpha' \neq -1$, such a subgroup is given by

$$\left\{\left(s, \alpha' \log|s|, \frac{\beta'}{\alpha' + 1}(|s|^{\alpha'+1} - 1\right), \frac{\gamma'}{\alpha'}(\mathrm{Sgn}(s)|s|^{\alpha'} - 1)) \mid s \in \mathbb{R}^*\right\}.$$

If $\alpha' = -1$, such a subgroup is given by

$$\left\{\left(s, -\log|s|, \beta' \log|s|, \gamma'\left(1 - \frac{1}{s}\right)\right) \mid s \in \mathbb{R}^*\right\}.$$

One can easily check that (G_1, G_2) is indeed a matched pair of Lie groups.

If next, $\alpha = 0$, we consider the action of $\mathbb{R}^*$ by automorphisms of H given by $\mu_s(a, h, x, y) = (sa, h, sx, y)$. Take $G = \mathbb{R}^* \ltimes H$ in which we consider the closed subgroup G_1 of elements $(s, 0, h, x, y)$, $s \in \mathbb{R}^*, h, x, y \in \mathbb{R}$. Conjugating with the element $(1, -\frac{1}{a}, 0, 0, 0)$, we have to find a good closed subgroup of $\mathbb{R}^* \ltimes G_0$ with tangent Lie algebra $\mathbb{R}(X_0 + \beta' X + \gamma' Y)$ and this is given by

$$\{(s, 0, \beta'(s - 1), \gamma' \log|s|) \mid s \in \mathbb{R}^*\}.$$

Again, we arrive at a matched pair of Lie groups. □

As it was promised, we now present a matched pair of dimension 5 + 1 which has no exponentiation.

Example 4.5. Consider the 6-dimensional Lie algebra $\mathfrak{g}$ with generators A, X_0, X_1, X_2, Y, Z and relations

$$[X_0, A] = A + Y, \quad [X_0, X_1] = \frac{1}{2}X_1 + X_2, \quad [X_0, X_2] = \frac{1}{2}X_2,$$
$$[X_0, Y] = Y, \quad [X_0, Z] = \frac{3}{2}Z,$$
$$[A, X_1] = [A, Y] = [A, Z] = 0, \quad [A, X_2] = Z, \quad [X_1, X_2] = Y,$$
$$[X_1, Y] = Z, \quad [X_1, Z] = [X_2, Y] = [X_2, Z] = [Y, Z] = 0.$$

Defining $\mathfrak{g}_1 = \langle X_0, X_1, X_2, Y, Z\rangle$ and $\mathfrak{g}_2 = \mathbb{R}(A + X_0)$, we have a matched pair of Lie algebras. There does not exist a matched pair of Lie groups which exponentiates the given matched pair of Lie algebras.

Proof. It is easy to check that $\mathfrak{g}$ is indeed a Lie algebra: first it is clear that $\mathfrak{g}_0 := \langle X_1, X_2, Y, Z\rangle$ is a Lie algebra. Next, we write $\rho = \mathrm{Ad}\, A$ and $\mu = \mathrm{Ad}\, X_0$ on $\mathfrak{g}_0$ and it is clear that μ and ρ are derivations of $\mathfrak{g}_0$ satisfying $[\mu, \rho] = \rho + \mathrm{Ad}\, Y$. The connected, simply connected Lie group G_0 of $\mathfrak{g}_0$ has $\mathbb{R}^4$ as an underlying space, with product

$$(x_1, x_2, y, z)(x_1', x_2', y', z') = \Big(x_1 + x_1', x_2 + x_2', y + y' + x_1 x_2', z + z' + \frac{x_1^2}{2} x_2' + x_1 y'\Big).$$

The derivation ρ gives rise to an action (ρ_a) of $\mathbb{R}$ on G_0 given by $\rho_a(x_1, x_2, y, z) = (x_1, x_2, y + a x_2, z)$ and we can define the Lie group $H := \mathbb{R}\,{}_\rho\!\ltimes G_0$ on the space $\mathbb{R}^5$. Finally, μ gives rise to an action (μ_x) of $\mathbb{R}$ on H given by

$$\begin{aligned}\mu_x(a, x_1, x_2, y, z) = \Big(&\exp(x)a, \exp\Big(\frac{1}{2}x\Big)x_1, \exp\Big(\frac{1}{2}x\Big)(x_2 + x x_1), \\ &\exp(x)\Big(y + xa + \frac{1}{2}x x_1^2\Big), \exp\Big(\frac{3}{2}x\Big)\Big(z + x a x_1 + \frac{6}{2}x x_1^3\Big)\Big).\end{aligned}$$

Defining $G = \mathbb{R}\,{}_\mu\!\ltimes H$ on the space $\mathbb{R}^6$, we obtain the connected, simply connected Lie group of $\mathfrak{g}$. One observes easily that the center of G is trivial. Hence, G is the only connected Lie group with Lie algebra $\mathfrak{g}$.

We claim the following: any automorphism α of G leaves the closed normal subgroup G_0 invariant and modulo an inner perturbation by $\operatorname{Ad} g$ ($g \in G$), we have $\alpha = \iota$ on G/G_0. Indeed, let α be an automorphism of G. Denote $\theta = d\alpha$, the corresponding automorphism of $\mathfrak{g}$. It is straightforward to check that $\mathfrak{g}$ has only two ideals of dimension 4: $\mathfrak{g}_0$ and $\langle A, X_2, Y, Z\rangle$. As a Lie algebra, the first one has rank 2 and the second one rank 1. Because $\theta(\mathfrak{g}_0)$ is an ideal of dimension 4 of $\mathfrak{g}$ isomorphic with $\mathfrak{g}_0$, we get $\theta(\mathfrak{g}_0) = \mathfrak{g}_0$. Then also $\alpha(G_0) = G_0$. Because $[X_0, A] = A \mod \mathfrak{g}_0$, a perturbation of α by $\operatorname{Ad}(x, a, 0, 0, 0, 0)$ for certain $x, a \in \mathbb{R}$, leaves two possibilities for θ: $\theta = \iota \mod \mathfrak{g}_0$ or $\theta(X_0) = X_0 \mod \mathfrak{g}_0$ and $\theta(A) = -A \mod \mathfrak{g}_0$. We have to prove that the second option is impossible. Hence, suppose that $\theta(X_0) = X_0 + C$ and $\theta(A) = -A + D$ with $C, D \in \mathfrak{g}_0$. Because θ is an automorphism of $\mathfrak{g}_0$ it has the following form

$$\begin{aligned}\theta(X_1) &= bX_1 + gX_2 + hY + kZ, \quad & \theta(X_2) &= dX_2 + eY + fZ, \\ \theta(Y) &= bdY + beZ, & \theta(Z) &= b^2 dZ,\end{aligned}$$

for $e, f, g, h, k \in \mathbb{R}$ and $b, d \in \mathbb{R}^*$. Because θ is an automorphism of $\mathfrak{g}$, we get $\theta\rho = (-\rho + \operatorname{Ad} D)\theta$. Suppose that $D = x_1 X_1 + x_2 X_2 + yY + zZ$. Verifying the equality on Y gives $x_1 bdZ = 0$ and, hence, $x_1 = 0$. Next, we verify on X_2 and conclude that $b^2 dZ = -dZ$, which yields the required contradiction $b^2 = -1$.

Suppose now that $\mathcal{G}$ is a Lie group with Lie algebra $\mathfrak{g}$ and with two closed subgroups $\mathcal{G}_1, \mathcal{G}_2$ whose tangent Lie algebras are $\mathfrak{g}_1, \mathfrak{g}_2$, respectively and such that $(\mathcal{G}_1, \mathcal{G}_2)$ is a matched pair of Lie groups. Because G is the only connected Lie group with Lie algebra $\mathfrak{g}$, we can identify G and $\mathcal{G}^{(e)}$, the connected component of e in $\mathcal{G}$.

Because any automorphism of G leaves G_0 invariant, we find that G_0 is a normal subgroup of $\mathcal{G}$. We can naturally identify $(\mathcal{G}/G_0)^{(e)}$ and G/G_0. Define $H = \mathcal{G}/G_0$. We claim that for any $\eta \in H/H^{(e)}$, there exists a unique representative $u(\eta) \in H$ such that $\operatorname{Ad} u(\eta) = \iota$ on $H^{(e)}$. Denote by π_1 the quotient map from $\mathcal{G}$ to H and by π_2 the quotient map from H to $H/H^{(e)}$. Let $\eta \in H/H^{(e)}$. Take $g \in G$ such that $\pi_2(\pi_1(g)) = \eta$. Then, $\operatorname{Ad} g$ defines an automorphism of $\mathcal{G}^{(e)} = G$. Using our claim, we can take an $h \in G$, such that $\operatorname{Ad}(hg)$ is trivial on G/G_0. This means that $\operatorname{Ad} \pi_1(hg)$ is trivial on $H^{(e)}$. But $\pi_2(\pi_1(hg)) = \eta$. So, we have proven the existence of the required representative. The uniqueness is trivial because $H^{(e)} = G/G_0$ has trivial center. Then, the map $\Psi : H/H^{(e)} \oplus H^{(e)} \to H : (\eta, g) \mapsto u(\eta)g$ is an isomorphism of Lie groups. (Recall that $H/H^{(e)}$ has dimension zero and the discrete topology.)

The only closed subgroup of G with tangent Lie algebra $\mathfrak{g}_1$ is the group consisting of the elements $(x, 0, x_1, x_2, y, z)$. Hence, this last subgroup coincides with $\mathcal{G}_1 \cap G$. On the other hand, the only closed subgroup of G with tangent Lie algebra $\mathfrak{g}_2$ consists of the elements $(x, \exp(x) - 1, 0, 0, x\exp(x), 0)$. Hence, this last subgroup coincides with $\mathcal{G}_2 \cap G$. Identifying $H^{(e)}$ with the group $K := \mathbb{R}^2$ with product $(x, a)(x', a') = (x + x', a + \exp(x)a')$, we get that $K_1 := \pi_1(\mathcal{G}_1) \cap H^{(e)} = \{(x, 0) \mid x \in \mathbb{R}\}$ and $K_2 := \pi_1(\mathcal{G}_2) \cap H^{(e)} = \{(x, \exp(x) - 1) \mid x \in \mathbb{R}\}$. Combining this with the fact that H is isomorphic with $H/H^{(e)} \oplus H^{(e)}$ and with the fact that the only elements of K normalizing K_1, resp. K_2, belong to K_1, resp. K_2, we conclude that $\pi_1(\mathcal{G}_i) = Q_i \oplus K_i$ for some subgroups $Q_i \subset H/H^{(e)}$ and $i = 1, 2$. Because $\mathcal{G}_1\mathcal{G}_2$ is dense in $\mathcal{G}$, it follows that K_1K_2 should be dense in K. This is clearly not the case. □

Case 3. Now we have $\mathfrak{g}_0 := \operatorname{Ker}\chi \cap \operatorname{Ker}\chi\beta$ and this is still an ideal of $\mathfrak{g}$. We can take $X, Y \in \mathfrak{g}_1$ such that

$$\chi(X) = 1, \quad \chi(Y) = 0, \quad \chi(\beta(X)) = 0, \quad \chi(\beta(Y)) = 1.$$

Using Equation (4.2), we get $\chi(\beta([X, Y])) = -1$ and $\chi([X, Y]) = 0$. Hence, $[X, Y] = -Y \mod \mathfrak{g}_0$. Because $\chi(\beta(X)) = 0$ and $\chi(\beta(Y)) = 1$, we get

$$\beta(X) = aY \mod \mathfrak{g}_0, \quad \beta(Y) = X + bY \mod \mathfrak{g}_0, \quad a, b \in k.$$

Checking Equation (4.1), we arrive at $b = 0$. Since the quotient Lie algebra $\mathfrak{g}/\mathfrak{g}_0$ is 3-dimensional and of rank 3, it is isomorphic to $\mathfrak{sl}_2(k)$ [14] (if $k = \mathbb{R}$, one must analyse also $\mathfrak{su}_2(\mathbb{R})$, but it has no 2-dimensional Lie subalgebras). Then, by the Levy–Maltsev theorem [39], Chapter X, we can find a Lie subalgebra $\tilde{\mathfrak{g}}$ of $\mathfrak{g}$ isomorphic to $\mathfrak{sl}_2(k)$ and such that $\mathfrak{g} = \tilde{\mathfrak{g}} \oplus \mathfrak{g}_0$ as vector spaces. This means that we are always in the following situation: $\mathfrak{sl}_2(k)$, with the generators A, X, Y satisfying

$$[X, A] = aY + A, \quad [Y, A] = X, \quad [X, Y] = -Y,$$

is represented by derivations of $\mathfrak{g}_0$ and in the semi-direct product $\mathfrak{g} := \mathfrak{sl}_2(k) \ltimes \mathfrak{g}_0$ we have the matched pair $\mathfrak{g}_1 = \langle X, Y \rangle + \mathfrak{g}_0$, $\mathfrak{g}_2 = k(A + Z)$ for some element $Z \in \mathfrak{g}_0$.

We first analyse the case $k = \mathbb{R}$. If we replace Y by rY and A by $(1/r)A$ for $r \neq 0$, then a changes to a/r^2. So, we only have to consider three cases: one with $a > 0$, one with $a = 0$ and one with $a > 0$.

In terms of the standard generators $H = e_{11} - e_{22}$, $K = e_{12}$ and $L = e_{21}$, we realize the relations above by putting $X = \frac{1}{2}H$, $Y = L$ and $A = -2aL - \frac{1}{2}K$. This means that we consider the matched pair $\mathfrak{g}_1 = \langle H, L\rangle + \mathfrak{g}_0$, $\mathfrak{g}_2 = \mathbb{R}(K + 4aL + Z)$ for some $Z \in \mathfrak{g}_0$.

Proposition 4.6. *If $k = \mathbb{R}$ and $a \leq 0$, the matched pair has an exponentiation. When $a < 0$, G, G_1, G_2 can be taken connected. When $a = 0$, we need two connected components for G_1.*

If $k = \mathbb{R}$ and $a > 0$, the matched pair has an exponentiation in at least the following cases:

1. *$\mathfrak{g}_0$ is abelian.*

2. *$\dim \mathfrak{g}_0 = 2$.*

We can take a matched pair of Lie groups where G, G_1 have at most two connected components and G_2 has at most four connected components.

In particular, for $k = \mathbb{R}$ the exponentiation exists for all matched pairs of dimension $n + 1, n \leq 4$.

Proof. We start off with the case $a \leq 0$ and we first suppose that $\mathfrak{g}_0 = \{0\}$. An exponentiation of the case $a = 0$ is given by $F = \mathrm{SL}_2(\mathbb{R})$ with subgroups

$$F_1 := \left\{ \begin{pmatrix} a & 0 \\ x & \frac{1}{a} \end{pmatrix} \mid a \neq 0, x \in \mathbb{R} \right\}, \quad F_2 := \left\{ \begin{pmatrix} 1 & b \\ 0 & 1 \end{pmatrix} \mid b \in \mathbb{R} \right\}. \tag{4.7}$$

An exponentiation of the case $a = -\frac{1}{4}$ (as we saw above, it is sufficient to consider one value of $a < 0$) is given by $F = \mathrm{SL}_2(\mathbb{R})$ with subgroups

$$F_1 := \left\{ \begin{pmatrix} a & 0 \\ x & \frac{1}{a} \end{pmatrix} \mid a > 0, x \in \mathbb{R} \right\}, \quad F_2 := \left\{ \begin{pmatrix} \cos t & \sin t \\ -\sin t & \cos t \end{pmatrix} \mid t \in \mathbb{R} \right\}.$$

For this last case, there is another exponentiation which is at least as important. We observe that $F = F_1 F_2$ and the multiplication map is a diffeomorphism of $F_1 \times F_2$ onto F. This means that the mutual actions $\alpha_g(s)$ and $\beta_s(g)$, $g \in F_1$, $s \in F_2$ are everywhere defined and smooth. Identifying F_2 with $\mathbb{T}$, we observe that for all $g \in F_1$, α_g is a diffeomorphism of $\mathbb{T}$ satisfying $\alpha_g(1) = 1$. Hence, there exists, for every $g \in F_1$ a unique diffeomorphism $\tilde{\alpha}_g$ of $\mathbb{R}$, such that $\tilde{\alpha}_g(0) = 0$ and $p(\tilde{\alpha}_g(t)) = \alpha_g(p(t))$ for all $t \in \mathbb{R}$, where $p(t) = \cos t + i \sin t$. Defining $\tilde{\beta}_t(g) = \beta_{p(t)}(g)$, we obtain a matched pair $(F_1, \mathbb{R})$, in which both actions are everywhere defined and smooth. Its corresponding big Lie group F_{sc} is the connected, simply connected Lie group of $\mathfrak{sl}_2(\mathbb{R})$.

Suppose now that $\mathfrak{g}_0$ is arbitrary and $Z \in \mathfrak{g}_0$. First, take $a = 0$. Because any finite-dimensional representation of $\mathfrak{sl}_2(\mathbb{R})$ can be exponentiated to $\mathrm{SL}_2(\mathbb{R})$ (although $\mathrm{SL}_2(\mathbb{R})$ is not simply connected, see e.g. [7], Chapitre VIII, par. 1, Théorème 2), we get an action μ of $F := \mathrm{SL}_2(\mathbb{R})$ with automorphisms of G_0, the connected, simply connected Lie group of $\mathfrak{g}_0$. We define $G := F \,{}_\mu\!\ltimes G_0$ and we denote by $\mathrm{Exp}_{\mathfrak{g}}$ its exponential mapping. Because

$$\mathrm{Exp}_{\mathfrak{g}}(b(K+Z)) = \left(\begin{pmatrix} 1 & b \\ 0 & 1 \end{pmatrix}, \dots \right),$$

it is clear that, defining

$$G_1 := \{(g,k) \mid g \in F_1, k \in G_0\}, \quad G_2 := \{\mathrm{Exp}_{\mathfrak{g}}(b(K+Z)) \mid b \in \mathbb{R}\},$$

where F_1 is as in Equation (4.7), we get the required matched pair of Lie groups.

When $a = -\frac{1}{4}$, we proceed similarly, but now with the simply connected Lie group F_{sc} of $\mathfrak{sl}_2(\mathbb{R})$. We get an action μ of F_{sc} on G_0 by automorphisms and we define $G := F_{\mathrm{sc}} \,{}_\mu\!\ltimes G_0$. As we explained above, we can find in F_{sc} a matched pair (F_1, F_2) with tangent Lie algebras $\langle X, Y \rangle$ and $\mathbb{R}(K-L)$, such that F_2 can be identified with $\mathbb{R}$. If $\mathrm{Exp}_{\mathfrak{g}}$ denotes the exponential mapping of $\mathfrak{g}$, it follows that

$$\mathrm{Exp}_{\mathfrak{g}}(t(K-L+Z)) = (t, \dots) \in F_2 \times G_0 \subset F_{\mathrm{sc}} \,{}_\mu\!\ltimes G_0.$$

So, we can define in the same way as above the required matched pair of Lie groups. Observe that this argument would not work with $SL_2(\mathbb{R})$ instead of F_{sc}, because then $\mathrm{Exp}_{\mathfrak{g}}(2\pi n(K-L+Z)) \in G_0 \subset G_1$ when $n \in \mathbb{Z}$ and hence, we no longer have $G_1 \cap G_2 = \{e\}$.

Next, we turn to $a > 0$ and we choose the value $a = \frac{1}{4}$. Then, $\mathfrak{g}_2 = \mathbb{R}(K+L+Z)$ for some $Z \in \mathfrak{g}_0$. When $\mathfrak{g}_0 = \{0\}$, we can exponentiate as follows. Write $F = \{T \in M_2(\mathbb{R}) \mid \det T = \pm 1\}$. Define

$$F_1 := \left\{ \begin{pmatrix} a & 0 \\ x & \frac{s}{a} \end{pmatrix} \mid a > 0, s = \pm 1, x \in \mathbb{R} \right\}, \quad F_2 := \left\{ \begin{pmatrix} b & c \\ c & b \end{pmatrix} \mid b^2 - c^2 = \pm 1 \right\}. \tag{4.8}$$

It is an easy exercise to check that we indeed get a matched pair of Lie groups. For general $\mathfrak{g}_0$, we want to proceed as in the case $a = 0$. We first get an action μ of $SL_2(\mathbb{R})$ on G_0. We now run into the same kind of problems as in the proof of Proposition 4.2. We should first extend the action μ to an action of F, by adding an involutive automorphism of G_0 corresponding to the action of $\left(\begin{smallmatrix} 1 & 0 \\ 0 & -1 \end{smallmatrix}\right)$ on G_0 and next, we should find the good closed subgroup with tangent Lie algebra $\mathbb{R}(K+L+Z)$ and with elements whose first components are precisely the matrices $\left(\begin{smallmatrix} b & c \\ c & b \end{smallmatrix}\right)$ with $b^2 - c^2 = \pm 1$.

First, take $\mathfrak{g}_0$ abelian. Write μ_H, μ_K and μ_L for the derivations of $\mathfrak{g}_0$ corresponding to the generators H, K and L of $\mathfrak{sl}_2(\mathbb{R})$. From [7] (Chapitre VIII, par. 1, no. 2, Corollaire), we can write

$$\mathfrak{g}_0 = \bigoplus_{-N \le n \le N} E_n,$$

where $N \in \mathbb{N}$, n only takes values in $\mathbb{Z}$ and where

$$\mu_H(X) = nX \quad \text{for } X \in E_n, \quad \mu_K(E_n) \subset E_{n+2}, \quad \mu_L(E_n) \subset E_{n-2}$$

with $E_m = \{0\}$ if $m < -N$ or $m > N$. We exponentiate μ to a homomorphism $\mu : \mathrm{SL}_2(\mathbb{R}) \to \mathrm{GL}(\mathfrak{g}_0)$. If we define the involution θ of $\mathfrak{g}_0$ by putting $\theta(X) = X$ if $X \in E_n$, $n = 0, 1 \bmod 4$ and $\theta(X) = -X$ if $X \in E_n$ and $n = 2, 3 \bmod 4$, then, θ commutes with μ_H and anti-commutes with μ_K and μ_L. So, we extend μ to the group F of matrices with determinant ± 1 by writing $\mu\left(\begin{smallmatrix} 1 & 0 \\ 0 & -1 \end{smallmatrix}\right) = \theta$. We define, on the space $F \times \mathfrak{g}_0$, the Lie group G with product $(P, X)(Q, Y) = (PQ, X + \mu(P)Y)$ for $P, Q \in F$ and $X, Y \in \mathfrak{g}_0$. Define G_1 consisting of the pairs (P, X) for $P \in F_1$ and $X \in \mathfrak{g}_0$, with F_1 as in Equation (4.8). Finally, we have to find the good closed subgroup with tangent Lie algebra $\mathbb{R}(K + L + Z)$ for $Z \in \mathfrak{g}_0$. This procedure is described in great detail in the proof of Proposition 4.2. After a well chosen conjugation, we have to find a closed subgroup of G with tangent Lie algebra $\mathbb{R}(H + C)$, for some $C \in \mathfrak{g}_0$, consisting of the elements

$$\left\{ \left(\begin{pmatrix} a & 0 \\ 0 & b \end{pmatrix}, v(a, b) \right) \mid ab = \pm 1 \right\}.$$

As we see from the proof of Lemma 4.3, the only point is to define $v(-1, -1)$ and $v(1, -1)$. It follows from [7], Chapitre VIII, par. 1, no. 5, Corollaire, that $\mu\left(\begin{smallmatrix} -1 & 0 \\ 0 & -1 \end{smallmatrix}\right) X = (-1)^n X$ for $X \in E_n$. Denote by ρ this involution of $\mathfrak{g}_0$. The proof of Lemma 4.3 suggests us to define $v(-1, -1) = \mu_H^{-1}(\rho(C) - C)$ and $v(1, -1) = \mu_H^{-1}(\theta(C) - C)$. Both are well-defined, because

$$\rho(C) - C \in \bigoplus_{n=1 \bmod 2} E_n, \quad \theta(C) - C \in \bigoplus_{n=2,3 \bmod 4} E_n$$

and μ_H is invertible on both subspaces. Also, $v(-1, -1)$ and $v(1, -1)$ should be compatible, in the sense that

$$v(-1, -1) + \rho(v(1, -1)) = v(1, -1) + \theta(v(-1, -1)),$$

or equivalently

$$(\rho - \iota)\mu_H^{-1}(\theta - \iota)(C) = (\theta - \iota)\mu_H^{-1}(\rho - \iota)(C),$$

which is the case because the factors commute.

To finish the proof of the proposition, it remains to consider the unique non-abelian 2-dimensional $\mathfrak{g}_0$. So, $\mathfrak{g}_0$ has generators X, Y with relation $[X, Y] = Y$. The Lie algebra of derivations of $\mathfrak{g}_0$ is now isomorphic to $\mathfrak{g}_0$. Any homomorphism of $\mathfrak{sl}_2(\mathbb{R})$ into $\mathfrak{g}_0$ must be trivial, because its kernel is a non-zero ideal of $\mathfrak{sl}_2(\mathbb{R})$. So, any action of $\mathfrak{sl}_2(\mathbb{R})$ on $\mathfrak{g}_0$ is necessarily trivial and we take $G := F \oplus G_0$, where G_0 is the $ax + b$-group. We take $G_1 = F_1 \oplus G_0$ and for any $Z \in \mathfrak{g}_0$ we can define

$$G_2 = \left\{ \left(\begin{pmatrix} b & c \\ c & b \end{pmatrix}, \mathrm{Exp}_{\mathfrak{g}_0}((\log|b + c|)Z) \right) \mid b^2 - c^2 = \pm 1 \right\}.$$

This again provides us with a matched pair of Lie groups. □

Remark 4.7. For the case $\mathfrak{g}_0 = \{0\}$, we will give more connected exponentiations below. In the proof of the previous proposition they are not so interesting, because we will then rather have $G = \mathrm{PSL}_2(\mathbb{R})$ and not every representation of $\mathrm{SL}_2(\mathbb{R})$ factors through $\mathrm{PSL}_2(\mathbb{R})$. Hence, we cannot make the right semi-direct products with $\mathrm{PSL}_2(\mathbb{R})$ acting.

Remark 4.8. In case 3, for $k = \mathbb{R}$ and $n \geq 5$, there are indications that there again exist matched pairs of Lie algebras that cannot be exponentiated. Their explicit description remains however open.

Next, we analyse the case $k = \mathbb{C}$. First, let us note, that now there are only two non-isomorphic cases: with $a = 0$ and with $a \neq 0$.

Proposition 4.9. *In case 3, any matched pair of complex Lie algebras can be exponentiated to a matched pair of connected complex Lie groups if $n \leq 3$.*

Proof. First, consider the case $\mathfrak{g}_0 = \{0\}$. If $a = 0$, we proceed exactly as in the case of $k = \mathbb{R}$ and just replace $\mathbb{R}$ by $\mathbb{C}$. If $a \neq 0$, we consider $\mathfrak{g}_1 = \langle H, L \rangle$ and $\mathfrak{g}_2 = \mathbb{C}(K - L)$. Define $F = \mathrm{PSL}_2(\mathbb{C})$ and define

$$F_1 := \left\{ \begin{pmatrix} a & 0 \\ x & \frac{1}{a} \end{pmatrix} \mod \{\pm 1\} \mid a \neq 0, x \in \mathbb{C} \right\}, \tag{4.9}$$

$$F_2 := \left\{ \begin{pmatrix} \cos z & \sin z \\ -\sin z & \cos z \end{pmatrix} \mod \{\pm 1\} \mid z \in \mathbb{C} \right\}.$$

Some care is needed in checking that we do get a matched pair of Lie groups. Writing the product of an element in F_1 and an element in F_2, we have to find a unique solution in F_1, F_2 of the equation

$$\begin{pmatrix} a \cos z & a \sin z \\ x \cos z - \frac{1}{a} \sin z & x \sin z + \frac{1}{a} \cos z \end{pmatrix} = \begin{pmatrix} u & v \\ w & r \end{pmatrix} \mod \{\pm 1\}$$

whenever $ur - vw = 1$. Given $u, v, w, r \in \mathbb{C}$ with $ur - vw = 1$, we proceed as follows: choose $a \in \mathbb{C}$ such that $a^2 = u^2 + v^2$ and define $\cos z = \frac{u}{a}$, $\sin z = \frac{v}{a}$ and $x = \frac{uw+vr}{a}$. Then, the required equation holds. If we choose the other square root of $u^2 + v^2$, then $a, x, \cos z$ and $\sin z$ change sign and hence, their projections mod$\{\pm 1\}$ do not change. Because clearly $F_1 \cap F_2 = \{e\}$, we have a matched pair of Lie groups.

If next, $\mathfrak{g}_0 = \mathbb{C}$, the action of $\mathfrak{sl}_2(\mathbb{C})$ on $\mathfrak{g}_0$ is necessarily trivial. Our matched pair has the form $\mathfrak{g} = \mathfrak{sl}_2(\mathbb{C}) \oplus \mathbb{C}$, $\mathfrak{g}_1 = \langle H, L, Z \rangle$, $\mathfrak{g}_2 = \mathbb{C}(K + 4aL + \lambda Z)$, where Z is the generator of $\mathfrak{g}_0 = \mathbb{C}$ and $\lambda \in \mathbb{C}$. If $\lambda = 0$, it is clear how to exponentiate, just adding a copy of $\mathbb{C}$ to F and F_1 above. If $\lambda \neq 0$, we change the generator Z and we may suppose that $\lambda = 1$. If $a = 0$, exponentiation is again easy. If we take $a = -\frac{1}{4}$, we denote by $\mathbb{T}$ the complex torus, consisting of the pairs $(\cos z, \sin z) \in \mathbb{C}^2$, we define $G = \mathrm{PSL}_2(\mathbb{C}) \oplus \mathbb{T}$ with subgroups $G_1 := F_1 \oplus \mathbb{T}$ (with F_1 as in Equation (4.9)) and

$$G_2 = \left\{ \left(\begin{pmatrix} \cos z & \sin z \\ -\sin z & \cos z \end{pmatrix}, (\cos z, \sin z) \right) \mid z \in \mathbb{C} \right\}.$$

We indeed have a matched pair of Lie groups. □

We conclude with the promised counterexample in dimension $4 + 1$.

Example 4.10. Define $\mathfrak{g}$ to be the complex Lie algebra $\mathfrak{g} := \mathfrak{sl}_2(\mathbb{C}) \oplus \mathfrak{g}_0$, where $\mathfrak{g}_0$ has the generators X, Y satisfying $[X, Y] = Y$. Use the canonical generators H, K, L of $\mathfrak{sl}_2(\mathbb{C})$ and define the matched pair

$$\mathfrak{g}_1 := \langle H, L, X, Y \rangle, \quad \mathfrak{g}_2 := \mathbb{C}(K - L + Y).$$

There does not exist an exponentiation of this matched pair of Lie algebras.

Proof. The connected, simply connected Lie group of $\mathfrak{g}$ is given by $\mathrm{SL}_2(\mathbb{C}) \oplus G_0$, where G_0 lives on the space $\mathbb{C}^2$ with product $(x, y)(x', y') = (x + x', y + \exp(x)y')$. Its center consists of the elements $(\pm 1, 2\pi n, 0)$, where $n \in \mathbb{Z}$. Hence, the only connected Lie groups with Lie algebra $\mathfrak{g}$ are $G := \mathrm{SL}_2(\mathbb{C}) \oplus G_0/H_N$ and $G' := \mathrm{PSL}_2(\mathbb{C}) \oplus G_0/H_N$, where H_N consists of the elements $(2\pi n N, 0)$, $n \in \mathbb{Z}$. If we take $G := \mathrm{SL}_2(\mathbb{C}) \oplus G_0$, the connected closed subgroup of G with tangent Lie algebra $\mathfrak{g}_1$ consists of the elements

$$\left(\begin{pmatrix} a & 0 \\ z & \frac{1}{a} \end{pmatrix}, x, y \right), \quad a \neq 0, \ z, x, y \in \mathbb{C}.$$

The connected closed subgroup of G with tangent Lie algebra $\mathfrak{g}_2$ consists of the elements

$$\left(\begin{pmatrix} \cos z & \sin z \\ -\sin z & \cos z \end{pmatrix}, 0, z \right), \quad z \in \mathbb{C}.$$

The intersection of both subgroups is non-trivial, because it contains the elements $(1, 0, 2\pi n)$ with $n \in \mathbb{Z}$. This intersection is not annihilated by any of the central subgroups of G. So, with the same kind of reasoning as in Example 3.6, we conclude that the matched pair cannot be exponentiated. □

5 Matched pairs of real Lie algebras of dimension $1 + 1$ and $2 + 1$ and their exponentiation

Next we classify, up to isomorphism, all matched pairs of real Lie algebras of dimension $1 + 1$ and $2 + 1$ and compute explicitly their exponentiation.

From now on, all Lie algebras and Lie groups are understood to be real.

Theorem 5.1. *In dimension $1 + 1$, there exist, up to isomorphism, the following non-isomorphic matched pairs of Lie algebras. We choose a generator X for $\mathfrak{g}_1$.*

1. $\chi = 0$ *and* $\beta = 0$.
2. $\chi = 0$ *and* $\beta(X) = X$.

3. $\chi(X) = 1$ *and* $\beta = 0$.

4. $\chi(X) = 1$ *and* $\beta(X) = X$.

In dimension $2+1$, *there exist, up to isomorphism, the following non-isomorphic matched pairs of Lie algebras. We choose generators* X, Y *for* $\mathfrak{g}_1$.

1. $\chi = 0$ *and* $\beta = 0$.

   1.1 $[X, Y] = 0$.

   1.2 $[X, Y] = Y$.

2. $\chi = 0$ *and* $\beta \neq 0$:

   2.1 $[X, Y] = 0, \beta(X) = X, \beta(Y) = rY, -1 \leq r \leq 1$.

   2.2 $[X, Y] = 0, \beta(X) = X + Y, \beta(Y) = Y$.

   2.3 $[X, Y] = 0, \beta(X) = Y, \beta(Y) = 0$.

   2.4 $[X, Y] = Y, \beta(X) = Y, \beta(Y) = 0$.

   2.5 $[X, Y] = Y, \beta(X) = 0, \beta(Y) = Y$.

3. $\chi \neq 0$ *and* $\beta = 0$: $[X, Y] = aY, \chi(X) = 1, \chi(Y) = 0, a \in \mathbb{R}$.

4. $\chi \neq 0$ *and* $\beta \neq 0$: $\chi(X) = 1, \chi(Y) = 0$ *and*

   4.1 $[X, Y] = dY, \beta(X) = X + bY, \beta(Y) = dY$, *either* $d = 1$ *and* $b \in \mathbb{R}$, *or* $d \neq 1$ *and* $b = 0$.

   4.2 $[X, Y] = dY, \beta(X) = Y, \beta(Y) = 0, d \in \mathbb{R}$.

   4.3 $[X, Y] = -Y, \beta(X) = aY, \beta(Y) = X, a = 1, 0, -1$.

Every matched pair above can be exponentiated to a matched pair of Lie groups, having at most 2 connected components.

Proof. In dimension 1+1 the classification is obvious. If either $\beta = 0$ or $\chi = 0$, an exponentiation can be given using the semi-direct product of the corresponding connected simply connected Lie groups. In the remaining case, the exponentiation was explicitly described in Remark 3.5.

In dimension $2+1$, it is again natural to separate the cases 1, 2, 3 and 4. In case 1, the classification follows from the classification of 2-dimensional Lie groups. In case 2, we observe that $\beta([X, Y]) = [\beta(X), Y] + [X, \beta(Y)]$, i.e., β is an action. If $[X, Y] = 0$, any linear map β defines an action. We have either β diagonalizable (case 2.1), either β not diagonalizable and not nilpotent (case 2.2), or β nilpotent (case 2.3). Multiplying A by a scalar, one can scale β. Hence, cases 2.2 and 2.3 cover all the non-diagonalizable β. In case 2.1, we not only scale β, but also interchange X and Y, so that we can limit ourselves to $-1 \leq r \leq 1$.

If $[X, Y] = Y$, we see that $\beta(Y) \in \mathbb{R}Y$, and then $[\beta(X), Y] = 0$. Hence, $\beta(X) \in \mathbb{R}Y$. Replacing, if necessary, X by $X - rY$ and rescaling A, we obtain two different cases: 2.4 and 2.5.

Next, suppose that $\chi \neq 0$ and $\beta = 0$. Then χ is a character. Take X, Y such that $\chi(X) = 1$ and $\chi(Y) = 0$. Hence $[X, Y] = aY$ for some $a \in \mathbb{R}$. These are all non-isomorphic: to pass from one $a \in \mathbb{R}$ to another, we have to multiply X by a scalar, but then $\chi(X) = 1$ is violated.

An exponentiation of all the above matched pairs of Lie algebras again can be given using semi-direct products of the corresponding connected simply connected Lie groups.

Finally, the general discussion before this theorem shows that in case 4, there are two special situations. First, if $\chi\beta$ is a multiple of χ, we separate the cases $\chi\beta \neq 0$ and $\chi\beta = 0$. If $\chi\beta \neq 0$, we rescale A, so that $\chi\beta = \chi$. We can take generators X, Y for $\mathfrak{g}_1$ such that $\chi(X) = 1$ and $\chi(Y) = 0$. Then, there exists $b, c \in \mathbb{R}$, such that

$$\beta(X) = X + bY, \quad \beta(Y) = cY.$$

Because $\chi([X, Y]) = 0$, we get $[X, Y] = dY$ for $d \in \mathbb{R}$. Checking Equation (4.1), we get $c = d$. If we replace X by $X + rY$, then b changes to $b + r(d - 1)$. Hence, we get two non-isomorphic families: $d = 1, b \in \mathbb{R}$ and $d \neq 1, b = 0$, as stated in case 4.1.

If $\chi\beta = 0$, an analogous reasoning gives generators X, Y for $\mathfrak{g}_1$ such that $[X, Y] = dY$, $\chi(X) = 1$, $\chi(Y) = 0$, $\beta(X) = bY$, $\beta(Y) = 0$, for $b, d \in \mathbb{R}$. Because $\beta \neq 0$, we get $b \neq 0$. Rescaling Y, we can assume that $b = 1$. This gives case 4.2.

The existence of the exponentiation of these matched pairs has been proven in Corollary 4.4, it will be described explicitly below.

The classification in case 4.3 as well as the exponentiation follows from Proposition 4.6. To reduce the number of connected components, we modify the exponentiation slightly.

If $a > 0$, in order to exponentiate, we define

$$G = \frac{\{X \in M_2(\mathbb{R}) \mid \det(X) = \pm 1\}}{\{\pm 1\}},$$

$$G_1 = \{(a, x) \mid a \neq 0, x \in \mathbb{R}\}, \ (a, x)(b, y) = (ab, x + ay), \ G_2 = (\mathbb{R} \setminus \{0\}, \cdot).$$

Making use of the function $\mathrm{Sq}(a) := \mathrm{Sgn}(a)\sqrt{|a|}$, we define

$$i(a, x) = \begin{pmatrix} \frac{1}{\sqrt{|a|}} & 0 \\ 2\frac{x}{\sqrt{|a|}} & \mathrm{Sq}(a) \end{pmatrix} \mod \{\pm 1\}, \tag{5.1}$$

$$j(s) = \begin{pmatrix} \sqrt{|s|} & \frac{1}{2}\left(\sqrt{|s|} - \frac{1}{\mathrm{Sq}(s)}\right) \\ 0 & \frac{1}{\mathrm{Sq}(s)} \end{pmatrix} \mod \{\pm 1\}.$$

It is clear that the tangent Lie algebras of $i(G_1)$ and $j(G_2)$ are given by span$\{\tilde{H}, \tilde{Y}\}$ and $\mathbb{R}(\tilde{H} + \tilde{X})$, respectively. It is not hard to check that we indeed get a matched pair of Lie groups.

If $a < 0$, the easiest way to exponentiate this matched pair goes as follows (as we saw in the proof of Proposition 4.6, there is also another way of doing so).

Define

$$G = \mathrm{PSL}_2(\mathbb{R}) = \frac{\mathrm{SL}_2(\mathbb{R})}{\{\pm 1\}}, \quad G_2 = (\mathbb{T} = \{z \in \mathbb{C} \mid |z| = 1\}, \cdot)$$
$$G_1 = \{(a, x) \mid a > 0, x \in \mathbb{R}\}.$$

Then define

$$i(a, x) = \begin{pmatrix} \frac{1}{\sqrt{a}} & 0 \\ \frac{x}{\sqrt{a}} & \sqrt{a} \end{pmatrix} \bmod \{\pm 1\}, \tag{5.2}$$
$$j(\cos t, \sin t) = \begin{pmatrix} \cos\frac{t}{2} & \sin\frac{t}{2} \\ -\sin\frac{t}{2} & \cos\frac{t}{2} \end{pmatrix} \bmod \{\pm 1\}.$$

One can check that the tangent Lie subalgebras of $i(G_1)$ and $j(G_2)$ agree with span$\{\tilde{H}, \tilde{Y}\}$ and $\mathbb{R}(\tilde{X} - \tilde{Y})$, respectively and that we get a matched pair of Lie groups.

The final case $a = 0$ has been exponentiated in [51], Section 5.4, but we recall it for completeness. We take again $G = \mathrm{PSL}_2(\mathbb{R})$. G_1 consists of pairs (a, x) with $a > 0$ and $x \in \mathbb{R}$ with product $(a, x)(b, y) = (ab, ay + \frac{x}{b})$. Putting

$$i(a, x) = \begin{pmatrix} a & x \\ 0 & \frac{1}{a} \end{pmatrix} \bmod \{\pm 1\}, \quad j(s) = \begin{pmatrix} 1 & 0 \\ s & 1 \end{pmatrix} \bmod \{\pm 1\}, \tag{5.3}$$

we get the required exponentiation to a matched pair of Lie groups. □

For any of the obtained matched pairs of Lie groups, we can now perform the bicrossed product construction in order to get a l.c. quantum group. Whenever one of the corresponding actions is trivial, we obtain a Kac algebra (see Corollary 2.5). When both actions are non-trivial, we find a lot of l.c. quantum groups which are not Kac algebras. To take a closer look at them, we need explicit forms for the corresponding mutual actions, and we use the formulas

$$\chi(X) = X_e[g \mapsto \frac{d}{ds}\alpha_g(s)|_{s=0}], \quad \beta(X) = \frac{d}{ds}((d\beta_s)(X))|_{s=0},$$

where X_e is the partial derivative in e in the direction of an arbitrary generator $X \in \mathfrak{g}_1$ and $d\beta_s$ is the canonical action of G_2 on $\mathfrak{g}_1$ coming from β_s.

The only case of dimension $1 + 1$ with both non-trivial actions has already been presented in Remark 3.5. It is easy to check that we do not get a Kac algebra, that $\delta_M \neq 1$, and that the corresponding l.c. quantum group is self-dual. For the details see [51], 5.3.

In dimension $2 + 1$, we analyse cases 4.1, 4.2 and 4.3.

In case 4.1, following the approach of Proposition 4.2, we define the Lie group G on the space $\mathbb{R} \setminus \{0\} \times \mathbb{R}^2$ with multiplication

$$(s, x, y)(s', x', y') = (ss', x + sx', y + bu_d(s)x' + s^d y'),$$

where

$$u_d(s) = \begin{cases} \frac{s^d - s}{d-1}, & \text{if } d \neq 1, \\ s \log |s|, & \text{if } d = 1, \end{cases} \quad \text{where } s^d = \operatorname{Sgn}(s)|s|^d,$$

and G_1 on the space $\mathbb{R}\setminus\{0\}\times\mathbb{R}$ with multiplication $(a, x)(a', x') = (aa', x+a^d x')$ and $i(a, x) = (a, a - 1, x + bu_d(a))$. Further, we put $G_2 = \mathbb{R} \setminus \{0\}$ and $j(s) = (s, 0, 0)$. Then, the mutual actions are given by

$$\alpha_{(a,x)}(s) = a(s-1) + 1, \tag{5.4}$$
$$\beta_s(a, x) = \left(\frac{sa}{a(s-1)+1}, \frac{x + b(u_d(a) + u_d(a(s-1)+1) - u_d(as))}{(a(s-1)+1)^d} \right).$$

One can check that the corresponding matched pair of Lie algebras is isomorphic to the initial one. Indeed, using the obvious generators, we have:

$$[X, Y] = dY, \ \chi(X) = 1, \ \chi(Y) = 0, \ \beta(X) = -X - bdY, \ \beta(Y) = -dY.$$

The needed isomorphism is given by $A \mapsto -A$, $Y \mapsto dY$ (if $d \neq 0$) ; if $d = 0$, $A \mapsto -A$ establishes an isomorphism with the special case of the initial matched pair: $b = d = 0$.

Because the modular functions of the groups G_1 and G are given by $\delta_1(a, x) = |a|^{-d}$ and $\delta(s, x, y) = |s|^{-d-1}$, we compute that the first equality of Proposition 2.4 does not hold and

$$\delta_M(a, x, s) = |a(s-1) + 1|^{d+1}, \quad \delta_{\hat{M}}(a, x, s) = \left| \frac{as}{a(s-1)+1} \right|^{d-1}.$$

So, both the l.c. quantum group and its dual are not Kac algebras, and are non-unimodular.

In case 4.2, the Lie groups G and G_1 are defined on $\mathbb{R}^+ \setminus \{0\} \times \mathbb{R}^2$ and $\mathbb{R}^+ \setminus \{0\} \times \mathbb{R}$, respectively, with the same multiplication as in case 4.1, but with parameter $b = 1$. We consider G_2 to be $\mathbb{R}$ with addition and define $i(a, x) = (a, 0, x)$ and $j(s) = (1, s, 0)$. Then, the mutual actions are

$$\alpha_{(a,x)}(s) = as, \quad \beta_s(a, x) = (a, x + u_d(a)s). \tag{5.5}$$

One can check that the corresponding matched pair of Lie algebras coincides with the initial one, that the first equality of Proposition 2.4 holds and that finally $\delta_M = 1$, $\delta_{\hat{M}}(a, x, s) = a^{d-1}$. Hence, (M, Δ) is a unimodular Kac algebra, and $(\hat{M}, \hat{\Delta})$ is unimodular if and only if $d = 1$.

Finally, the exponentiations of case 4.3 with $a > 0, < 0, = 0$, are determined by Equations (5.1), (5.2) and (5.3), respectively. In the case $a > 0$, the mutual actions are given by

$$\alpha_{(a,x)}(s) = \frac{(x+1)s + a - x - 1}{xs + a - x}, \tag{5.6}$$

$$\beta_s(a,x) = \left(\frac{((x+1)s + a - x - 1)\,(xs + a - x)}{as}, \; \frac{x\,((x+1)s + a - x - 1)}{a} \right).$$

The corresponding matched pair of Lie algebras

$$[X,Y] = Y,\ \chi(X) = -1,\ \chi(Y) = 0,\ \beta(X) = -X,\ \beta(Y) = 2X + Y$$

is isomorphic to the initial one: $X \mapsto -X - \frac{Y}{2},\ Y \mapsto -\frac{Y}{4},\ A \mapsto 2A$.

In the case $a < 0$, the mutual actions are

$$\alpha_{(a,x)}(\cos t, \sin t) = \tag{5.7}$$
$$\frac{\big((a^2 + x^2 - 1) + (a^2 - x^2 + 1)\cos t + 2ax \sin t,\ 2x - 2x\cos t + 2a \sin t\big)}{(x^2 + a^2 + 1) + (-x^2 + a^2 - 1)\cos t + 2ax \sin t},$$

$$\beta_{(\cos t, \sin t)}(a,x) = \left(\frac{1}{2a}\big((x^2 + a^2 + 1) + (-x^2 + a^2 - 1)\cos t + 2ax \sin t\big), \right.$$
$$\left. \frac{1}{2a}\big((x^2 - a^2 + 1)\sin t + 2ax \cos t\big) \right).$$

Observe that these actions are everywhere defined and continuous. The corresponding matched pair of Lie algebras

$$[X,Y] = Y,\ \chi(X) = -1,\ \chi(Y) = 0,\ \beta(X) = -Y,\ \beta(Y) = X$$

is isomorphic to the initial one: $X \mapsto -X,\ Y \mapsto \frac{Y}{2},\ A \mapsto -2A$.

In the case $a = 0$, the mutual actions are

$$\alpha_{(a,x)}(s) = \frac{s}{a(a+xs)}, \qquad \beta_s(a,x) = (|a + sx|, \operatorname{Sgn}(a+sx)x). \tag{5.8}$$

The corresponding matched pair of Lie algebras

$$[X,Y] = 2Y,\ \chi(X) = -2,\ \chi(Y) = 0,\ \beta(X) = 0,\ \beta(Y) = X$$

is isomorphic to the initial one: $X \mapsto -\frac{X}{2},\ Y \mapsto -\frac{Y}{2}$.

In all three cases $a > 0, < 0, = 0$, one verifies that the first equality of Proposition 2.4 does not hold, $\delta_M = 1$, while $\delta_{\hat{M}} \neq 1$. In particular, we do not get Kac algebras.

6 Cocycle matched pairs of Lie groups and Lie algebras in low dimensions

So far, we have explained how to construct l.c. quantum groups which are bicrossed products of low-dimensional Lie groups without 2-cocycles. The usage of 2-cocycles gives much more concrete examples and, what is more important, gives a more complete picture of low-dimensional l.c. quantum groups. Again, we first explain the infinitesimal picture, i.e. how 2-cocycles for matched pairs of Lie algebras look like, how they are related to the problem of extensions and then show how to exponentiate them.

The first thing that we need here is the notion of a Lie bialgebra, due to V.G. Drinfeld [10]. A Lie bialgebra is a Lie algebra $\mathfrak{g}$ equipped with a Lie bracket $[\cdot, \cdot]$ and a Lie cobracket δ, i.e., a linear map $\delta : \mathfrak{g} \to \mathfrak{g} \otimes \mathfrak{g}$ satisfying the co-anticommutativity and the co-Jacobi identity, that is:

$$(\iota - \tau)\delta = 0, \quad (\iota + \zeta + \zeta^2)(\iota \otimes \delta)\delta = 0,$$

where

$$\tau(u \otimes v) = v \otimes u, \quad \zeta(u \otimes v \otimes w) = v \otimes w \otimes u \quad \text{(for all } u, v, w \in \mathfrak{g})$$

are the flip maps, and these Lie bracket and cobracket are compatible in the following sense:

$$\delta[u, v] = [u, v_{[1]}] \otimes v_{[2]} + v_{[1]} \otimes [u, v_{[2]}] + [u_{[1]}, v] \otimes u_{[2]} + u_{[1]} \otimes [u_{[2]}, v].$$

Any Lie algebra (respectively, Lie coalgebra, i.e., vector space dual to a Lie algebra) is a Lie bialgebra with zero Lie cobracket (respectively, zero Lie bracket). The definition of a morphism of Lie bialgebras is obvious.

Given a pair of Lie algebras $(\mathfrak{g}_1, \mathfrak{g}_2)$, let us ask if there exists a Lie bialgebra $\mathfrak{g}$ such that

$$\mathfrak{g}_2^* \longrightarrow \mathfrak{g} \longrightarrow \mathfrak{g}_1$$

is a short exact sequence in the category of Lie bialgebras. This means precisely that $\mathfrak{g}$ has a sub-bialgebra with trivial bracket, which is an ideal and such that the quotient is a Lie bialgebra with trivial cobracket.

The theory of extensions in this framework has been developed in [36] and is quite similar to the theory of extensions of l.c. groups that we have recalled above. Namely, for the existence of an extension $\mathfrak{g}$ it is necessary and sufficient that $(\mathfrak{g}_1, \mathfrak{g}_2)$ form a matched pair, and all extensions are bicrossed products with cocycles. We consider this theory as an infinitesimal version of the theory of extensions of Lie groups.

As we remember, for any matched pair of Lie algebras $(\mathfrak{g}_1, \mathfrak{g}_2)$, there are mutual actions $\triangleright : \mathfrak{g}_2 \otimes \mathfrak{g}_1 \to \mathfrak{g}_1$ and $\triangleleft : \mathfrak{g}_2 \otimes \mathfrak{g}_1 \to \mathfrak{g}_2$, compatible in a way explained in Section 3 and such that for all $a, b \in \mathfrak{g}_1$, $x, y \in \mathfrak{g}_2$ we have

$$[a \oplus x, b \oplus y] = ([a, b] + x \triangleright b - y \triangleright a) \oplus ([x, y] + x \triangleleft b - y \triangleleft a).$$

For the general definition of a pair of 2-cocycles on such a matched pair, we refer to [33], [36]. For our needs, it suffices to understand that these 2-cocycles are linear maps

$$\mathcal{U} : \mathfrak{g}_1 \wedge \mathfrak{g}_1 \to \mathfrak{g}_2^*, \quad \mathcal{V} : \mathfrak{g}_2 \wedge \mathfrak{g}_2 \to \mathfrak{g}_1^*$$

verifying certain 2-cocycle equations and compatibility equations that are infinitesimal forms of Equations (2.1). For the case of dimension $n + 1$, we give these equations explicitly below. Let us formulate the link between 2-cocycles on matched pairs of Lie algebras and those of Lie groups as a proposition whose proof is straightforward.

Proposition 6.1. *Let (G_1, G_2) be a matched pair of Lie groups equipped with cocycles $\mathcal{U}$ and $\mathcal{V}$, which are differentiable around the unit elements, and let $(\mathfrak{g}_1, \mathfrak{g}_2)$ be the corresponding matched pair of Lie algebras. Defining*

$$\langle \mathcal{U}(X, Y), A\rangle = -i(X_e \otimes Y_e \otimes A_e - Y_e \otimes X_e \otimes A_e)(\mathcal{U}) \quad \textit{and}$$
$$\langle \mathcal{V}(A, B), X\rangle = -i(A_e \otimes B_e \otimes X_e - B_e \otimes A_e \otimes X_e)(\mathcal{V}),$$

for $X, Y \in \mathfrak{g}_1$ and $A, B \in \mathfrak{g}_2$, we get a pair of cocycles on $(\mathfrak{g}_1, \mathfrak{g}_2)$.

Here $\langle \cdot, \cdot \rangle$ denotes the duality between $\mathfrak{g}_i$ and $\mathfrak{g}_i^*$ and X_e, Y_e, A_e, B_e denote the partial derivatives at e in the direction of the corresponding generator. The factor $-i$ appears because for Lie groups $\mathcal{U}$ and $\mathcal{V}$ take values in $\mathbb{T}$, and for real Lie algebras we consider 2-cocycles as real linear maps.

In dimension $n + 1$, $\mathcal{V}$ is necessarily trivial (if also $n = 1$, then also $\mathcal{U}$ is trivial, so there are no non-trivial cocycles in dimension $1 + 1$). Returning to arbitrary n, we choose a generator A for $\mathfrak{g}_2$ and define maps β and χ as above. Then $\mathcal{U}$ can be regarded as an antisymmetric, bilinear form on $\mathfrak{g}_1$, and the 2-cocycle equations of [33],[36] reduce to the equation

$$\mathcal{U}([X, Y], Z) + \chi(X)\mathcal{U}(Y, Z) + \text{ cyclic permutation } = 0 \quad \text{for all } X, Y, Z \in \mathfrak{g}_1.$$

It is clear that these 2-cocycles $\mathcal{U}$ form a real vector space.

In Section 2, we defined the notion of the group of extensions for a matched pair of Lie groups (G_1, G_2) using the notion of cohomologous 2-cocycles. The same can be done for a matched pair of Lie algebras [36]. In particular, for the dimension $n + 1$, a 2-cocycle $\mathcal{U}$ is called *cohomologous to trivial*, if there exists a linear form ρ in $\mathfrak{g}_1^*$ such that

$$\mathcal{U}(X, Y) = \rho([X, Y]) + \chi(X)\rho(Y) - \chi(Y)\rho(X).$$

Two cocycles $\mathcal{U}_1$ and $\mathcal{U}_2$ are called *cohomologous* if $\mathcal{U}_1 - \mathcal{U}_2$ is cohomologous to trivial.

The quotient space of 2-cocycles modulo 2-cocycles cohomologous to trivial, with addition as the group operation, is called the *group of extensions* of the matched pair $(\mathfrak{g}_1, \mathfrak{g}_2)$.

Now let us describe all 2-cocycles on the matched pairs of real Lie algebras of dimension $2 + 1$.

Proposition 6.2. *Referring to the classification of matched pairs of Lie algebras of dimension* $2+1$ *given in Theorem* 5.1, *the following holds: the group of extensions is* $\mathbb{R}$ *in the cases* 1.1, 2.1, 2.2, 2.3, 3 $(a=-1)$, 4.1 $(d=-1)$, 4.2 $(d=-1)$ *and* 4.3. *The cocycles are defined by* $\mathcal{U}(X,Y)=\lambda$, *for* $\lambda\in\mathbb{R}$. *In the other cases, the group of extensions is trivial.*

Proof. Since dim $\mathfrak{g}_1=2$, any antisymmetric bilinear form $\mathcal{U}$ on $\mathfrak{g}_1$ is a cocycle. Suppose that X, Y are generators of $\mathfrak{g}_1$ with $[X,Y]=dY$. A cocycle $\mathcal{U}$ is entirely determined by $\mathcal{U}(X,Y)=\lambda$ for $\lambda\in\mathbb{R}$. If $\chi=0$ and $d\neq 0$, we take $\rho(Y)=\lambda/d$ and get that $\mathcal{U}$ is cohomologous to trivial. If $\chi=0$ and $d=0$, it is clear that $\mathcal{U}$ is not cohomologous to trivial if $\lambda\neq 0$.

If $\chi\neq 0$, we may suppose that $\chi(X)=1$ and $\chi(Y)=0$. If $d\neq -1$, we take $\rho(Y)=\lambda/(1+d)$ and get that $\mathcal{U}$ is cohomologous to trivial. If $d=-1$, it is clear that $\mathcal{U}$ is not cohomologous to trivial if $\lambda\neq 0$. □

Next, we want to exponentiate a cocycle on a matched pair of real Lie algebras, i.e., to construct a measurable map $\mathcal{U}: G_1\times G_1\times G_2\to U(1)$, with values in the unit circle of $\mathbb{C}$, satisfying

$$\mathcal{U}(g,h,\alpha_k(s))\,\mathcal{U}(gh,k,s)=\mathcal{U}(h,k,s)\,\mathcal{U}(g,hk,s),$$
$$\mathcal{U}(g,h,s)\,\mathcal{U}(\beta_{\alpha_h(s)}(g),\beta_s(h),t)=\mathcal{U}(g,h,ts)$$

almost everywhere. Let us define a function $A(\cdot)$ by

$$\mathcal{U}(g,h,s)=\exp(iA(g,h,s)).$$

So $A(\cdot)$ should satisfy

$$A(g,h,\alpha_k(s))+A(gh,k,s)=A(h,k,s)+A(g,hk,s)\mod 2\pi,$$
$$A(g,h,s)+A(\beta_{\alpha_h(s)}(g),\beta_s(h),t)=A(g,h,ts)\mod 2\pi$$

almost everywhere.

Proposition 6.3. *If the group of extensions of a matched pair of Lie algebras of dimension* $2+1$ *is non-trivial, there exists an exponentiation of this matched pair with cocycles. These cocycles are labeled by* $\mathbb{R}$ *in all the cases, except case* 4.3 $(a=0,1)$, *where they are labeled by* $\mathbb{Z}$.

Proof. Following [51], Section 5.5, we look for the above function A in the form

$$A(g,h,s)=P\int_0^s f(\phi_r(g,h))\,dr,$$

where $\phi_r(g,h):=(\beta_{\alpha_h(r)}(g),\beta_r(h))$ and where the function f on $G_1\times G_1$ is such that for almost all $g,h\in G_1$ the function $r\mapsto f(\phi_r(g,h))$ has a principal value integral over any interval in $\mathbb{R}$ (dr is the Haar measure on the 1-dimensional Lie group $(\mathbb{R},+)$ or on $\mathbb{R}\backslash\{0\}$, in which case we integrate from 1 to s). A necessary condition to be satisfied by f is

$$\frac{d}{dt}\alpha_k(t)\Big|_{t=0} f(g,h) + f(gh,k) = f(h,k) + f(g,hk).$$

Finally, having found such an f, we have to check if it really gives rise to a 2-cocycle.

In those cases where the actions α and β are everywhere defined and smooth, one can check that any smooth solution of this equation gives indeed rise to a 2-cocycle (for the details see [51], Section 5.5). In this way, it is easy to find 2-cocycles in the cases 1.1, 2.1, 2.2, 2.3, 3 ($a = -1$), and 4.2 ($d = -1$), namely: in the cases 1.1, 2.1, 2.2 and 2.3, the action α is trivial, and $G_1 = \mathbb{R}^2$ with addition. So, we can take $f(x_1, x_2; y_1, y_2) = \lambda(x_1 y_2 - x_2 y_1)$, for any $\lambda \in \mathbb{R}$. In the cases 3 ($a = -1$) and 4.2 ($d = -1$), we observe that $G_1 = \{(a, x) \mid a > 0, x \in \mathbb{R}\}$ with $(a,x)(b,y) = (ab, x + y/a)$. Because $\chi(X) = 1$, the character $g \mapsto \frac{d}{dt}\alpha_g(t)\big|_{t=0}$ is given by $(a,x) \mapsto a$, and we can take $f(a,x;b,y) = \lambda abx \log b$, for any $\lambda \in \mathbb{R}$.

The case 4.3 ($a = 0$) has been studied in [51], Section 5.5: $f(a,x;b,y) = \lambda\frac{x\log b}{ab^2}$. Checking if we really get 2-cocycles, observe that

$$f(\phi_r(a,x;b,y)) = \frac{\lambda x}{(b+ry)(ab+r(ay+\frac{x}{b}))}\log|c+dr|,$$

then

$$P\int_{-\infty}^{\infty} f(\phi_r(a,x;b,y))\,dr = \frac{\lambda}{2}\pi^2\,\mathrm{Sgn}\big(\frac{y}{x}(ay+\frac{x}{b})\big).$$

From this, it follows that we do get 2-cocycles if and only if $\lambda = \frac{4n}{\pi}$, with $n \in \mathbb{Z}$.

The same phenomenon happens in the cases 4.1 ($d = -1$) and 4.3 ($a > 0$): although the above principal value integral is well defined, we do not always get a 2-cocycle $\mathcal{U}$, as explained in [51], after Proposition 5.6. In case 4.1 ($d = -1$), we use the matched pair explicitly described in Equation (5.4) with $d = -1$ and $b = 0$. We take again $f(a,x;b,y) = \lambda abx\log|b|$, for any $\lambda \in \mathbb{R}$, and we can explicitly perform the integration, to obtain 2-cocycles:

$$A(a,x;b,y;s) = \lambda ax\big(-(b(s-1)+1)\log|b(s-1)+1| + bs\log|bs| - b\log|b|\big).$$

On the contrary, the situation of case 4.3 ($a > 0$) is more delicate. We use the matched pair explicitly described in Equation (5.6). We can take $f(a,x;b,y) = \lambda\frac{y}{b}\log|a|$ with $\lambda \in \mathbb{R}$, and our candidate for $A(\cdot)$ becomes:

$$A(a,x;b,y;s) = \lambda\ P\int_1^s \frac{y}{yr+b-y}$$
$$\log\left|\frac{\big((x+ay+1)r+ab-x-ay-1\big)\big((x+ay)r+ab-x-ay\big)}{a\big((y+1)r+b-y-1\big)\big(yr+b-y\big)}\right|\,dr.$$

Because

$$P\int_{-\infty}^{+\infty} \frac{c}{cr+d} \log|ar+b|\, dr = \frac{\pi^2}{2} \operatorname{Sgn}\left(\frac{b}{a}-\frac{d}{c}\right),$$

the same reasoning as in Section 5.5 of [51] implies that we do get a 2-cocycle if $\lambda = \frac{4n}{\pi}$ for $n \in \mathbb{Z}$.

Finally, in case 4.3 ($a > 0$), with the explicit exponentiation given in Equation (5.7), the mutual actions are defined everywhere and are smooth, but $G_1 = \mathbb{T}$. Taking $f(a,x;b,y) = \lambda \frac{y}{b} \log a$ with $\lambda \in \mathbb{R}$, it is natural to use

$$A(a,x;b,y;\cos t, \sin t) = \lambda \int_0^t f\big(\phi_{(\cos s, \sin s)}(a,x,b,y)\big)\, ds.$$

To have a 2-cocycle, we need that

$$\lambda \int_0^{2\pi} f\big(\phi_{(\cos s, \sin s)}(a,x,b,y)\big)\, ds = 0 \mod 2\pi.$$

Denote the left-hand side of this expression by $I_\lambda(a,x,b,y)$. Then, one can compute that

$$I_\lambda(a,x,b,y) = \lambda\big(H(a,x) + H(b,y) - H(ab, x+ay)\big),$$

where $H(a,x) := -4\pi \arctan\left(\frac{x}{1+a}\right)$. Hence, there is no λ which gives us a 2-cocycle.

We can, however, find 2-cocycles, using the other exponentiation of the same matched pair of Lie algebras, as explained in the proof of Proposition 4.6. We obtain a matched pair $(G_1, \mathbb{R})$, in which both actions are everywhere defined and smooth. Hence, we obtain cocycles labeled by $\mathbb{R}$, following the procedure described in the beginning of the proof. □

7 Infinitesimal objects for low-dimensional l.c. quantum groups

The case of cocycle bicrossed product l.c. quantum groups. Given a cocycle matched pair of Lie groups (G_1, G_2) whose 2-cocycles $\mathcal{U}$ and $\mathcal{V}$ are differentiable around the unit elements, we can construct the corresponding l.c. quantum group using the cocycle bicrossed product construction.

But, in this situation, we can also construct two other intimately related algebraic structures which can be viewed as infinitesimal objects of this l.c. quantum group: a Lie bialgebra and a Hopf *-algebra, as in [51], Section 5.2. The precise mathematical link between these three structures is not completely clear at the moment (it is tempting to consider it as a kind of a Lie theory for our cocycle bicrossed product l.c. quantum groups). We will discuss it mainly on the level of examples.

Let us recall the construction of infinitesimal Lie bialgebras and Hopf *-algebras in the special case of dimension $n+1$.

Let (G_1, G_2) be a cocycle matched pair of Lie groups with $G_2 = \mathbb{R}$ (the case $\mathbb{R} \setminus \{0\}$ is completely analogous, replacing differentials in 0 by differentials in 1) and let $\mathcal{U}$ be a 2-cocycle differentiable around the unit elements. Denote by $\alpha_g(s)$ and $\beta_s(g)$ the corresponding mutual actions. Then the cocycle matched pair of Lie algebras is determined (see Section 4 and Proposition 6.1) by

$$\chi(X) = X_e[g \mapsto \frac{d}{ds}\alpha_g(s)|_{s=0}], \quad X \in \mathfrak{g}_1,$$
$$\beta(X) = \frac{d}{ds}((d\beta_s)(X))|_{s=0}, \quad X \in \mathfrak{g}_1,$$
$$\mathcal{U}(X, Y) = -i\frac{d}{ds}((X_e \otimes Y_e - Y_e \otimes X_e)(\mathcal{U}(\cdot, \cdot, s)))|_{s=0}]\,\tilde{A}, \quad X, Y \in \mathfrak{g}_1.$$

The infinitesimal Lie bialgebra is precisely the corresponding cocycle bicrossed product Lie bialgebra and has generators $\tilde{A} \in \mathfrak{g}_2$ and $X \in \mathfrak{g}_1$, subject to the relations

$$[\tilde{A}, X] = \chi(X)\tilde{A},$$
$$[X, Y] = [X, Y]_1 + \mathcal{U}(X, Y)\tilde{A},$$
$$\delta(\tilde{A}) = 0,$$
$$\delta(X) = \beta(X) \wedge \tilde{A}.$$

The dual infinitesimal Lie bialgebra has generators A and $\tilde{X}_i$, subject to the relations

$$[\tilde{X}_i, A] = \sum_j \beta(X_i)_j \tilde{X}_j,$$
$$[\tilde{X}_i, \tilde{X}_j] = 0,$$
$$\langle \delta(\tilde{X}_i), X \otimes Y\rangle = \langle \tilde{X}_i, [X, Y]_1\rangle,$$
$$\delta(A) = A \wedge \Big(\sum_i \chi(X_i)\tilde{X}_i\Big) + \sum_{i<j} \mathcal{U}(X_i, X_j)\tilde{X}_i \wedge \tilde{X}_j.$$

Following [51], Section 5.2 and in order to construct the infinitesimal Hopf *-algebra, which is an algebraic cocycle bicrossed product in the sense of [33], we denote by $\tilde{A}$ the function $\tilde{A}(x) = x$ on G_2. Then, our Hopf *-algebra has generators $\tilde{A}$ and $\{X \mid X \in \mathfrak{g}_1\}$, where $\tilde{A}^* = \tilde{A}$ and $X^* = -X$, with relations

$$[\tilde{A}, X] = X_e[g \mapsto \alpha_g(s)], \quad X \in \mathfrak{g}_1,$$
$$[X, Y] = [X, Y]_1 + (X_e \otimes Y_e - Y_e \otimes X_e)(\mathcal{U}(\cdot, \cdot, s)), \quad X, Y \in \mathfrak{g}_1,$$
$$\Delta(\tilde{A}) = \tilde{A} \otimes 1 + 1 \otimes \tilde{A}, \quad (\tilde{A} \otimes \tilde{A} \text{ when } G_2 = \mathbb{R} \setminus \{0\}),$$
$$\Delta(X) = 1 \otimes X + \sum_i X_i \otimes (d\beta_s)(X)_i, \quad X \in \mathfrak{g}_1.$$

This needs some explanation: several of the used expressions are functions of $s \in G_2$, and it is understood that these functions belong to some algebra of functions in $\tilde{A}$ (in the simplest case - to the algebra of polynomials in $\tilde{A}$). Next, (X_i) is a basis for $\mathfrak{g}_1$ and $(d\beta_s)(X)_i$ is the i-th component of $(d\beta_s)(X)$ in this basis, and is, therefore, again a function of $s \in G_2$. Finally, $[X, Y]_1$ denotes the Lie bracket in $\mathfrak{g}_1$. When $G_2 = (\mathbb{R}, +)$, co-unit and antipode are given by $\varepsilon(\tilde{A}) = \varepsilon(X) = 0$ $(X \in \mathfrak{g}_1)$, $S(\tilde{A}) = -\tilde{A}$ and

$$S(X) = -\sum_i X_i \, (d\beta_{-s})(X)_i, \quad X \in \mathfrak{g}_1.$$

When $G_2 = (\mathbb{R} \setminus \{0\}, \cdot)$, we rather have $\varepsilon(\tilde{A}) = 1$, $\varepsilon(X) = 0$ $(X \in \mathfrak{g}_1)$, $S(\tilde{A}) = \tilde{A}^{-1}$ and

$$S(X) = -\sum_i X_i \, (d\beta_{1/s})(X)_i, \quad X \in \mathfrak{g}_1.$$

To describe the dual infinitesimal Hopf *-algebra, suppose that we have coordinate functions $(\tilde{X}_i)$ on G_1, dual to the basis (X_i) of $\mathfrak{g}_1$. Then we can write down, with the same kind of conventions, the dual Hopf *-algebra, with generators $(\tilde{X}_i)$ and A, such that $\tilde{X}_i^* = \tilde{X}_i$ and $A^* = -A$, and with relations:

$$\begin{aligned} [\tilde{X}_i, A] &= \frac{d}{ds}(\tilde{X}_i(\beta_s(g)))|_{s=0}, \\ [\tilde{X}_i, \tilde{X}_j] &= 0, \\ \Delta(\tilde{X}_i)(g, h) &= \tilde{X}_i(gh), \\ \Delta(A) &= 1 \otimes A + A \otimes \frac{d}{ds}(\alpha_g(s))|_{s=0} + \frac{d}{ds}\mathcal{U}(g, h, s)|_{s=0}. \end{aligned}$$

Again, we write functions of $g, h \in G_1$ and they are tacitly assumed to belong to some algebra of functions in $\tilde{X}_i$. Co-unit and antipode are given by $\varepsilon(\tilde{X}_i) = \tilde{X}_i(e)$, $\varepsilon(A) = 0$, $S(\tilde{X}_i)(g) = \tilde{X}_i(g^{-1})$ and

$$S(A) = -A \, \frac{d}{ds}(\alpha_{g^{-1}}(s))|_{s=0} - \frac{d}{ds}\mathcal{U}(g, g^{-1}, s)|_{s=0}.$$

Remark 7.1. Observe that in order to pass from the infinitesimal Hopf *-algebra to the infinitesimal Lie bialgebra, one replaces functions on G_2 by their first-order approximations, which are multiples of $\tilde{A}$, one takes anti-selfadjoint generators $i\tilde{A}$ and $X \in \mathfrak{g}_1$ and takes as δ the linear part of $i(\Delta - \Delta^{\mathrm{op}})$, where Δ^{op} denotes the opposite comultiplication.

Let us turn now to concrete examples. For the matched pair of dimension 1+1 (see Remark 3.5) we do not have cocycles. The infinitesimal Hopf *-algebra is determined by

$$\tilde{A} = \tilde{A}^*, \quad X^* = -X, \quad [\tilde{A}, X] = \tilde{A} - 1,$$
$$\Delta(\tilde{A}) = \tilde{A} \otimes \tilde{A}, \quad \Delta(X) = X \otimes \tilde{A}^{-1} + 1 \otimes X.$$

Co-unit and antipode are $\varepsilon(\tilde{A}) = 1, \varepsilon(X) = 0, S(\tilde{A}) = \tilde{A}^{-1}, S(X) = -X\tilde{A}$. The infinitesimal Lie bialgebra is given by

$$[\tilde{A}, X] = \tilde{A}, \quad \delta(\tilde{A}) = 0, \quad \delta(X) = \tilde{A} \wedge X.$$

We will now make the link between the Hopf *-algebra and the Lie bialgebra slightly more explicit. For a smooth function f in $\tilde{A}$ we get $[f(\tilde{A}), X] = f'(\tilde{A})(\tilde{A} - 1)$. If we take $A_1 := i \log \tilde{A}$ as a new generator, we observe that

$$[A_1, X] = A_1 + \frac{i}{2}A_1^2 + \cdots, \quad \Delta(A_1) = A_1 \otimes 1 - 1 \otimes A_1,$$
$$\Delta(X) = X \otimes \left(1 + iA_1 - \frac{1}{2}A_1^2 + \cdots\right) + 1 \otimes X.$$

Linearizing now the above bracket and $i(\Delta - \Delta^{\mathrm{op}})$, we precisely get our Lie bialgebra.

Since this example is self-dual, the dual infinitesimal Hopf *-algebra and Lie bialgebra are the same as the original ones.

Remark 7.2. It is easy to show that the only non-trivial Lie bialgebras of dimension 2 non-isomorphic to the above mentioned Lie bialgebra, are defined by the relations

$$[A, X] = X, \quad \delta(A) = 0, \quad \delta(X) = qX \wedge A.$$

On the other hand, the Hopf *-algebras corresponding to the l.c. quantum "$ax + b$"-group considered by S. L. Woronowicz and S. Zakrzevski [66] and by A. Van Daele [55] are defined by the relations

$$a^* = a, \; x^* = x, \quad ax = \exp(iq)xa, \quad \Delta(a) = a \otimes a, \quad \Delta(x) = x \otimes a + 1 \otimes x.$$

Although this quantum group cannot be obtained by the bicrossed product construction (see [51], 5.4.d), we can apply to it the same formal procedure to associate with it a Lie bialgebra. Namely, observing that $[\log a, x] = iqx$, we put $A = \frac{-i}{q} \log a$ and $X = ix$. Linearizing as above, we find precisely the previous Lie bialgebra which then can be considered as an infinitesimal Lie bialgebra of the l.c. quantum "$ax + b$"-group (we mentioned already that the exact relation between these structures is not completely clear).

Let us now list the infinitesimal Hopf *-algebras and Lie bialgebras of all the cocycle matched pairs of Lie algebras of dimension $2+1$, with two non-trivial actions (referring to Theorem 5.1, case 4).

Case 4.1 has been exponentiated concretely in Equation (5.4). Taking the obvious generators X and Y for $\mathfrak{g}_1$ (differentiating to a and x, respectively), satisfying $[X, Y] = dY$, and A for $\mathfrak{g}_2$, we get the infinitesimal form $\chi(X) = 1$, $\chi(Y) = 0$, $\beta(X) =$

$-X - bdY$ and $\beta(Y) = -dY$. When $d \neq -1$, we do not have non-trivial cocycles and we get the Hopf *-algebra, with $\tilde{A}^* = \tilde{A}$, $X^* = -X$, $Y^* = -Y$ and

$$\begin{aligned}
&[\tilde{A}, X] = \tilde{A} - 1, \quad [\tilde{A}, Y] = 0, \quad [X, Y] = dY, \\
&\Delta(\tilde{A}) = \tilde{A} \otimes \tilde{A}, \\
&\Delta(X) = X \otimes \tilde{A}^{-1} + 1 \otimes X + bY \otimes \tilde{A}^{-d}(1 - u'_d(\tilde{A})), \\
&\Delta(Y) = Y \otimes \tilde{A}^{-d} + 1 \otimes Y.
\end{aligned}$$

Co-unit and antipode are $\varepsilon(\tilde{A}) = 1, \varepsilon(X) = \varepsilon(Y) = 0, S(\tilde{A}) = \tilde{A}^{-1}, S(X) = -X\tilde{A} - bdYu_d(\tilde{A}), S(Y) = -Y\tilde{A}^d$. The corresponding Lie bialgebra is

$$\begin{aligned}
&[\tilde{A}, X] = \tilde{A}, \quad [\tilde{A}, Y] = 0, \qquad [X, Y] = dY, \\
&\delta(\tilde{A}) = 0, \quad \delta(X) = \tilde{A} \wedge (X + bdY), \quad \delta(Y) = d\tilde{A} \wedge Y.
\end{aligned}$$

When $d = -1$, we have found non-trivial cocycles, which are infinitesimally given by $\mathcal{U}(X, Y) = -\lambda\tilde{A}$. For the Hopf *-algebra, there changes $[X, Y] = -Y - i\lambda \log \tilde{A}$, and on the Lie bialgebra level, this becomes $[X, Y] = -Y - \lambda\tilde{A}$. Co-unit and antipode remain unchanged.

The dual Hopf *-algebra for $d \neq -1$ has anti-self-adjoint generators $\tilde{X}, \tilde{Y}$, and a self-adjoint generator A, subject to the relations

$$\begin{aligned}
&[\tilde{X}, A] = \tilde{X}(1 - \tilde{X}), \quad [\tilde{Y}, A] = b\tilde{X}(1 - u'_d(\tilde{X})) - d\tilde{X}\tilde{Y}, \quad [\tilde{X}, \tilde{Y}] = 0, \\
&\Delta(\tilde{X}) = \tilde{X} \otimes \tilde{X}, \\
&\Delta(\tilde{Y}) = \tilde{X}^d \otimes \tilde{Y} + \tilde{Y} \otimes 1, \\
&\Delta(A) = A \otimes \tilde{X} + 1 \otimes A.
\end{aligned}$$

Co-unit and antipode are $\varepsilon(\tilde{X}) = 1, \varepsilon(\tilde{Y}) = \varepsilon(A) = 0, S(\tilde{X}) = \tilde{X}^{-1}, S(\tilde{Y}) = -\tilde{X}^{-d}\tilde{Y}, S(A) = -A\tilde{X}^{-1}$. The corresponding infinitesimal Lie bialgebra is determined by

$$\begin{aligned}
&[\tilde{X}, A] = -\tilde{X}, \quad [\tilde{Y}, A] = -bd\tilde{X} - d\tilde{Y}, \quad [\tilde{X}, \tilde{Y}] = 0, \\
&\delta(\tilde{X}) = 0, \quad \delta(\tilde{Y}) = d\tilde{X} \wedge \tilde{Y}, \quad \delta(A) = A \wedge \tilde{X}.
\end{aligned}$$

In the case $d = -1$ and with the same cocycle as above, we should change $\Delta(A) = A \otimes \tilde{X} + 1 \otimes A + i\lambda\tilde{X}\tilde{Y} \otimes \tilde{X} \log \tilde{X}$ in the definition of the Hopf *-algebra and $\delta(A) = A \wedge \tilde{X} - \lambda\tilde{X} \wedge \tilde{Y}$ in the definition of the Lie bialgebra. For the antipode, we change $S(A) = -A\tilde{X} + i\lambda\tilde{Y} \log \tilde{X}$.

Case 4.2 has been exponentiated in Equation (5.5). Taking again the obvious generators, we have the infinitesimal form $[X, Y] = dY$, $\beta(X) = Y$, $\beta(Y) = 0$, $\chi(X) = 1$ and $\chi(Y) = 0$. When $d \neq -1$, there are no non-trivial cocycles. We find the Hopf *-algebra with generators $\tilde{A} = \tilde{A}^*$ and X, Y anti-self-adjoint, satisfying

$$\begin{aligned} &[\tilde{A}, X] = \tilde{A}, \quad [\tilde{A}, Y] = 0, \quad [X, Y] = dY, \\ &\Delta(\tilde{A}) = \tilde{A} \otimes \tilde{A}, \\ &\Delta(X) = X \otimes 1 + 1 \otimes X + Y \otimes \tilde{A}, \\ &\Delta(Y) = Y \otimes 1 + 1 \otimes Y. \end{aligned}$$

Co-unit and antipode are $\varepsilon(\tilde{A}) = 1, \varepsilon(X) = \varepsilon(Y) = 0, S(\tilde{A}) = \tilde{A}^{-1}, S(X) = -X - Y\tilde{A}^{-1}, S(Y) = -Y$. The corresponding Lie bialgebra is

$$\begin{aligned} &[\tilde{A}, X] = \tilde{A}, \quad [\tilde{A}, Y] = 0, \qquad [X, Y] = dY, \\ &\delta(\tilde{A}) = 0, \qquad \delta(X) = Y \wedge \tilde{A}, \qquad \delta(Y) = 0. \end{aligned}$$

When $d = -1$, we have found non-trivial cocycles, which are infinitesimally given by $\mathcal{U}(X, Y) = -\lambda\tilde{A}$. For the Hopf *-algebra, there changes $[X, Y] = -Y - i\lambda\tilde{A}$, and on the Lie bialgebra level, this becomes $[X, Y] = -Y - \lambda\tilde{A}$. Co-unit and antipode remain unchanged.

The dual Hopf *-algebra for $d \neq -1$ has anti-self-adjoint generators $\tilde{X}, \tilde{Y}$ and a self-adjoint generator A, subject to the relations

$$\begin{aligned} &[\tilde{X}, A] = 0, \quad [\tilde{Y}, A] = u_d(\tilde{X}), \quad [\tilde{X}, \tilde{Y}] = 0, \\ &\Delta(\tilde{X}) = \tilde{X} \otimes \tilde{X}, \\ &\Delta(\tilde{Y}) = \tilde{X}^d \otimes \tilde{Y} + \tilde{Y} \otimes 1, \\ &\Delta(A) = A \otimes \tilde{X} + 1 \otimes A. \end{aligned}$$

Co-unit and antipode are $\varepsilon(\tilde{X}) = 1, \varepsilon(\tilde{Y}) = \varepsilon(A) = 0, S(\tilde{X}) = \tilde{X}^{-1}, S(\tilde{Y}) = -\tilde{X}^{-d}\tilde{Y}, S(A) = -A\tilde{X}^{-1}$. The corresponding infinitesimal Lie bialgebra is determined by

$$\begin{aligned} &[\tilde{X}, A] = 0, \quad [\tilde{Y}, A] = \tilde{X}, \qquad [\tilde{X}, \tilde{Y}] = 0, \\ &\delta(\tilde{X}) = 0, \qquad \delta(\tilde{Y}) = d\tilde{X} \wedge \tilde{Y}, \qquad \delta(A) = A \wedge \tilde{X}. \end{aligned}$$

In the case $d = -1$ and with the same cocycle as above, we should change $\Delta(A) = A \otimes \tilde{X} + 1 \otimes A + i\lambda\tilde{X}\tilde{Y} \otimes \tilde{X}\log\tilde{X}$ in the definition of the Hopf *-algebra and $\delta(A) = A \wedge \tilde{X} - \lambda\tilde{X} \wedge \tilde{Y}$ in the definition of the Lie bialgebra. For the antipode, we change $S(A) = -A\tilde{X}^{-1} + i\lambda\tilde{Y}\log\tilde{X}$.

Case 4.3 ($a < 0$). For the exponentiation given in Equation (5.6), we have natural generators X, Y and A, giving the infinitesimal form $[X, Y] = Y$, $\chi(X) = -1$, $\chi(Y) = 0$, $\beta(X) = -X$ and $\beta(Y) = 2X + Y$. There are cocycles, infinitesimally given by $\mathcal{U}(X, Y) = -\lambda\tilde{A}$. The Hopf *-algebra has generators $\tilde{A} = \tilde{A}^*$ and X, Y, anti-self-adjoint, satisfying

$$
\begin{aligned}
&[\tilde{A}, X] = 1 - \tilde{A}, \quad [\tilde{A}, Y] = -(1-\tilde{A})^2, \quad [X, Y] = Y - i\lambda \log \tilde{A},\\
&\Delta(\tilde{A}) = \tilde{A} \otimes \tilde{A},\\
&\Delta(X) = X \otimes \tilde{A}^{-1} + 1 \otimes X,\\
&\Delta(Y) = Y \otimes \tilde{A} + 1 \otimes Y + X \otimes (\tilde{A} - \tilde{A}^{-1}).
\end{aligned}
$$

Co-unit and antipode are $\varepsilon(\tilde{A}) = 1, \varepsilon(X) = \varepsilon(Y) = 0, S(\tilde{A}) = \tilde{A}^{-1}, S(X) = -X\tilde{A}, S(Y) = -Y\tilde{A}^{-1} + X(\tilde{A} - \tilde{A}^{-1})$. The corresponding Lie bialgebra is

$$
\begin{aligned}
&[\tilde{A}, X] = -\tilde{A}, \quad [\tilde{A}, Y] = 0, \qquad [X, Y] = Y - \lambda\tilde{A},\\
&\delta(\tilde{A}) = 0, \qquad \delta(X) = \tilde{A} \wedge X, \quad \delta(Y) = (2X + Y) \wedge \tilde{A}.
\end{aligned}
$$

The dual Hopf *-algebra has self-adjoint generators $\tilde{X}, \tilde{Y}$ and an anti-self-adjoint generator A, subject to the relations

$$
\begin{aligned}
&[\tilde{X}, A] = 1 - \tilde{X} + 2\tilde{Y}, \quad [\tilde{Y}, A] = \tilde{X}^{-1}\tilde{Y}(1 + \tilde{Y}), \quad [\tilde{X}, \tilde{Y}] = 0,\\
&\Delta(\tilde{X}) = \tilde{X} \otimes \tilde{X},\\
&\Delta(\tilde{Y}) = \tilde{X} \otimes \tilde{Y} + \tilde{Y} \otimes 1,\\
&\Delta(A) = A \otimes \tilde{X}^{-1} + 1 \otimes A + i\lambda \log \tilde{X} \otimes \tilde{X}^{-1}\tilde{Y}.
\end{aligned}
$$

Co-unit and antipode are $\varepsilon(\tilde{X}) = 1, \varepsilon(\tilde{Y}) = \varepsilon(A) = 0, S(\tilde{X}) = \tilde{X}^{-1}, S(\tilde{Y}) = -\tilde{X}^{-1}\tilde{Y}, S(A) = -A\tilde{X} + i\lambda\tilde{Y} \log \tilde{X}$. The corresponding Lie bialgebra is determined by the equations

$$
\begin{aligned}
&[\tilde{X}, A] = -\tilde{X} + 2\tilde{Y}, \quad [\tilde{Y}, A] = \tilde{Y}, \qquad [\tilde{X}, \tilde{Y}] = 0,\\
&\delta(\tilde{X}) = 0, \qquad \delta(\tilde{Y}) = \tilde{X} \wedge \tilde{Y}, \quad \delta(A) = \tilde{X} \wedge A - \lambda\tilde{X} \wedge \tilde{Y}.
\end{aligned}
$$

For the exponentiation of the case 4.3 ($a = 0$) described in Equation (5.8) with infinitesimal form $[X, Y] = 2Y, \chi(X) = -2, \chi(Y) = 0, \beta(X) = 0$ and $\beta(Y) = X$, there are non-trivial cocycles determined by $\mathcal{U}(X, Y) = -\lambda\tilde{A}$, giving rise to the following Hopf *-algebra, with a self-adjoint generator $\tilde{A}$ and anti-self-adjoint generators X, Y, subject to the relations

$$
\begin{aligned}
&[\tilde{A}, X] = -2\tilde{A}, \quad [\tilde{A}, Y] = -\tilde{A}^2, \quad [X, Y] = 2Y - i\lambda\tilde{A},\\
&\Delta(\tilde{A}) = \tilde{A} \otimes 1 + 1 \otimes \tilde{A},\\
&\Delta(X) = X \otimes 1 + 1 \otimes X,\\
&\Delta(Y) = Y \otimes 1 + X \otimes \tilde{A} + 1 \otimes Y.
\end{aligned}
$$

Co-unit and antipode are $\varepsilon(\tilde{A}) = \varepsilon(X) = \varepsilon(Y) = 0$, $S(\tilde{A}) = -\tilde{A}$, $S(X) = -X$, $S(Y) = -Y + X\tilde{A}$. The corresponding Lie bialgebra is determined by

$$[\tilde{A}, X] = -2\tilde{A}, \quad [\tilde{A}, Y] = 0, \quad [X, Y] = 2Y - \lambda\tilde{A},$$
$$\delta(\tilde{A}) = 0, \qquad \delta(X) = 0, \qquad \delta(Y) = X \wedge \tilde{A}.$$

The dual Hopf *-algebra has self-adjoint generators $\tilde{X}$, $\tilde{Y}$ and an anti-self-adjoint generator A, subject to the relations

$$\begin{aligned}
&[\tilde{X}, A] = \tilde{Y}, \quad [\tilde{Y}, A] = 0, \quad [\tilde{X}, \tilde{Y}] = 0,\\
&\Delta(\tilde{X}) = \tilde{X} \otimes \tilde{X},\\
&\Delta(\tilde{Y}) = \tilde{X} \otimes \tilde{Y} + \tilde{Y} \otimes \tilde{X}^{-1},\\
&\Delta(A) = A \otimes \tilde{X}^{-2} + 1 \otimes A + i\lambda\tilde{X}^{-1}\tilde{Y} \otimes \tilde{X}^{-2} \log \tilde{X}.
\end{aligned}$$

Co-unit and antipode are $\varepsilon(\tilde{X}) = 1$, $\varepsilon(\tilde{Y}) = \varepsilon(A) = 0$, $S(\tilde{X}) = \tilde{X}^{-1}$, $S(\tilde{Y}) = -\tilde{Y}$, $S(A) = -A\tilde{X}^2 + i\lambda\tilde{X}\tilde{Y} \log \tilde{X}$. The corresponding Lie bialgebra is determined by

$$[\tilde{X}, A] = \tilde{Y}, \quad [\tilde{Y}, A] = 0, \qquad [\tilde{X}, \tilde{Y}] = 0,$$
$$\delta(\tilde{X}) = 0, \quad \delta(\tilde{Y}) = 2\tilde{X} \wedge \tilde{Y}, \quad \delta(A) = -2A \wedge \tilde{X} + \lambda\tilde{Y} \wedge \tilde{X}.$$

Finally, to treat case 4.3 ($a < 0$), we use the modification of the exponentiation (5.7) with the usage of the universal covering Lie group (see the very end of Section 5). This gives $[X, Y] = Y$, $\chi(X) = -1$, $\chi(Y) = 0$, $\beta(X) = -Y$ and $\beta(Y) = X$. There are non-trivial cocycles, given by $\mathcal{U}(X, Y) = -\lambda\tilde{A}$. Hence, we get a Hopf *-algebra with generators $\tilde{A} = \tilde{A}^*$, $X = -X^*$ and $Y = -Y^*$, subject to the relations

$$\begin{aligned}
&[\tilde{A}, X] = -\sin\tilde{A}, \quad [\tilde{A}, Y] = -1 + \cos\tilde{A}, \quad [X, Y] = Y - i\lambda\tilde{A},\\
&\Delta(\tilde{A}) = \tilde{A} \otimes 1 + 1 \otimes \tilde{A},\\
&\Delta(X) = X \otimes \cos\tilde{A} - Y \otimes \sin\tilde{A} + 1 \otimes X,\\
&\Delta(Y) = X \otimes \sin\tilde{A} + Y \otimes \cos\tilde{A} + 1 \otimes Y.
\end{aligned}$$

Co-unit and antipode are $\varepsilon(\tilde{A}) = \varepsilon(X) = \varepsilon(Y) = 0$, $S(\tilde{A}) = -\tilde{A}$, $S(X) = -X\cos\tilde{A} - Y\sin\tilde{A}$, $S(Y) = X\sin\tilde{A} - Y\cos\tilde{A}$. The corresponding Lie bialgebra is

$$[\tilde{A}, X] = -\tilde{A}, \quad [\tilde{A}, Y] = 0, \qquad [X, Y] = Y - \lambda\tilde{A},$$
$$\delta(\tilde{A}) = 0, \qquad \delta(X) = \tilde{A} \wedge Y, \qquad \delta(Y) = X \wedge \tilde{A}.$$

The dual Hopf $*$-algebra has self-adjoint generators $\tilde{X}$, $\tilde{Y}$ and the generator $A = -A^*$, subject to the relations

$$\begin{aligned}
&[\tilde{X}, A] = \tilde{Y}, \quad [\tilde{Y}, A] = \frac{1}{2}(\tilde{X}^{-1}\tilde{Y}^2 - \tilde{X} + \tilde{X}^{-1}), \quad [\tilde{X}, \tilde{Y}] = 0, \\
&\Delta(\tilde{X}) = \tilde{X} \otimes \tilde{X}, \\
&\Delta(\tilde{Y}) = \tilde{X} \otimes \tilde{Y} + \tilde{Y} \otimes 1, \\
&\Delta(A) = A \otimes \tilde{X}^{-1} + 1 \otimes A - i\lambda \log \tilde{X} \otimes \tilde{X}^{-1}\tilde{Y}.
\end{aligned}$$

Co-unit and antipode are $\varepsilon(\tilde{X}) = 1, \varepsilon(\tilde{Y}) = \varepsilon(A) = 0, S(\tilde{X}) = \tilde{X}^{-1}, S(\tilde{Y}) = -\tilde{X}^{-1}\tilde{Y}, S(A) = -A\tilde{X} - i\lambda\tilde{Y} \log \tilde{X}$. The corresponding Lie bialgebra is determined by

$$\begin{aligned}
&[\tilde{X}, A] = \tilde{Y}, \quad [\tilde{Y}, A] = -\tilde{X}, \qquad [\tilde{X}, \tilde{Y}] = 0, \\
&\delta(\tilde{X}) = 0, \qquad \delta(\tilde{Y}) = \tilde{X} \wedge \tilde{Y}, \qquad \delta(A) = \tilde{X} \wedge A - \lambda\tilde{X} \wedge \tilde{Y}.
\end{aligned}$$

The above list of Lie bialgebras $\mathfrak{g}$ covers all decomposable Lie bialgebras of dimension 3, i.e., those that are extensions of the form

$$\mathfrak{g}_2^* \longrightarrow \mathfrak{g} \longrightarrow \mathfrak{g}_1.$$

Naturally, their duals are extensions of the form

$$\mathfrak{g}_1^* \longrightarrow \mathfrak{g} \longrightarrow \mathfrak{g}_2.$$

Here $\mathfrak{g}_1$ and $\mathfrak{g}_2$ are the Lie algebras forming the cocycle matched pair.

The case of indecomposable low-dimensional l.c. quantum groups. Comparing the above list with the full list of all mutually non-isomorphic Lie bialgebras $\mathfrak{g}$ of dimension 3 obtained by X. Gomez [15], one can observe that there are also indecomposable Lie bialgebras of dimension 3. Below we list them and describe the corresponding Hopf $*$-algebras and, when they are known, the corresponding locally compact quantum groups.

We start with the Hopf $*$-algebras $U_q(\mathfrak{su}_2)$, $U_q(\mathfrak{su}_{1,1})$, $U_q(\mathfrak{sl}_2(\mathbb{R}))$, which are real forms of the Hopf algebra $U_q(\mathfrak{sl}_2(\mathbb{C}))$, and their duals $\mathrm{SU}_q(2)$, $\mathrm{SU}_q(1,1)$ and $\mathrm{SL}_q(2,\mathbb{R})$. Hopf algebraically, they are studied by several authors, see e.g. [13]. Operator algebraically, $\mathrm{SU}_q(2)$ and its dual are studied in [52, 61, 62] as an example of a compact (dually, discrete) quantum group. $\mathrm{SL}_q(2,\mathbb{R})$ is studied as a l.c. quantum group recently by W. Pusz and S. L. Woronowicz, see [42] for a related example. $\mathrm{SU}_q(1,1)$ and its dual is treated as a l.c. quantum group in [24, 25, 65].

The Hopf algebra $U_q(\mathfrak{sl}_2(\mathbb{C}))$ is defined by 3 generators a, x, y and the following relations (q is a complex parameter):

$$ax = qxa, \quad ay = q^{-1}ya, \quad [x, y] = \frac{1}{q - q^{-1}}(a^2 - a^{-2}),$$
$$\Delta(a) = a \otimes a,$$
$$\Delta(x) = x \otimes a + a^{-1} \otimes x,$$
$$\Delta(y) = y \otimes a + a^{-1} \otimes y.$$

It is well known that there exist three different Hopf *-algebra structures on $U_q(\mathfrak{sl}_2(\mathbb{C}))$: we get $U_q(\mathfrak{su}_2)$ by putting $q > 0$, $q \neq 1$, $a = a^*$ and $y = x^*$, we get $U_q(\mathfrak{su}_{1,1})$ for the same values of q, $a = a^*$ and $x = -y^*$ and we finally get $U_q(\mathfrak{sl}_2(\mathbb{R}))$ for $|q| = 1$, $q \neq \pm 1$, $a = a^*$, $x^* = -x$, $y^* = -y$.

One can construct the corresponding Lie bialgebras using the above mentioned formal procedure of linearization. For $U_q(\mathfrak{su}_2)$, we put $H = -i\frac{1}{\log q}\log a$, $X = i(x + y)$ and $Y = x - y$, and we arrive at the Lie bialgebra

$$[H, X] = Y, \quad [H, Y] = -X, \qquad [X, Y] = \frac{8\log q}{q - q^{-1}}H,$$
$$\delta(H) = 0, \qquad \delta(X) = 2\log q \; H \wedge X, \qquad \delta(Y) = 2\log q \; H \wedge Y.$$

For $U_q(\mathfrak{su}_{1,1})$, we put $H = -i\frac{1}{\log q}\log a$, $X = x + y$ and $Y = i(x - y)$ to obtain

$$[H, X] = -Y, \quad [H, Y] = X, \qquad [X, Y] = \frac{8\log q}{q - q^{-1}}H,$$
$$\delta(H) = 0, \qquad \delta(X) = 2\log q \; H \wedge X, \qquad \delta(Y) = 2\log q \; H \wedge Y.$$

Finally, for $U_q(\mathfrak{sl}_2)$, we put $q = \exp(ir)$, $H = -i\frac{2}{r}\log a$, $X = x$ and $Y = y$ to arrive at

$$[H, X] = 2X, \quad [H, Y] = -2Y, \qquad [X, Y] = \frac{r}{\sin r}H,$$
$$\delta(H) = 0, \qquad \delta(X) = rH \wedge X, \qquad \delta(Y) = rH \wedge Y.$$

For the dual Hopf *-algebras, we take the following versions. $SU_q(2)$ has generators a, a^*, b, b^*, parameter $q > 0$ and relations

$$ab = qba, \quad ab^* = qb^*a, \quad bb^* = b^*b,$$
$$aa^* - a^*a = (q^{-2} - 1)bb^*, \quad aa^* + bb^* = 1,$$
$$\Delta(a) = a \otimes a - q^{-1}b \otimes b^*,$$
$$\Delta(b) = a \otimes b + b \otimes a^*.$$

Suppose $a = \exp(A)$. We first formally calculate that

$$[A, b] = \log q \; b, \quad [A^*, b] = -\log q \; b.$$

So, we put $H = A - A^*$, $X = i(b + b^*)$ and $Y = b - b^*$ to obtain the linearization

$$[H, X] = 2\log q X, \quad [H, Y] = 2\log q Y, \quad [X, Y] = 0,$$
$$\delta(H) = q^{-1} X \wedge Y, \quad \delta(X) = Y \wedge H, \quad \delta(Y) = H \wedge X.$$

Next, we write $SL_q(2, \mathbb{R})$ with $|q| = 1$ and self-adjoint generators a, b, c, d, subject to the relations

$$ab = qba, \quad ac = qca, \quad bd = qdb, \quad cd = qdc,$$
$$bc = cb, \quad [a, d] = (q - q^{-1})bc, \quad ad - qbc = 1,$$
$$\Delta(a) = a \otimes a + b \otimes c, \quad \Delta(b) = a \otimes b + b \otimes d,$$
$$\Delta(c) = c \otimes a + d \otimes c, \quad \Delta(d) = c \otimes b + d \otimes d.$$

Again, writing $a = \exp(A)$, $q = \exp(ir)$, $H = iA$, $X = ib$ and $Y = ic$, we get the Lie bialgebra

$$[H, X] = -rX, \quad [H, Y] = -rY, \quad [X, Y] = 0,$$
$$\delta(H) = X \wedge Y, \quad \delta(X) = 2H \wedge X, \quad \delta(Y) = 2Y \wedge H.$$

Finally, we present $SU_q(1, 1)$ with $q > 0$, generators a, b and relations

$$ab = qba, \quad ab^* = qb^*a, \quad bb^* = b^*b,$$
$$aa^* - a^*a = (1 - q^{-2})bb^*, \quad aa^* - bb^* = 1,$$
$$\Delta(a) = a \otimes a + q^{-1} b \otimes b^*,$$
$$\Delta(b) = a \otimes b + b \otimes a^*.$$

Following the same road as for $SU_q(2)$, we get the Lie bialgebra

$$[H, X] = 2\log q \; X, \quad [H, Y] = 2\log q \; Y, \quad [X, Y] = 0,$$
$$\delta(H) = q^{-1} Y \wedge X, \quad \delta(X) = Y \wedge H, \quad \delta(Y) = H \wedge X.$$

The Hopf *-algebras corresponding to the l.c. quantum group of motions of the plane and its dual was considered in e.g. [22]. They are treated as l.c. quantum groups in [3], [56], [57] and [63].

Take $\mu > 0$ and consider the Hopf algebra defined by

$$ax = \mu xa, \quad \Delta(a) = a \otimes a, \quad \Delta(x) = a \otimes x + x \otimes a^{-1}.$$

We can put two different Hopf *-algebra structures. First, we get $U_\mu(\mathfrak{e}_2)$ by taking a self-adjoint and x normal. Next, we get $E_\mu(2)$ by supposing that a is unitary and x is normal. We linearize $U_\mu(\mathfrak{e}_2)$ by writing $H = i \log a$, $X = i(x + x^*)$ and $Y = x - x^*$. This gives us the Lie bialgebra

$$[H, X] = -\log \mu \; Y, \quad [H, Y] = \log \mu \; X, \quad [X, Y] = 0,$$
$$\delta(H) = 0, \quad \delta(X) = 2H \wedge X, \quad \delta(Y) = 2H \wedge Y.$$

For $E_\mu(2)$, we write $H = \log a$ (which is indeed anti-self-adjoint), $X = i(x + x^*)$ and $Y = x - x^*$, to arrive at the Lie bialgebra

$$\begin{aligned} &[H, X] = \log\mu\ X, \quad [H, Y] = \log\mu\ Y, \quad [X, Y] = 0, \\ &\delta(H) = 0, \qquad \delta(X) = 2Y \wedge H, \qquad \delta(Y) = 2H \wedge X. \end{aligned}$$

Observe that the list of 3-dimensional Lie bialgebras [15] contains some more objects, and we now want to present the corresponding Hopf *-algebras, which are less known. As far as we know, they have not yet been considered on the level of l.c. quantum groups.

Let $\mu \in \mathbb{C}$, $\mu \neq 0$ and let $\rho > 0$. Put $\lambda = -\frac{\log \rho}{\overline{\mu}}$. Then, we can define a Hopf *-algebra with relations

$$\begin{aligned} &[a, x] = \mu x, \quad xx^* = \rho x^* x, \quad a = a^*, \\ &\Delta(a) = a \otimes 1 + 1 \otimes a, \\ &\Delta(x) = x \otimes \exp(\lambda a) + 1 \otimes x. \end{aligned}$$

Co-unit and antipode are given by $\varepsilon(a) = \varepsilon(x) = 0$, $S(a) = -a$, $S(x) = -x\exp(-\lambda a)$. The specific form of λ is needed to ensure that Δ respects the relation $xx^* = \rho x^* x$. Then, putting $H = ia$, $X = i(x + x^*)$ and $Y = x - x^*$, and observing that X and Y commute in a first order approximation, we get the corresponding Lie bialgebra

$$\begin{aligned} &[H, X] = -\operatorname{Im}\mu\ X - \operatorname{Re}\mu\ Y, \quad [H, Y] = \operatorname{Re}\mu\ X - \operatorname{Im}\mu\ Y, \quad [X, Y] = 0, \\ &\delta(H) = 0, \quad \delta(X) = (\operatorname{Re}\lambda\ X - \operatorname{Im}\lambda\ Y) \wedge H, \quad \delta(Y) = (\operatorname{Im}\lambda\ X + \operatorname{Re}\lambda\ Y) \wedge H. \end{aligned}$$

One can check that δ respects the relation $[X, Y] = 0$, because $\operatorname{Im}(\lambda\overline{\mu}) = 0$. Also, one can check that this family of Lie bialgebras is self-dual, i.e., the dual of any Lie bialgebra with specific values of μ and ρ belongs again to this family (but with different values of μ and ρ). So, the dual Hopf *-algebras are of the same form as above.

Next, we take real numbers α and β, and we write the Hopf *-algebra with self-adjoint generators a, x, y and relations:

$$\begin{aligned} &[a, x] = -ix, \quad [a, y] = -i\alpha y, \quad xy = \exp(-i\alpha\beta)yx, \\ &\Delta(a) = a \otimes 1 + 1 \otimes a, \\ &\Delta(x) = x \otimes \exp(\beta a) + 1 \otimes x, \\ &\Delta(y) = y \otimes \exp(-\alpha\beta a) + 1 \otimes y. \end{aligned}$$

Co-unit and antipode are given by $S(x) = -x\exp(-\beta a)$, $S(y) = -y\exp(\alpha\beta a)$, $\varepsilon(a) = \varepsilon(x) = \varepsilon(y) = 0$. To linearize, we write $H = ia$, $X = ix$ and $Y = iy$. Observing again that X and Y commute in a first order approximation, we obtain the corresponding Lie bialgebra

$$\begin{aligned} &[H, X] = X, \quad [H, Y] = \alpha Y, \qquad [X, Y] = 0, \\ &\delta(H) = 0, \qquad \delta(X) = \beta X \wedge H, \qquad \delta(Y) = \alpha\beta H \wedge Y. \end{aligned}$$

In the same sense as in the previous paragraph, this family of Lie bialgebras is self-dual, so the dual Hopf *-algebras are of the same form.

Finally, there is one isolated Lie bialgebra, which is defined by

$$[H,X]=2X,\qquad [H,Y]=-2Y,\qquad [X,Y]=H,$$
$$\delta(H)=H\wedge Y,\qquad \delta(X)=X\wedge Y,\qquad \delta(Y)=0.$$

We can write the following Hopf *-algebra, which appears in [9], Section 6.4.F and which has generators $h^*=-h$, $x=x^*$, $y=y^*$ and relations

$$[h,x]=2x-\frac{1}{2}h^2,\quad [h,y]=2(1-\exp(y)),\quad [x,y]=h,$$
$$\Delta(h)=h\otimes\exp(y)+1\otimes h,$$
$$\Delta(x)=x\otimes\exp(y)+1\otimes x,$$
$$\Delta(y)=y\otimes 1+1\otimes y.$$

One can check that $[x,\exp(y)]=\exp(y)h+\exp(y)(1-\exp(y))$, so taking $H=h$, $X=-ix$ and $Y=iy$, and linearizing we get the above Lie bialgebra. For the dual Lie bialgebra, we cannot construct at the moment a corresponding Hopf *-algebra. The main problem to construct this exponentiation is the fact that the dual Lie bialgebra has no non-trivial Lie sub-bialgebra.

References

[1] N. Andruskiewitsch, Notes on extensions of Hopf algebras, *Canad. J. Math.* 48 (1) (1996), 3–42.

[2] N. Andruskiewitsch and J. Devoto, Extensions of Hopf algebras, *Algebra Anal.* 7 (1995), 22–61 and *St. Petersburg Math. J.* 7 (1996), 17–52.

[3] S. Baaj, Représentation régulière du groupe quantique des déplacements de Woronowicz, *Astérisque* 232 (1995), 11-48.

[4] S. Baaj and G. Skandalis, Unitaires multiplicatifs et dualité pour les produits croisés de C*-algèbres, *Ann. Scient. Ec. Norm. Sup.* (4) 26 (1993), 425–488.

[5] S. Baaj and G. Skandalis, Transformations pentagonales, *C. R. Acad. Sci. Paris Sér.* I *Math.* 327 (1998), 623–628.

[6] S. Baaj, G. Skandalis and S. Vaes, Non-semi-regular quantum groups coming from number theory, preprint, math.OA/0209069 (2002).

[7] N. Bourbaki, *Groupes et algèbres de Lie*, Chapitres 7 et 8, Masson, Paris 1990.

[8] J. De Canniére, Produit croisé d'une algèbre de Kac par un groupe localement compact, *Bull. Soc. Math. France* 107 (1979), 337–372.

[9] V. Chari and A. Pressley, *A guide to quantum groups*, Cambridge University Press, Cambridge 1994.

[10] V. G. Drinfeld, Hamiltonian structures on Lie groups, Lie bialgebras and the geometric meaning of the classical Yang-Baxter equation, *Soviet Math. Dokl.* 27 (1983), 68–72.

[11] M. Enock, Sous-facteurs intermédiaires et groupes quantiques mesurés, *J. Operator Theory* 42 (1999), 305–330.

[12] M. Enock and J.-M. Schwartz, *Kac Algebras and Duality of Locally Compact Groups*, Springer-Verlag, Berlin 1992.

[13] L. D. Fadeev, N. Yu. Reshetikhin and L.A. Takhtajan, Quantization of Lie groups and Lie algebras, *Leningrad Math. J.* 1 (1990), 193–225.

[14] W. Fulton and J. Harris, *Representation Theory*, Grad. Texts in Math. 129, Springer-Verlag, New York 1991.

[15] X. Gomez, Classification of three-dimensional Lie bialgebras, *J. Math. Phys.* 41 (7) (2000), 4939–4956.

[16] S. Helgason, *Groups and Geometric Analysis*, Pure Appl. Math. 113, Academic Press, Inc., Orlando 1984.

[17] G. I. Kac, Extensions of groups to ring groups, *Math. USSR Sb.* 5 (1968), 451–474.

[18] G. I. Kac, Ring groups and the principle of duality, I, II, *Trans. Moscow Math. Soc.* (1963), 291–339; ibid. (1965), 94–126.

[19] G. I. Kac and V.G. Paljutkin, Finite ring groups, *Trans. Moscow Math. Soc.* 5 (1966), 251–294.

[20] G. I. Kac and V. G. Paljutkin, Example of a ring group generated by Lie groups, *Ukraïn. Mat. Zh.* 16 (1964), 99–105 (in Russian).

[21] G. I. Kac and L. I. Vainerman, Nonunimodular ring groups and Hopf-von Neumann algebras, *Math. USSR. Sb.* 23 (1974), 185–214.

[22] H. Koelink, On quantum groups and q-special functions, Ph.D. Thesis, Rijksuniversiteit Leiden (1991).

[23] J. Kustermans, Locally compact quantum groups in the universal setting, *Internat. J. Math.* 12 (2000), 289–338.

[24] E. Koelink and J. Kustermans, A locally compact quantum group analogue of the normalizer of $SU(1, 1)$ in $SL(2, C)$, to appear in *Comm. Math. Phys.*, preprint `math.QA/0105117` (2001).

[25] E. Koelink and J. Kustermans, Quantum SU(1, 1) and its Pontryagin dual, in *Locally compact quantum groups and groupoids* (L. Vainerman, ed.), IRMA Lectures in Math. Theoret. Phys. 2, Walter de Gruyter, Berlin–New York 2003, 49–77.

[26] J. Kustermans and S. Vaes, Locally compact quantum groups, *Ann. Sci. École Norm. Sup.* (4) 33 (2000), 837–934.

[27] J. Kustermans and S. Vaes, Locally compact quantum groups in the von Neumann algebraic setting, to appear in *Math. Scand.*.

[28] J. Kustermans and S. Vaes, A simple definition for locally compact quantum groups, *C. R. Acad. Sci. Paris Sér.* I *Math.* 328 (10) (1999), 871–876.

[29] J. Kustermans, S. Vaes, L. Vainerman, A. Van Daele and S. L. Woronowicz, Locally Compact Quantum Groups, in *Lecture Notes for the school on Noncommutative Geometry and Quantum Groups in Warsaw* (17–29 September 2001), Banach Centre Publ., to appear.

[30] S. Majid, Physics for algebraists: Non-commutative and non-cocommutative Hopf algebras by a bicrossproduct construction, *J. Algebra* 130 (1990), 17–64.

[31] S. Majid, Hopf-von Neumann algebra bicrossproducts, Kac algebra bicrossproducts, and the classical Yang-Baxter equations, *J. Funct. Anal.* 95 (1991), 291–319.

[32] S. Majid, More examples of bicrossproduct and double cross product Hopf algebras, *Israel J. Math.* 72 (1990), 133–148.

[33] S. Majid, *Foundations of quantum group theory*, Cambridge University Press, Cambridge 1995.

[34] T. Masuda and Y. Nakagami, A von Neumann algebraic framework for the duality of the quantum groups, *Publ. Res. Inst. Math. Sci.* 30 (1994), 799–850.

[35] A. Masuoka, Calculations of Some Groups of Hopf algebra Extensions, *J. Algebra* 191 (1997), 568–588.

[36] A. Masuoka, Extensions of Hopf Algebras and Lie Bialgebras, *Trans. Amer. Math. Soc.* 352 (2000), 3837–3879.

[37] A. Masuoka, Cohomology and Coquasi-Bialgebra Extensions Associated to a Matched Pair of Bialgebras, preprint, Institute of Mathematics, University of Tsukuba (2001).

[38] S. Montgomery, *Hopf Algebras and Their Actions on Rings*, CBMS Regional Conf. Ser. Math. 82, Amer. Math. Soc., Providence, RI, 1993.

[39] M. A. Naimark and A. I. Stern, *Theory of Group Representations*, Grundlehren Math. Wiss. 246, Springer-Verlag, New York–Heidelberg–Berlin 1982.

[40] J. A. Packer and I. Raeburn, Twisted crossed products of C*-algebras, *Math. Proc. Cambridge Philos. Soc.* 106 (1989), 293–311.

[41] P. Podleś and S. L. Woronowicz, Quantum deformation of Lorenz group, *Comm. Math. Phys.* 130 (1990), 381–431.

[42] W. Pusz and S. L. Woronowicz, A quantum $GL(2, \mathbb{C})$ group at roots of unity, *Rep. Math. Phys.* 47 (2001), 431–462.

[43] H.-J. Schneider, Normal basis and transitivity of crossed products for Hopf algebras, *J. Algebra* 152 (1992), 289–312.

[44] G. Skandalis, Duality for locally compact 'quantum groups' (joint work with S. Baaj), Mathematisches Forschungsinstitut Oberwolfach, Tagungsbericht 46/1991, C*-algebren, 20.10–26.10.1991, pp. 20.

[45] S. Stratila, *Modular Theory in Operator Algebras*, Editura Academiei/Abacus Press, Bucuresti/Tunbridge Wells, Kent, 1981.

[46] M. Takesaki and N. Tatsuuma, Duality and subgroups, *Ann. of Math.* 93 (1971), 344–364.

[47] M. Takeuchi, Matched pairs of groups and bismash products of Hopf algebras, *Comm. Algebra* 9 (1981), 841–882.

[48] S. Vaes, The unitary implementation of a locally compact quantum group action, *J. Funct. Anal.* 180 (2001), 426–480.

[49] S. Vaes, Examples of locally compact quantum groups through the bicrossed product construction, in *Proceedings of the XIIIth Int. Conf. Math. Phys., London* 2000 (A. Grigoryan, A. Fokas, T. Kibble and B. Zegarlinski, eds.), International Press of Boston, Somerville, MA, 2001, 341–348.

[50] S. Vaes, A Radon-Nikodym theorem for von Neumann algebras, *J. Operator Theory* 46 (3) (2001), 477–489.

[51] S. Vaes and L. Vainerman, Extensions of locally compact quantum groups and the bicrossed product construction, to appear in *Adv. Math.*, preprint `math.QA/0101133`.

[52] L. L. Vaksman and Ya.S. Soibelman, Algebra of functions on the quantum group $SU(2)$, *Funct. Anal. Appl.* 22 (1988),170–181.

[53] A. Van Daele, Quantum deformation of the Heisenberg group, in *Current topics in operator algebras* (H. Araki et al., eds.), World Scientific, Singapore 1991, 314–325.

[54] A. Van Daele, An algebraic framework for group duality, *Adv. Math.* 140 (1998), 323–366.

[55] A. Van Daele, Dual pairs of Hopf $*$-algebras, *Bull. London Math. Soc.* 25 (1993), 209–230.

[56] A. Van Daele and S. L. Woronowicz, Duality for the quantum $E(2)$ group, *Pacific J. Math.* 173 (2) (1996), 375–385.

[57] L. Vaksman and L. Korogodsky, Algebra of bounded functions on the quantum group of motions of the plane and q-analogues of Bessel functions, *Soviet Math. Dokl.* 39 (1989), 173–177.

[58] S. L. Woronowicz, Compact quantum groups, in *Symétries quantiques* (A. Connes et al., eds.), North-Holland, Amsterdam 1998, 845–884.

[59] S. L. Woronowicz, From multiplicative unitaries to quantum groups, *Internat. J. Math.* 7 (1996), 127–149.

[60] S. L. Woronowicz, Compact matrix pseudogroups, *Comm. Math. Phys.* 111 (1987), 613–665.

[61] S. L. Woronowicz, Twisted $SU(2)$ group. An example of a non-commutative differential calculus, *Publ. Res. Inst. Math. Sci.* 23 (1987), 117–181.

[62] S. L. Woronowicz, Tannaka-Krein duality for compact matrix pseudogroups. Twisted $SU(N)$ groups, *Invent. Math.* 93 (1988), 35–76.

[63] S. L. Woronowicz, Quantum $E(2)$ group and its Pontryagin dual, *Lett. Math. Phys.* 23 (1991), 251–263.

[64] S. L. Woronowicz, Quantum $'az + b'$ group on complex plane, *Internat. J. Math.* 12 (2001), 461–503.

[65] S. L. Woronowicz, Extended $SU(1,1)$ quantum group. Hilbert space level, preprint KMMF (in preparation).

[66] S. L. Woronowicz and S. Zakrzevski, Quantum $'ax + b'$ group, preprint KMMF (1999).

Multiplicative partial isometries and finite quantum groupoids

Jean-Michel Vallin

UMR CNRS 6628
Université d'Orleans
and
Institut de Mathématiques de Chevaleret
Plateau 7D, 175 rue du Chevaleret, 75013 Paris
email: `jmva@math.jussieu.fr`

Abstract. In this work we continue, after [Val1], the study of multiplicative partial isometries over a finite dimensional Hilbert space. We prove that, after an ampliation and a reduction, any regular multiplicative partial isometry is isomorphic to an irreducible one. For this irreducible multiplicative partial isometry we prove quantum Markov properties. Namely, both normalized Haar measures of the quantum groupoids associated to a multiplicative partial isometry can be extended to a unique faithful positive linear form on the involutive algebra generated by these groupoids (the Weyl algebra). Using this Markov extension a multiplicative partial isometry can be expressed as a composition of two very simple partial isometries. The two Haar conditional expectations of the quantum groupoids with values in the intersection of their algebras can be, in a unique way, extended to a multiplicative conditional expectation on the Weyl algebra; moreover, this extension is invariant with respect to the Markov extension of the Haar measures. We prove that a multiplicative partial isometry is completely determined by the two quantum groupoids in duality which it generates and the spaces of fixed and cofixed vectors. Finally, we give a complete characterization of quantum groupoids in duality acting on the same Hilbert space in the irreducible situation.

1 Introduction

Multiplicative partial isometries (*mpi*) generalize Baaj and Skandalis multiplicative unitaries in a finite dimension [BS], [BBS]. They are the finite-dimensional version of so-called pseudo-multiplicative unitaries, which appeared first in a commutative context dealing with locally compact groupoids [Val0], and then in the general case for a very large class of depth two inclusions of von Neumann algebras [EV].

Any regular *mpi* generates two involutive subalgebras of the algebra of all bounded linear operators on the corresponding Hilbert space. Due to canonical pairing, these two algebras have structures generalizing involutive Hopf algebras. The first examples of these new structures were discovered by the theoretical physicists Böhm, Szlachanyi and Nill [BoSz] [BoSzNi], who called them weak Hopf C^*-algebras.

D. Nikshych and L. Vainerman, using general inclusions of depth two factors of type II_1 with finite index $M_0 \subset M_1$, and a special pairing between relative commutants $M_0' \cap M_2$ and $M_1' \cap M_3$, gave explicit formulas for weak Hopf C^*-algebra structures in duality for these two last involutive algebras [NV2] and found a Galois correspondence between intermediate subfactors and involutive coideals for $M_1' \cap M_3$ [NV4].

The aim of this article is to continue the study of *mpi's*, after [Val1], in the spirit of [BBS], and to express all these structures and relations between them in terms of *mpi's*. In particular, we prove the close relation between weak Hopf C^*-algebras in duality acting on the same Hilbert space and a multiplicativity condition for a natural conditional expectation with values in their intersection.

In the second chapter we recall the definition of multiplicative partial isometries with a base, their relation with quantum groupoids (or weak Hopf C^*-algebras), the meaning of fixed and cofixed vectors. Finally, we prove that any regular multiplicative partial isometry can be reduced, in a certain sense, to a canonical one (irreducibility).

In the third chapter we study this irreducible situation. We prove quantum Markov properties: both normalized Haar measures of the quantum groupoids S and $\hat{S}$, associated to a *mpi*, can be extended to the Weyl algebra $S\hat{S}$ and also both Haar conditional expectations of S and $\hat{S}$ with values in $S \cap \hat{S}$ can be, in a unique way, extended to a multiplicative conditional expectation on the Weyl algebra; moreover, this extension is invariant with respect to the Markov extension of the Haar measures.

In the last chapter we study pairs of involutive subalgebras of $\mathcal{L}(H)$ for which the conditional expectation associated to the canonical trace of $\mathcal{L}(H)$ on their intersection has a multiplicativity property with a coefficient. We completely characterize quantum groupoids in duality acting on the same Hilbert space. As a consequence of the demonstration we obtain, using the Markov extension of the two Haar measures, that any irreducible *mpi* can be expressed as a composition of two very simple partial isometries.

I want to thank M. Enock, S. Baaj, L. Vainerman and M. C. David for the numerous discussions we had on this topic.

2 Multiplicative partial isometries and quantum groupoids

In what follows, N is a finite dimensional von Neumann algebra, so N is isomorphic to a sum of matrix algebras $\underset{\gamma}{\oplus} M_{n_\gamma}$. We denote the family of minimal central projections of N by $\{p^\gamma\}$, and we fix a family $\{e^\gamma_{i,j} / 1 \le i, j \le n_\gamma\}$ of matrix units of N. We denote the opposite von Neumann algebra of N by N^o, so N^o is N with the opposite multiplication and matrix units of N^o are given by $\{e^\gamma_{j,i} / 1 \le i, j \le n_\gamma\}$.

2.0.1 Lemma. *The element* $f = \sum_\gamma \sum_{i,j} \frac{1}{n_\gamma} e_{i,j}^{\gamma\ o} \otimes e_{j,i}^\gamma$ *is the only projection of* $N^o \otimes N$ *such that, for any* n *in* N,

1) $f(n^o \otimes 1) = f(1 \otimes n)$,

2) *if* $f(1 \otimes n) = 0$ *then* $n = 0$.

Proof. Obvious. □

2.1 Spatial theory in finite dimension

2.1.1 Notations. Let M_1, M_2 be two other finite dimensional von Neumann algebras, and let H_1 (resp., H_2) be a finite dimensional Hilbert space in which M_1 (resp., M_2) acts. Let s (resp., r) be a faithful non-degenerate antirepresentation (resp., a representation) from N to M_1 (resp., M_2). Then s can be viewed as a representation s^o of N^o. We define

$$e_{s,r} = (s^o \otimes r)(f) = \sum_\gamma \sum_{i,j} \frac{1}{n_\gamma} s(e_{i,j}^\gamma) \otimes r(e_{j,i}^\gamma).$$

For any Hilbert space, let us denote by tr_H the canonical trace with value 1 on minimal projections. The following result is obtained as a consequence of Lemma 2.0.1 (see also Lemmas 2.1.1 and 2.1.2 of [Val1]).

2.1.2 Lemma. *The element* $e_{s,r}$ *is a projection in* $s(N) \otimes r(N)$ *such that*

1) $(i \otimes \mathrm{tr}_{H_2})(e_{s,r})$ *is positive and invertible in the center of* $s(N)$;

2) $e_{s,r}$ *is the only projection* e *in* $M_1 \otimes r(N)$ *which satisfies the following two conditions:*

a) *for every* m_1,m_2 *in* M_1 *and* M_2 *respectively, the relation* $e(m_1 \otimes 1) = 0$ *implies* $m_1 = 0$ *and the relation* $e(1 \otimes m_2) = 0$ *implies* $m_2 = 0$.

b) *for every* n *in* N: $e(s(n) \otimes 1) = e(1 \otimes r(n))$.

The representation r can be decomposed into the direct product of the representations $r^\gamma = r \mid r(p^\gamma N)H_2$ and each r^γ can be viewed as a direct sum of faithful irreducible representations r_p^γ of M_{n_γ} for $p = 1, \ldots, q_\gamma$ $\left(= \frac{\mathrm{tr}_{H_2} \circ r(p^\gamma)}{\dim H_2}\right)$. Hence for every $\delta = (\gamma, p)$, where $p \in \{1, \ldots, q_\gamma\}$, one can find a unitary cyclic separating vector e_δ for r_p^γ, so $r_p^\gamma(p^\gamma N))H_2$ is given the orthonormal base $r(e_{i,1}^\gamma)e_\delta$ for $i = 1, \ldots, n_\gamma$. The antirepresentation s is a representation of N^o, so one can construct a similar family f_j for $j = (\gamma, p)$, where $p \in \left\{1, \ldots, m_\gamma \left(= \frac{\mathrm{tr}_{H_1} \circ s(p^\gamma)}{\dim H_1}\right)\right\}$; and $(s(e_{1,i})f_j)_{ij}$, for any j and for $i = 1, \ldots, n_\gamma$, is an orthonormal base of H_1.

2.1.3 Definition. The families e_δ and f_j will be called an $r(N)$-base of H_2 and an $s(N)$-base of H_1, respectively.

2.1.4 Lemma. *Every vector in $e_{s,r}(H_1 \otimes H_2)$ is the sum of terms of the form $e_{s,r}(f_j \otimes r(e^{\gamma}_{i,1})\eta)$, for any $j = (\gamma, p)$, where $p \in \{1, \dots, m_\gamma)\}$, and for any $i = 1, \dots, n_\gamma$.*

Proof. Any vector ξ in H_1 can be written as $\xi = \sum_j \sum_i \lambda^i_j s(e^{\gamma}_{1,i}) f_j$, where the λ^i_j's are scalars. So for any η in H_2, one has, due to Lemma 2.1.2 2),

$$\begin{aligned} e_{s,r}(\xi \otimes \eta) &= e_{s,r}\Big(\sum_j \sum_i \lambda^i_j s(e^{\gamma}_{1,i}) f_j \otimes \eta\Big) \\ &= \sum_j \sum_i e_{s,r}(s(e^{\gamma}_{1,i}) f_j \otimes \lambda^i_j \eta) \\ &= \sum_j \sum_i e_{s,r}(f_j \otimes r(e^{\gamma}_{1,i})(\lambda^i_j \eta)). \qquad \square \end{aligned}$$

2.1.5 Lemma. *We have, for every $j = (\gamma, p)$, $j' = (\gamma', p')$:*

$$\frac{d(\omega_{f_j, f_{j'}} \circ s)}{d(\mathrm{tr}_{H_1} \circ s)} = \delta_{j,j'} m_\gamma^{-1} e^{\gamma}_{1,1}.$$

Proof. The formula is obvious if $j \neq j'$. One can notice that $s(e^{\gamma}_{1,1}) f_{(\gamma',k)} = f_{(\gamma',k)}$ and $\omega_{f_{(\gamma,k)}, f_{(\gamma,k)}} \circ s$ does not depend on k. Therefore, for any n in N,

$$\begin{aligned} \mathrm{tr}(s(m_\gamma^{-1} e^{\gamma}_{1,1}) s(n)) &= \mathrm{tr}(s(m_\gamma^{-1} e^{\gamma}_{1,1}) s(n) s(e^{\gamma}_{1,1})) \\ &= \sum_{(\gamma',k)} \sum_i (s(m_\gamma^{-1} e^{\gamma}_{1,1}) s(n) s(e^{\gamma}_{1,1}) s(e^{\gamma'}_{1,i}) f_{(\gamma',k)}, s(e^{\gamma'}_{1,i}) f_{(\gamma',k)}) \\ &= \sum_{(\gamma',k)} \sum_i (s(m_\gamma^{-1} e^{\gamma}_{1,1} e^{\gamma'}_{i,1}) s(n) s(e^{\gamma'}_{1,i} e^{\gamma}_{1,1}) f_{(\gamma',k)}, f_{(\gamma',k)}) \\ &= m_\gamma^{-1} \sum_k (s(e^{\gamma}_{1,1}) s(n) s(e^{\gamma}_{1,1}) f_{(\gamma,k)}, f_{(\gamma,k)}) \\ &= (s(n)) f_{(\gamma,p)}, f_{(\gamma,p)}). \qquad \square \end{aligned}$$

2.2 Multiplicative partial isometries

2.2.1 Notations. Let H be a finite dimensional Hilbert space, and let α (resp., β, $\widehat{\beta}$) be an injective non-degenerate representation (resp., two injective non-degenerate antirepresentations) which commute two by two pointwise. We also suppose that $\mathrm{tr} \circ \alpha = \mathrm{tr} \circ \beta = \mathrm{tr} \circ \hat{\beta}$, where tr is the canonical trace on H. Let us denote $\tau = \mathrm{tr} \circ \alpha$. One must keep in mind that β and α are a representation and an antirepresentation of N^o.

2.2.2 Definition. We call a multiplicative partial isometry with the base $(N, \alpha, \beta, \hat{\beta})$ a partial isometry I on $H \otimes H$, if

0) its initial (resp., final) support is $e_{\hat{\beta},\alpha}$ (resp., $e_{\alpha,\beta}$).

1) I commutes with $\beta(N) \otimes \widehat{\beta}(N)$;

2) for every n, n' in N, one has $I(\alpha(n) \otimes \beta(n')) = (\widehat{\beta}(n') \otimes \alpha(n))I$;

3) I verifies the pentagonal relation $I_{12}I_{13}I_{23} = I_{23}I_{12}$.

By Lemmas 2.3.1, 2.4.2, 2.4.6 in [Val1], one has:

2.2.3 Notations and Lemma. *Let I be a multiplicative partial isometry with the base $(N, \alpha, \beta, \hat{\beta})$, let us denote the set $\{(\omega \otimes i)(I)/\omega$ is a linear form on $\mathcal{L}(H)\}$ by S and the set $\{(i \otimes \omega)(I)/\omega$ is a linear form on $\mathcal{L}(H)\}$ by $\widehat{S}$. Then S and $\widehat{S}$ are non-degenerate subalgebras of $\mathcal{L}(H)$ and their adjoints S^* and $\widehat{S}^*$, too. If M (resp., $\widehat{M}$) is the involutive algebra (i.e., von Neumann algebra) generated by S (resp.,$\widehat{S}$), then the algebras $\beta(N)$ et $\alpha(N)$ (resp., $\hat{\beta}(N)$ and $\alpha(N)$) are included in the multipliers of S and in M (resp., $\widehat{S}$ and in $\widehat{M}$).*

According to Lemma 2.3.4 of [Val1] and its corollary, one can be more precise for the non-degeneracy of S and $\widehat{S}$ in $\mathcal{L}(H)$.

2.2.4 Lemma. *For every γ the following assertions are equivalent:*

1) $\beta(p^{\gamma})S\beta(p^{\gamma}) = 0$,

2) *the canonical trace is identically equal to zero on $\beta(p^{\gamma})S\beta(p^{\gamma})$,*

3) *the algebra $\beta(p^{\gamma})S\beta(p^{\gamma})$ is degenerated in $\mathcal{L}(\beta(p^{\gamma})H)$.*

4) $\alpha(p^{\gamma})\hat{\beta}(p^{\gamma}) = 0$.

If for every $\gamma, \alpha(p^{\gamma})\hat{\beta}(p^{\gamma})$ is not equal to zero, we will say that the base $(N, \alpha, \beta, \hat{\beta})$ is non-degenerate.

2.3 Fixed and cofixed vectors for multiplicative partial isometries

Let I be some multiplicative partial isometry with the base $(N, \alpha, \beta, \hat{\beta})$.

2.3.1 Definition. 1) A vector e of H is said to be *fixed* if $I(e \otimes \eta) = e_{\alpha,\beta}(e \otimes \eta)$ for every vector η in H (or, equivalently, $I^*(e \otimes \eta) = e_{\hat{\beta},\alpha}(e \otimes \eta)$). A vector e of H is said to be cofixed if and only if every vector η in H verifies the condition $I(\eta \otimes e') = e_{\alpha,\beta}(\eta \otimes e')$ (or, equivalently, $I^*(\eta \otimes e') = e_{\hat{\beta},\alpha}(\eta \otimes e')$).

2) By Lemma 2.1.2, there exists a unique positive invertible central element n_0 of N satisfying the relations $\alpha(n_0) = (i \otimes \mathrm{tr})(e_{\alpha,\beta}) = (\mathrm{tr} \otimes i)(e_{\hat{\beta},\alpha})$, $\hat{\beta}(n_0) = (i \otimes \mathrm{tr})(e_{\hat{\beta},\alpha})$, $\beta(n_0) = (\mathrm{tr} \otimes i)(e_{\alpha,\beta})$. This element will be called the *index* of I.

It is easy to see that the set of fixed vectors is a $\beta(N)$ Hilbert module (even on $\hat{M}'$). This set will be denoted by $\mathcal{F}$. A characterisation of fixed vectors is given by Lemma 2.5.13 of [Val1]:

2.3.2 Lemma. *Let us define a faithful positive form on $\mathcal{L}(H)$ by*

$$h(\cdot) = \mathrm{tr}(\beta(n_0)^{-1}\cdot).$$

Then a vector e of H is fixed for I if and only if $(i \otimes h)(I)e = e$.

2.3.3 Definition. A fixed (resp., cofixed) vector e is said to be *normalized* if it satisfies one of the following equivalent conditions:

1) $(i \otimes \omega_e)(e_{\hat{\beta},\alpha}) = 1$ in $\mathcal{L}(H)$,

2) $(\omega_e \otimes i)(e_{\alpha,\beta}) = 1$ in $\mathcal{L}(H)$,

3) $(\omega_e \otimes i)(e_{\hat{\beta},\alpha}) = 1$ (resp., $(i \otimes \omega_e)(e_{\alpha,\beta}) = 1$) in $\mathcal{L}(H)$,

4) $\dfrac{d(\omega_e \circ \alpha)}{d\tau} = n_0^{-1}$ in N,

5) $\dfrac{d(\omega_e \circ \hat{\beta})}{d\tau} = n_0^{-1}$ (resp., $\dfrac{d(\omega_e \circ \beta)}{d\tau} = n_0^{-1}$) in N.

Now one can give a criterion of the existence of normalized fixed vectors.

2.3.4 Proposition. *The algebra S has a unit if and only if there exists a normalized fixed vector for I.*

Proof. If there exists a normalized fixed vector e for I, then by Proposition 2.5.11 of [Val1] the operator $(\omega_e \otimes i)(I)$ is the unit of S. Conversely, if S has a unit, then this unit is that of $\mathcal{L}(H)$ by non-degeneracy. By 2.2.3, there exists a linear form θ on $\mathcal{L}(H)$ such that $(\theta \otimes i)(I) = \beta(n_0)$. Let h be the form defined by $h(x) = \mathrm{tr}(\beta(n_0)^{-1}x)$, and let h^m be the form defined by iteration using the formula $h^{m+1} = (h^m \otimes h)(I(x \otimes 1)I^*)$.

Let k_m be the sequence of the Cesaro averages of h, i.e., $k_m = \frac{1}{m}\sum_{i=1,\dots,m} h^i$. For every n in N and every natural number m one has

$$\begin{aligned} h^{m+1}(\beta(n)) &= (h^m \otimes h)(I(\beta(n) \otimes 1)I^*) = h^m((i \otimes h)(e_{\alpha,\beta}(\beta(n) \otimes 1))) \\ &= h^m((i \otimes h)(e_{\alpha,\beta})(\beta(n)) = h^m((i \otimes \mathrm{tr})(e_{\alpha,\beta}(1 \otimes \beta(n_0)^{-1}))(\beta(n)) \\ &= h^m((i \otimes \mathrm{tr})(e_{\alpha,\beta})\alpha(n_0)^{-1}\beta(n)) \\ &= h^m(\beta(n)). \end{aligned}$$

So $h^m(\beta(n)) = h(\beta(n))$. The same is true for k_m, in particular, the sequences $(\|h^m\|)_{m\in\mathbb{N}}$ and $(\|k_m\|)_{m\in\mathbb{N}}$ are bounded.

Let $p = (i \otimes h)(I)$. Then, using the same argument that in Propodition 1.4 of [BS], for every natural number m one has $p^m = (i \otimes h^m)(I)$, so the sequence $\|p^m\|$ is also bounded. One has

$$(1-p)((i \otimes k_m)(I) = (1-p)\frac{1}{m}\sum_{1,\dots,m} p^i = \frac{1}{n}(p - p^{m+1}),$$

hence this sequence tends to 0 as $\frac{1}{n}$. If $(1-p)$ was injective, then the sequence $(i \otimes k_m)(I)$ should also tend to 0. But $\theta(i \otimes k_m)(I) = k_m((\theta \otimes i)(I)) = k_m((\beta(n_0)) = h(\beta(n_0))$, which is a contradiction. Thus, 1 is an eigen value for p, and fixed vectors exist.

Let k_∞ be an accumulation point for the sequence $(k_m)_{m\in\mathbb{N}}$. If $\star$ is the convolution product of linear forms on $\mathcal{L}(H)$ defined by the formula $\omega \star \omega'(x) = (\omega \otimes \omega')(I(x \otimes 1)I^*)$, then it is obvious that $h \star k_\infty = k_\infty \star k = k_\infty$. Then $k_\infty \star k_\infty = k_\infty$; so the operator $p_\infty = (i \otimes k_\infty)(I)$ is an algebraic projection on the set of fixed vectors. Let q be an orthogonal projection of N. Then

$$\begin{aligned}\theta(\alpha(q)p_\infty) &= (\theta \otimes k_\infty)((\alpha(p) \otimes 1)I) = k_\infty((\theta \otimes i)((\alpha(p) \otimes 1)I)) \\ &= k_\infty((\theta \otimes i)((1 \otimes \beta(p))I)) = k_\infty(\beta(p)) = h(\beta(p)) \\ &\neq 0.\end{aligned}$$

Thus, with notations of Lemma 2.1, for every γ, $\alpha(p^\gamma)p_\infty \neq 0$, by Baire's theorem, there exists a vector η in H such that for every γ we have $\alpha(p^\gamma)p_\infty\eta \neq 0$.

Let $\xi = p_\infty\eta$, then, according to Lemma 2.1.5 5) and Lemma 2.5.8 of [Val1], we have

$$\frac{d(\omega_{\xi,\xi} \circ \alpha)}{d\tau} = \sum_\gamma \frac{1}{m_\gamma^2} \langle \alpha(p^\gamma)\xi, \xi\rangle p^\gamma,$$

and so $t = \dfrac{d(\omega_{\xi,\xi} \circ \alpha)}{d\tau}$ is an invertible element of N such that, by Lemma 2.5.8 of [Val1], $\alpha(t) = \beta(t)$. It is then easy to prove that $e = \beta(t^{-1})\xi$ is a normalized fixed vector. □

One can characterize, in terms of a regularity condition, the fact that S is an involutive subalgebra of $\mathcal{L}(H)$.

2.4 Regular multiplicative partial isometries and quantum groupoids

2.4.1 Lemma (cf. Lemma 2.6.2 of [Val1]). *S and $\hat{S}$ are sub-C^*-algebras of $\mathcal{L}(H)$ if and only if $\{(i \otimes \omega)(\sum I)/\omega$ is a linear form on $\mathcal{L}(H)\}$ is equal to $\alpha(N)'$; in this case one says that I is regular.*

Let us recall the definition of a quantum groupoid (or a weak Hopf C^*-algebra).

2.4.2 Definition (G. Böhm, K. Szlachányi, F. Nill, [BoSzNi]). A weak Hopf C^*-algebra is a collection $(A, \Gamma, \kappa, \varepsilon)$, where A is a finite-dimensional C^*-algebra, $\Gamma : A \mapsto A \otimes A$ is a generalized coproduct (which means that $(\Gamma \otimes i)\Gamma = (i \otimes \Gamma)\Gamma$), κ is an antipode on A, i.e., a linear application from A to A such that $(\kappa \circ *)^2 = i$ (where $*$ is the involution on A), $\kappa(xy) = \kappa(y)\kappa(x)$ for every x, y in A with $(\kappa \otimes \kappa)\Gamma = \varsigma\Gamma\kappa$ (where ς is the usual flip on $A \otimes A$).

We also suppose that $(m(\kappa \otimes i) \otimes i)(\Gamma \otimes i)\Gamma(x) = (1 \otimes x)\Gamma(1)$ (where m is the multiplication of tensors, i.e., $m(a \otimes b) = ab$), and that ε is a counit, i.e., a positive linear form on A such that $(\varepsilon \otimes i)\Gamma = (i \otimes \varepsilon)\Gamma = i$, and, for every x, y in A, $(\varepsilon \otimes \varepsilon)((x \otimes 1)\Gamma(1)(1 \otimes y)) = \varepsilon(xy)$.

2.4.3 Results (cf. [NV1], [NV3], [BoSzNi]). *If $(A, \Gamma, \kappa, \varepsilon)$ is a weak Hopf C^*-algebra, then the following holds.*

0) *The sets*

$$S_t = \{x \in A / \Gamma(x) = \Gamma(1)(x \otimes 1) = (x \otimes 1)\Gamma(1)\},$$

$$S_s = \{x \in A / \Gamma(x) = \Gamma(1)(1 \otimes x) = (1 \otimes x)\Gamma(1)\}$$

are sub-C^*-algebras of A; we call them target and source Cartan subalgebra of (A, Γ), respectively.

1) *The applications $\varepsilon_t = m(i \otimes \kappa)\Gamma$ and $\varepsilon_s = m(\kappa \otimes i)\Gamma)$ take values in S_t and S_s, respectively. We will call them target and source counit applications, respectively.*

2) *There is a unique projection p in A satisfying the relations $\kappa(p) = p, \varepsilon_t(p) = 1$, and, for every a in A, $ap = \varepsilon_t(a)p$. This element is called the Haar projection of $(A, \Gamma, \kappa, \varepsilon)$.*

3) *There exists a unique faithful positive linear form ϕ on A, satisfying the following three properties:*

   $\phi \circ \kappa = \phi, \quad (i \otimes \phi)(\Gamma(1)) = 1,$ *and for every x, y in A*

   $$(i \otimes \phi)((1 \otimes y)\Gamma(x)) = \kappa((i \otimes \phi)(\Gamma(y)(1 \otimes x))).$$

   This form is called the normalized Haar measure of $(A, \Gamma, \kappa, \varepsilon)$.

4) *The application $E^s_\phi = (\phi \otimes i)\Gamma$ (resp., $E^t_\phi = (i \otimes \phi)\Gamma$) is the conditional expectation with values in the source (resp., target) Cartan subalgebra, such that $\phi \circ E^s_\phi = \phi$ (resp., $\phi \circ E^t_\phi = \phi$). It is called a source (resp., target) Haar conditional expectation. If $g_s = E^s_\phi(p)^{\frac{1}{2}}$ and $g_t = E^t_\phi(p)^{\frac{1}{2}}$, then one has $g_t = \kappa(g_s)$. For every a in A, $\kappa^2(a) = g_t g_s^{-1} a g_t^{-1} g_s$, and the modular group σ^ϕ_{-i} is given by $\sigma^\phi_{-i}(a) = g_t g_s a g_t^{-1} g_s^{-1}$; this leads to a polar decomposition $\kappa = j \circ \mathrm{Ad}\, g_t$, where j is the involutive anti-homomorphism of A (coinvolution) defined by $j(y) = g_s \kappa(x) g_s^{-1}$ for any x in A.*

5) *One says that the collection $(A, \Gamma, \kappa, \varepsilon)$ is a weak Kac algebra if it is a weak Hopf C^*-algebra for which κ is involutive. This is equivalent to the fact that ϕ is a trace.*

It is shown in [Val1], that if I is regular, then it generates two quantum groupoids in duality:

2.4.4 Proposition. *If I is regular, then one can define $\Gamma, \kappa, \varepsilon$ on S and $\widehat{\Gamma}, \widehat{\kappa}, \widehat{\varepsilon}$ on $\widehat{S}$ by the formulas $\Gamma(s) = I(s \otimes 1)I^*$, $\widehat{\Gamma}(\hat{s}) = I^*(1 \otimes \hat{s})I$, $\kappa((\omega \otimes i)(I)) = (\omega \otimes i)(I^*)$, $\widehat{\kappa}((i \otimes \omega)(I)) = (i \otimes \omega)(I^*)$, $\varepsilon((\omega \otimes i)(I)) = \omega(1)$ and $\widehat{\varepsilon}((i \otimes \omega)(I)) = \omega(1)$. Then $(S, \Gamma, \kappa, \varepsilon)$ and $(\widehat{S}, \widehat{\Gamma}, \widehat{\kappa}, \widehat{\varepsilon})$ are quantum groupoids in duality defined by the formula $\langle(\omega \otimes i)(I), (i \otimes \omega')(I)\rangle = (\omega \otimes \omega')(I)$.*

The Haar projection p is equal to $(\hat{\phi} \otimes i)(I)$, and it is also the orthogonal projection on the space of cofixed vectors; the operator $(i \otimes \phi)(I)$ is the Haar projection of $(\hat{S}, \hat{\Gamma}, \hat{\kappa}, \hat{\epsilon})$ and is also the orthogonal projection onto the space of fixed vectors. There exist non-zero fixed and non-zero cofixed vectors. For every normalized fixed vector the restriction of $\omega_e(\omega_e(x) = (xe, e))$ is equal to ϕ if and only if $\omega_e \circ \alpha$ is equal to $\omega_e \circ \beta$. For every normalized cofixed vector $\hat{e}$ the restriction of $\omega_{\hat{e}}$ to S is equal to ε.

2.4.5 Remark. In general, weak Hopf C^*-algebras obtained in Proposition 2.4.4 are not weak Kac algebras, but the restrictions of antipodes κ and $\hat{\kappa}$ are involutive on the Cartan subalgebras, since one has $\kappa \circ \alpha = \beta$, $\kappa \circ \beta = \alpha$, $\hat{\kappa} \circ \alpha = \hat{\beta}$ and $\hat{\kappa} \circ \hat{\beta} = \alpha$.

2.4.6 Lemma. *There exists a unique n_1 in the center of N such that:*
$\alpha(n_1) = g_s = \hat{g}_t$ *(related to $\hat{S}$),*
$\beta(n_1) = g_t$,
$\hat{\beta}(n_1) = \hat{g}_s$ *(related to $\hat{S}$, which will be also denoted by $g_{\hat{\beta}}$).*

Proof. One has

$$\begin{aligned} g_s^2 &= (\phi \otimes i)(I((\hat{\phi} \otimes I)(I) \otimes 1)I) = (\hat{\phi} \otimes \phi \otimes i)(I_{23} I_{12} I_{23}^*) \\ &= (\hat{\phi} \otimes \phi \otimes i)(I_{12} I_{13}) = (\hat{\phi} \otimes i)((1 \otimes \phi)(I) \otimes 1)I) \\ &= (\hat{\phi} \otimes i)((\hat{p} \otimes 1)I) = (\hat{\phi} \otimes i)((\hat{p} \otimes 1)e_{\hat{\beta},\alpha}) \\ &= (\hat{\phi} \otimes i)((E^s_{\hat{\phi}}(\hat{p}) \otimes 1)e_{\hat{\beta},\alpha}) = (\hat{\phi} \otimes i)((1 \otimes \hat{\kappa}(E^s_{\hat{\phi}}(\hat{p})))e_{\hat{\beta},\alpha}) \\ &= \hat{\kappa}(E^s_{\hat{\phi}}(\hat{p}))(\hat{\phi} \otimes i)(\hat{e}_{\hat{\beta},\alpha}) = E^t_{\hat{\phi}}(\hat{\kappa}(\hat{p})) = E^t_{\hat{\phi}}(\hat{p}) = \hat{g}_t^2. \end{aligned}$$

This leads to the result. □

Now let I be a regular multiplicative partial isometry and ϕ the normalized Haar measure of S. Using the GNS representation (H_ϕ, Λ, π) of ϕ, by the same arguments as in [Val 1], Theorem 3.2.3, one can define a multiplicative partial isometry I_ϕ on $H_\phi \otimes H_\phi$ by the formula

$$I_\phi(\Lambda x \otimes \Lambda y) = (\Lambda \otimes \Lambda)(\Gamma(x)(1 \otimes y))$$

with the base $(N, \pi \circ \alpha, \pi \circ \beta, \mathrm{Ad}\, J_\phi \circ \pi \circ \alpha \circ \star)$, where $\star$ stands for the usual passage to adjoint. We will denote this base by $(N, \alpha_\phi, \beta_\phi, \hat{\beta}_\phi)$, and the application $\mathrm{Ad}\, J_\phi \circ \pi \circ \beta \circ \star$ by $\hat{\alpha}_\phi$. Then it is easy to see that $\hat{\alpha}_\phi(N) \subset S' \cap \hat{S}'$.

2.5 Binormalized fixed and cofixed vectors

In general, for a normalized fixed vector e of a multiplicative partial isometry we have $\omega_e \circ \alpha \neq \omega_e \circ \beta$. Let us give the following definition.

2.5.1 Definition. A fixed (resp., cofixed) vector e in H is said to be binormalized if and only if it is a normalized fixed (resp., cofixed) vector for which $\omega_e \circ \alpha = \omega_e \circ \beta$ (resp., $\omega_e \circ \alpha = \omega_e \circ \hat{\beta}$). If binormalized fixed (resp., cofixed) vectors exist, then I is said to be compact (resp., discrete).

2.5.2 Proposition. *If I is regular and e is a fixed binormalized vector for I, then $\hat{e} = pg_t^{-1}e$ is cofixed and binormalized, and such that $(i \otimes \omega_{e,\hat{e}})(e_{\alpha,\beta}) = g_s$ and $(\omega_{e,\hat{e}} \otimes i)(e_{\alpha,\beta}) = g_t$.*

Proof. The vector $\hat{e}$ is obviously cofixed, and for every n in N one has $\alpha(n)p = \beta(n)p$, and in particular $g_t p = g_s p$. Hence

$$\begin{aligned}
\omega_{\hat{e}}(\alpha(n)) &= \omega_e(g_t^{-1}p\alpha(n)pg_t^{-1}) = \phi(\alpha(n)pg_t^{-2}\sigma_{-i}^{\phi}(p)) \\
&= \phi(\alpha(n)pg_t^{-2}g_t g_s p g_t^{-1}g_s^{-1}) = \phi(\alpha(n)pg_s^{-2}) = \phi(\alpha(n)E_\phi^s(p)g_s^{-2}) \\
&= \phi(\alpha(n)) = \mathrm{tr}(\alpha(n_0)^{-1}\alpha(n)),
\end{aligned}$$

$$\begin{aligned}
\omega_{\hat{e}}(\hat{\beta}(n)) &= \omega_e(g_t^{-1}p\hat{\beta}(n)pg_t^{-1}) = \omega_e(g_t^{-1}pg_t^{-1}\hat{\beta}(n)) = \omega_e(g_t^{-1}pg_t^{-1}\alpha(n)) \\
&= \phi(g_s^{-1}pg_s^{-1}\alpha(n)) = \phi(g_s^{-1}E_\phi^s(p)g_s^{-1}\alpha(n)) = \phi(\alpha(n)) \\
&= \mathrm{tr}(\alpha(n_0)^{-1}\alpha(n)),
\end{aligned}$$

$$\begin{aligned}
(i \otimes \omega_{e,\hat{e}})(e_{\alpha,\beta}) &= (i \otimes \omega_{e,e})((1 \otimes g_t^{-1}pe_{\alpha,\beta}) = (i \otimes \phi)((1 \otimes g_t^{-1}pe_{\alpha,\beta}) \\
&= (i \otimes \phi)((1 \otimes g_t^{-1}E_\phi^t(p)e_{\alpha,\beta}) = (i \otimes \phi)((1 \otimes g_t)e_{\alpha,\beta}) \\
&= g_s(i \otimes \phi)(e_{\alpha,\beta}) = g_s.
\end{aligned}$$

The remaining equality can be proved by a similar calculation. □

2.6 The canonical decomposition of a regular multiplicative partial isometry

We will prove now that any regular multiplicative partial isometry can be decomposed in a simple way.

Let I be a multiplicative partial isometry on the Hilbert space H with the base $(N, \alpha, \beta, \hat{\beta})$ and let K be a finite dimensional Hilbert space; then the ampliation $I^K = I_{13}$ on $H \otimes K \otimes H \otimes K$ is clearly a multiplicative partial isometry on the Hilbert space $H \otimes K$ with the base $(N, \alpha \otimes 1_K, \beta \otimes 1_K, \hat{\beta} \otimes 1_K)$.

If U is a unitary from H on another Hilbert space H', then $(U \otimes U)I(U \otimes U)$ is a multiplicative partial isometry on the Hilbert space H' with the base $(N, \operatorname{Ad} U \circ \alpha, \operatorname{Ad} U \circ \beta, \operatorname{Ad} U \circ \hat{\beta})$; we say that these two objects are isomorphic.

We say that an orthogonal projection p on H reduces I if $(p \otimes 1)I = I(p \otimes 1)$, $(1 \otimes p)I = I(1 \otimes p)$ and if α, β, $\hat{\beta}$ restricted to pH are faithful. It is easy to see that $(p \otimes p)I(p \otimes p)$ is a multiplicative partial isometry on pH with the base $(N, \operatorname{Ad} p \circ \alpha, \operatorname{Ad} p \circ \beta, \operatorname{Ad} p \circ \hat{\beta})$. We call it the reduction of I by p.

Now let I be a regular multiplicative partial isometry together with the constructions of the previous chapter. Let f_j be an $\alpha(N)$-base of H in the sense of Definition 2.1.3

2.6.1 Lemma. *For any operator T in $\mathcal{L}(H \otimes H)$ such that its image is contained in $e_{\alpha,\beta}(H \otimes H)$, any $\alpha(N)$-base f_j of H and for all ξ, η in H, the following equality is true:*

$$T(\xi \otimes \eta) = e_{\alpha,\beta}\Big(\sum_j f_j \otimes n_\gamma(\omega_{\xi,f_j} \otimes i)(T)\eta\Big).$$

Proof. Due to Lemmas 2.1.4 and 2.1.5, for every natural j there exists a vector η_j in H such that $T(\xi \otimes \eta) = \sum_j e_{\alpha,\beta}(f_j \otimes \beta(e^\gamma_{i,1})\eta_j)$ (recall that β is a representation of N^o, so we have to transpose the matrix unit of N). Then, using the fact that $m_\gamma^{-1} n_0 e^\gamma_{i',1} = n_\gamma^{-1} e^\gamma_{i',1}$, we have for any $j' = (\gamma', i')$ and θ in H that

$$\begin{aligned}
\langle(\omega_{\xi,f_{j'}} \otimes i)(T)\eta, \theta\rangle &= \langle T(\xi \otimes \eta), e_{\alpha,\beta}(f_{j'} \otimes \theta)\rangle \\
&= \Big\langle \sum_j e_{\alpha,\beta}(f_j \otimes \beta(e^\gamma_{i,1})\eta_j), e_{\alpha,\beta}(f_{j'} \otimes \theta)\Big\rangle \\
&= \sum_j \Big\langle \beta\Big(n_0 \frac{d(\omega_{f_j,f_{j'}} \circ \alpha)}{d(\mathrm{tr}_{H_1} \circ \alpha)}\Big)\beta(e^\gamma_{i,1})\eta_j, \theta\Big\rangle \\
&= \langle \beta(n_0 m_\gamma^{-1} e^{\gamma'}_{1,1})\beta(e^{\gamma'}_{i',1})\eta_{j'}, \theta\rangle \\
&= \langle n_\gamma^{-1}\beta(e^{\gamma'}_{i',1})\eta_{j'}, \theta\rangle.
\end{aligned}$$

□

2.6.2 Lemma. *For any y in $\hat{S}'$, the operator $I(1 \otimes y)I^*$ belongs to $\mathcal{L}(H) \otimes \hat{S}'$.*

Proof. Let x be any element in $\hat{S}$. Then

$$\begin{aligned}
I(1 \otimes y)I^*(1 \otimes x) &= I(1 \otimes y)I^* e_{\alpha,\beta}(1 \otimes x) = I(1 \otimes y)I^*(1 \otimes x)e_{\alpha,\beta} \\
&= I(1 \otimes y)I^*(1 \otimes x)II^* = I(1 \otimes y)\hat{\Gamma}(x)I^* = I\hat{\Gamma}(x)(1 \otimes y)I^* \\
&= II^*(1 \otimes x)I(1 \otimes y)I^* = e_{\alpha,\beta}(1 \otimes x)I(1 \otimes y)I^* \\
&= (1 \otimes x)e_{\alpha,\beta}I(1 \otimes y)I^* = (1 \otimes x)I(1 \otimes y)I^*.
\end{aligned}$$

This equality proves the lemma. □

2.6.3 Lemma. *The von Neumann algebra generated by $\hat{S}'$ and S is equal to the vector space generated by the products xy for any x in S and y in $\hat{S}'$.*

Proof. To prove the lemma, it suffices to be able to write any product yx, for x in S and y in $\hat{S}'$, as a sum of elements of the form $x'y'$ for some x' in S and y' in $\hat{S}'$. For any ξ, ξ', η, η' in H and any y in $\hat{S}'$, due to Lemma 2.6.1 one has

$$\begin{aligned}(y(\omega_{\xi,\xi'} \otimes i)(I)^*\eta, \eta') &= (I^*(\xi' \otimes \eta), \xi \otimes y^*\eta') = (I^*(\xi' \otimes \eta), e_{\hat{\beta},\alpha}(\xi \otimes y^*\eta')) \\ &= ((1 \otimes y)I^*(\xi' \otimes \eta), e_{\hat{\beta},\alpha}(\xi \otimes \eta')) \\ &= (I(1 \otimes y)I^*(\xi' \otimes \eta), I(\xi \otimes \eta')) \\ &= (I(1 \otimes y)I^*(\xi' \otimes \eta), e_{\alpha,\beta}\Big(\sum_j f_j \otimes n_\gamma(\omega_{\xi,f_j} \otimes i)(I)\eta'\Big)) \\ &= \sum_j (I(1 \otimes y)I^*(\xi' \otimes \eta), f_j \otimes n_\gamma(\omega_{\xi,f_j} \otimes i)(I)\eta') \\ &= \sum_j ((\omega_{\xi',f_j} \otimes i)(I(1 \otimes y)I^*)\eta, n_\gamma(\omega_{\xi,f_j} \otimes i)(I)\eta') \\ &= \sum_j ((\omega_{\xi,f_j} \otimes i)(I)^* n_\gamma(\omega_{\xi',f_j} \otimes i)(I(1 \otimes y)I^*)\eta), \eta').\end{aligned}$$

This gives the equality

$$y(\omega_{\xi,\xi'} \otimes i)(I)^* = \sum_j (\omega_{\xi,f_j} \otimes i)(I)^* n_\gamma(\omega_{\xi',f_j} \otimes i)(I(1 \otimes y)I^*).$$

The result follows from the regularity of I and from Lemma 2.6.2. □

2.6.4 Proposition. *The set $S\mathcal{F} = \{xe/x \in S, e \in \mathcal{F}\}$ generates the Hilbert space H.*

Proof. Since I is regular, there exists a normalized fixed vector e. So e is separating for $\hat{\beta}(N)$ and cyclic for $\beta(N)'$, which, by Corollary 3.1.6 of [Val1], is equal to the algebra generated by S and $\hat{S}'$ and equal to the vector space generated by xy for x in S and y in $\hat{S}'$ by Lemma 2.6.3. But $\mathcal{F}$ is invariant by $\hat{S}'$, and the result follows. □

2.6.5 Theorem. *There is a unique isomorphism V from $e_{\hat{\alpha}_\phi,\beta}(H_\phi \otimes \mathcal{F})$ onto H such that*

$$V(e_{\hat{\alpha}_\phi,\beta}(\Lambda x \otimes e)) = xe$$

for every x in S and any fixed vector e. The projection $e_{\hat{\alpha}_\phi,\beta}$ reduces the ampliation $I_\phi^{\mathcal{F}}$, and U is an isomorphism of the reduction of $I_\phi^{\mathcal{F}}$ by $e_{\hat{\alpha}_\phi,\beta}$ and I.

Proof. Let x, y be in S and e, f be in $\mathcal{F}$, then

$$
\begin{aligned}
\langle e_{\hat{\alpha}_\phi,\beta}(\Lambda x\otimes e), e_{\hat{\alpha}_\phi,\beta}(\Lambda y\otimes f)\rangle &= \Big\langle \hat{\alpha}_\phi\Big(\frac{d(\omega_{e,f}\circ\beta)}{d\tau}n_0^{-1}\Big)\Lambda x, \Lambda y\Big\rangle\\
&= \Big\langle \hat{\alpha}_\phi\Big(\frac{d(\omega_{e,f}\circ\beta)}{d\phi\circ\beta}\Big)\Lambda x, \Lambda y\Big\rangle\\
&= \Big\langle \Lambda x\beta\Big(\frac{d(\omega_{e,f}\circ\beta)}{d\phi\circ\beta}\Big), \Lambda y\Big\rangle\\
&= \phi\Big(y^* x\beta\Big(\frac{d(\omega_{e,f}\circ\beta)}{d\phi\circ\beta}\Big)\Big)\\
&= \phi\Big(E_\phi^\beta(y^*x)\beta\Big(\frac{d(\omega_{e,f}\circ\beta)}{d\phi\circ\beta}\Big)\Big)\\
&= (\phi\circ\beta)\Big(\frac{d(\omega_{e,f}\circ\beta)}{d\phi\circ\beta}\beta^{-1}(E_\phi^\beta(y^*x))\Big)\\
&= \omega_{e,f}(E_\phi^\beta(y^*x)) = (\omega_{e,f}\otimes\phi)(\Gamma(y^*x))\\
&= \phi((\omega_{e,f}\otimes i)(\Gamma(y^*x))\\
&= \phi((\omega_{e,f}\otimes i)(e_{\hat{\beta},\alpha}((y^*x\otimes 1))\\
&= \omega_{e,f}(i\otimes\phi)(e_{\hat{\beta},\alpha}((y^*x\otimes 1))\\
&= \langle xe, yf\rangle.
\end{aligned}
$$

This proves the existence and uniqueness of the application V and the fact that it is an isometry; by Lemma 2.6.3, this is a unitary. Due to the inclusion of $\hat{\alpha}_\phi(N)$ in $S'\cap\hat{S}'$ and Lemma 2.1.2 2), it is easy to see that $e_{\hat{\alpha}_\phi,\beta}$ reduces $I_\phi^{\mathcal{F}}$.

Let x, x' be in S, and e, e' be in $\mathcal{F}$, then the following equalities, with Sweedler notations, are true:

$$
\begin{aligned}
&I(V\otimes V)(e_{\hat{\alpha}_\phi,\beta}(\Lambda x\otimes e)\otimes e_{\hat{\alpha}_\phi,\beta}(\Lambda x'\otimes e'))\\
&\quad= I(xe\otimes x'e') = Ie_{\hat{\beta},\alpha}(xe\otimes x'e')\\
&\quad= I(x\otimes 1)e_{\hat{\beta},\alpha}(e\otimes x'e') = I(x\otimes 1)I^*(e\otimes x'e')\\
&\quad= \Gamma(x)_1 e\otimes\Gamma(x)_2 x'e'\\
&\quad= (V\otimes V)(e_{\hat{\alpha}_\phi,\beta}\otimes e_{\hat{\alpha}_\phi,\beta})\pi_{\phi\otimes\phi}(\Gamma(x)(1\otimes x'))_{13}(\Lambda 1\otimes e\otimes\Lambda 1\otimes e')\\
&\quad= (V\otimes V)(e_{\hat{\alpha}_\phi,\beta}\otimes e_{\hat{\alpha}_\phi,\beta})I_\phi^{\mathcal{F}}(\Lambda x\otimes e\otimes\Lambda x'\otimes e').
\end{aligned}
$$

This completes the proof. □

3 Irreducible multiplicative partial isometries

Notations. In this section we consider a regular multiplicative partial isometry I, so by Theorem 2.6.5 one can suppose that there is a binormalized fixed vector e, the Haar measure for S is ω_e, e is cyclic and separating for S; there also exists a binormalized cofixed vector $\hat{e} = g_{\hat{\beta}}^{-1} pe$ (where p is the orthogonal projection on the space of fixed vectors) which is separating for $\hat{S}$ and such that $\hat{\phi} = \omega_{\hat{e}}$. In such a situation we will say that I is irreducible.

Let T be the linear bounded operator acting on H by $se \mapsto \kappa(s)e$, and let $T = U(T^*T)^{\frac{1}{2}}$ be its polar decomposition.

3.1 The quantum groupoid structures of the commutants

In this subsection we construct two *mpi's* related to the commutants of S and $\widehat{S}$ in $\mathcal{L}(H)$.

3.1.1 Proposition. *The operator U is an orthogonal symmetry. It satisfies* $\operatorname{Ad} U \circ \beta = \hat{\beta}, USU \subset S'$.

Proof. As the modular group of the Haar measure is $\sigma^{\phi}_{-i} = \operatorname{Ad} g_s g_t$ and $\phi \circ \kappa = \phi$, for every x, y in S one has

$$\begin{aligned}(T(xe), ye) &= \phi(y^*\kappa(x)) = \phi(x\kappa^{-1}(y^*)) = \phi(\kappa^{-1}(y^*)\sigma^{\phi}_{-i}(x)) \\ &= \phi(g_s^{-1}g_t^{-1}\kappa^{-1}(y^*)g_sg_tx)) = \phi((g_sg_t\kappa^{-1}(y^*)^*g_s^{-1}g_t^{-1})^*x)) \\ &= \phi((g_sg_t\kappa(y)g_s^{-1}g_t^{-1})^*x)).\end{aligned}$$

Therefore $T^*(ye) = g_sg_t\kappa(y)g_s^{-1}g_t^{-1}e$. As $\kappa^2(y) = g_tg_s^{-1}yg_t^{-1}g_s$, it holds that $T^*T(ye) = g_t^2yg_t^{-2}e$, hence

$$Uye = T(T^*T)^{-\frac{1}{2}}(ye) = T(g_t^{-1}yg_t)e = g_s\kappa(y)g_s^{-1}e, \tag{3.1}$$

and one concludes that

$$U^2(ye) = U(g_s\kappa(y)g_s^{-1}e) = g_s(g_t^{-1}\kappa^2(y)g_t)g_s^{-1}e = ye.$$

For any s, x in S one has

$$\begin{aligned}(\operatorname{Ad} U \circ s)xye &= Usg_s\kappa(xy)g_s^{-1}e = g_s\kappa(sg_s\kappa(xy)g_s^{-1})g_s^{-1}e \qquad (3.2)\\ &= g_sg_t^{-1}\kappa^2(xy)g_t\kappa(s)g_s^{-1}e = xyg_s\kappa(s)g_s^{-1}e. \qquad (3.3)\end{aligned}$$

In particular, for $x = 1$, one deduces that

$$(\operatorname{Ad} U \circ s)ye = yg_s\kappa(s)g_s^{-1}e, \tag{3.4}$$

so we get $(\operatorname{Ad} U \circ s)xye = x(\operatorname{Ad} U \circ s)ye$, hence $USU \subset S'$.

Applying the last numbered equality to $\beta(n)$ and using the fact that $\alpha(n)e = \hat{\beta}(n)e$, one gets

$$(\operatorname{Ad} U \circ \beta(n))ye = y\alpha(n)e = y\hat{\beta}(n)e = \hat{\beta}(n)ye.$$

Therefore $\operatorname{Ad} U \circ \beta = \hat{\beta}$. □

3.1.2 Definition and Corollary. *Let us define $\hat{\alpha}$ as $\operatorname{Ad} U \circ \alpha$. This is obviously a faithful non-degenerate representation of N; one has $Ue = e$ and $U\hat{e} = \hat{e}$, e and $\hat{e}$ satisfy the fourth normalization condition:*

$$\frac{d(\omega_e \circ \hat{\alpha})}{d\tau} = \frac{d(\omega_{\hat{e}} \circ \hat{\alpha})}{d\tau} = n_0^{-1}.$$

Proof. The formula $Ue = e$ is a particular case of the first equality in the proof of Proposition 3.1.1. Using Proposition 2.5.2 and the facts that $p = \kappa(p)$ is in S and $pg_s^{-1} = pg_t^{-1}$ implies $U\hat{e} = Upg_t^{-1}e = g_s g_s^{-1} pg_s^{-1}e = pg_s^{-1}e = \hat{e}$. □

3.1.3 Lemma. *For every n in N, $\hat{\alpha}(n)e = \beta(n)e$ and $\hat{\alpha}(n)\hat{e} = \hat{\beta}(n)\hat{e}$; $\hat{\alpha}$ commutes with α, β, $\hat{\beta}$; one has $\hat{\alpha}(N) \subset \hat{S}' \cap S'$ and I belongs to $\hat{\alpha}(N)' \otimes \hat{\alpha}(N)'$.*

Proof. From Corollary 3.1.2, for every n in N one has $\hat{\alpha}(n)e = U\alpha(n)e = \beta(n)e$. On the other hand,

$$\begin{aligned}\hat{\alpha}(n)\hat{e} &= U\alpha(n)\hat{e} = U\beta(n)\hat{e} = U\beta(n)pg_t^{-1}e \\ &= g_s^{-1}p\alpha(n)e = g_s^{-1}p\hat{\beta}(n)e \\ &= \hat{\beta}(n)g_s^{-1}pe = \hat{\beta}(n)\hat{e}.\end{aligned}$$

By Proposition 3.1.1, $\hat{\alpha}$ commutes with α and β; from the proof of Proposition 3.1.1 one gets, for every n in N and s in S, that the equality $\hat{\alpha}(n)se = s\beta(n)e$ is true. Hence, for every η in H,

$$\begin{aligned}I(\hat{\alpha}(n) \otimes 1)(se \otimes \eta) &= I(s\beta(n)e \otimes \eta) = \Gamma(s\beta(n))(e \otimes \eta) \\ &= \Gamma(s)(\beta(n) \otimes 1)(e \otimes \eta) = (\hat{\alpha}(n) \otimes 1)\Gamma(s)(e \otimes \eta) \\ &= (\hat{\alpha}(n) \otimes 1)I(se \otimes \eta),\end{aligned}$$

and one deduces that $I(\hat{\alpha}(n) \otimes 1) = (\hat{\alpha}(n) \otimes 1)I$ and $\hat{\beta}(n)se = s\alpha(n)e$. Now the remaining statements of the lemma follow. □

Let us define $\hat{I} = \sum(U \otimes 1)I(U \otimes 1)\sum$ and $\tilde{I} = \sum(1 \otimes U)I(1 \otimes U)\sum$ where $\sum$ is the flip for $H \otimes H$ (so $\hat{I} = (U \otimes U)\tilde{I}(U \otimes U)$); these are partial isometries. With the notations of Lemma 3.1.3, one can see that the initial (resp., final) support of $\tilde{I}$ is $e_{\hat{\alpha},\hat{\beta}}$ (resp., $e_{\hat{\beta},\alpha}$). The initial (resp., final) support of $\hat{I}$ is $e_{\alpha,\beta}$ (resp., $e_{\beta,\hat{\alpha}}$). Obviously, e (resp., $\hat{e}$) is a binormalized cofixed (resp., fixed) vector for both $\hat{I}$ and $\tilde{I}$.

3.1.4 Proposition. *$\tilde{I}$ and $\hat{I}$ are regular multiplicative partial isometries with the bases $(N^o, \hat{\beta}, \alpha, \hat{\alpha})$ and $(N^o, \beta, \hat{\alpha}, \alpha)$ respectively. $\tilde{I}$ belongs to $S' \otimes \hat{S}$ and $\hat{I}$ to $S \otimes \hat{S}'$. For every s in S and $\hat{s}$ in $\hat{S}$ we have $\Gamma(s) = \hat{I}^*(1 \otimes s)\hat{I}$, $\hat{\Gamma}(\hat{s}) = \tilde{I}(\hat{s} \otimes 1)\tilde{I}^*$.*

Proof. By Proposition 3.1.1, one has $USU \subset S'$, as I is in $\hat{S} \otimes S$, then $\tilde{I} = \sum(1 \otimes U)I(1 \otimes U)\sum$ is in $S' \otimes \hat{S}$. It is easy to compute, in Sweedler notations, that for every s, s' in S we have $\hat{I}(se \otimes s'e) = g_s\kappa(s'_1)g_s^{-1}se \otimes g_t s'_2 g_t^{-1}e$ and $\tilde{I}(se \otimes s'e) = sg_s\kappa(s'_2)g_s^{-1}e \otimes s'_1 e$. So we can deduce that, for every s, s' in S,

$$\begin{aligned}\hat{I}(se \otimes s'e) &= g_s\kappa(s'_1)g_s^{-1}se \otimes g_t s'_2 g_t^{-1}e = (\kappa \otimes i)(g_t^{-1}(s'_1)g_t \otimes g_t s'_2 g_t^{-1})(se \otimes e)\\ &= (\kappa \otimes i)((g_t^{-1} \otimes g_t)\Gamma(s')(g_t \otimes g_t^{-1})(se \otimes e)\\ &= (\kappa \otimes i)((g_s g_t^{-1} \otimes 1)\Gamma(s')(g_t g_s^{-1} \otimes 1))(se \otimes e)\\ &= (\kappa^{-1} \otimes i)(\Gamma(s'))(se \otimes e).\end{aligned}$$

Therefore, for every x, y in S,

$$\begin{aligned}(\hat{I}(se \otimes s'e), xe \otimes ye) &= (\phi \otimes \phi)((x^* \otimes y^*)(\kappa^{-1} \otimes i)(\Gamma(s'))(1 \otimes s))\\ &= \phi(x^*\kappa^{-1}((i \otimes \phi)((1 \otimes y^*)\Gamma(s'))s)\\ &= \phi(x^*\kappa^{-1}((i \otimes \phi)((1 \otimes y^*)\Gamma(s'))s)\\ &= \phi(x^*((i \otimes \phi)(\Gamma(y^*)(1 \otimes s')))s)\\ &= (\phi \otimes \phi)((x^* \otimes 1)\Gamma(y^*(s \otimes s'))\\ &= (se \otimes s'e, \Gamma(y)(x \otimes 1)(e \otimes e)).\end{aligned}$$

One deduces that $\hat{I}^*(xe \otimes ye) = \Gamma(y)(xe \otimes e)$ for any x, y in S. From this one easily proves that $\sum \hat{I}^* \sum$ (and so $\hat{I}$) satisfies the pentagonal equation. As $\tilde{I} = (U \otimes U)\hat{I}(U \otimes U)$, the same is true for $\tilde{I}$, and easy verification shows that these are multiplicative partial isometries with the bases $(N^o, \hat{\beta}, \alpha, \hat{\alpha})$ and $(N^o, \beta, \hat{\alpha}, \alpha)$ respectively. Since $(i \otimes \omega_{se,s'e})(\hat{I}) = (i \otimes \phi)(\Gamma(s'^*)(1 \otimes s))$, then $\hat{I}$ and $\tilde{I}$ are regular by [Val1], Corollary 3.2.4.

$\hat{I}$ obviously belongs to $S \otimes \mathcal{L}(H)$ and, for any x, y in S and η in H, we get

$$\begin{aligned}\hat{I}_{12}I_{23}(xe \otimes ye \otimes \eta) &= \hat{I}_{12}(xe \otimes y_1 e \otimes y_2\eta)\\ &= (\kappa^{-1} \otimes i \otimes i)(\Gamma \otimes i)\Gamma(y)(xe \otimes e \otimes \eta)\\ &= (\kappa^{-1} \otimes i \otimes i)(i \otimes \Gamma)\Gamma(y)(xe \otimes e \otimes \eta)\\ &= \kappa^{-1}(y_1)xe \otimes \Gamma(y_2)(e \otimes \eta)\\ &= I_{23}(\kappa^{-1}(y_1)xe \otimes y_2 e \otimes \eta) = I_{23}\hat{I}_{12}(xe \otimes ye \otimes \eta)\end{aligned}$$

which proves that $\hat{I}$ belongs to $\mathcal{L}(H) \otimes \hat{S}'$, so $\hat{I}$ belongs to $S \otimes \hat{S}'$. As in Proposition 6.1 3) of [BS], it is easy to compute: $\tilde{I}_{12}I_{13} = I_{13}I_{23}\tilde{I}_{12}$, this leads to $\tilde{I}_{12}I_{13}\tilde{I}_{12}^* = I_{13}I_{23}e_{\hat{\beta},\alpha} = I_{13}I_{23}$. Applying, for any linear form ω on $\mathcal{L}(H)$, the slice map $(i \otimes i \otimes \omega)$ to this last equality, one proves that $\hat{\Gamma}(\hat{s}) = \tilde{I}(\hat{s} \otimes 1)\tilde{I}^*$ for every $\hat{s}$ in $\hat{S}$. Following the same argument as in Proposition 6.1 2) of [BS], one has $\hat{I}_{23}I_{12}I_{13} = I_{12}\hat{I}_{23}$, so $(1 \otimes e_{\alpha,\beta})I_{12}I_{13} = \hat{I}^*_{23}I_{13}\hat{I}_{23} = I_{12}I_{13}$. Applying, for any linear form ω on $\mathcal{L}(H)$, the slice map $(\omega \otimes i \otimes i)$ to this latter equality, one proves that $\Gamma(s) = \hat{I}^*(1 \otimes \hat{s})\hat{I}$ for every s in S. □

3.1.5 Proposition. *The vector space generated by the products $\hat{s}s$ (resp., $s\hat{s}'$), for all s in S and $\hat{s}$ in $\hat{S}$ (resp., s in S and $\hat{s}'$ in $\hat{S}'$), is equal to $\hat{\alpha}(N)'$ (resp., $\hat{\beta}(N)'$).*

Proof. Using the same arguments as in Proposition 6.3. of [BS], we can prove that the vector space generated by $\{(i \otimes \omega)(\hat{I}I(s \otimes 1)/\omega \in \mathcal{L}(H)_*, s \in S\}$ is equal to $\hat{\alpha}(N)' = U\alpha(N)'U$. As the final support of I is equal to $e_{\alpha,\beta}$, then $\hat{\alpha}(N)'$ is also equal to the vector space generated by $\{(i \otimes \omega)(\hat{I}e_{\alpha,\beta}(s \otimes 1)/\omega \in \mathcal{L}(H)_*, s \in S\}$. But $e_{\alpha,\beta}$ is the initial support of $\hat{I}$, so $\hat{\alpha}(N)'$ is also generated by the products $(i \otimes \omega)(\hat{I})s = ((U\omega U) \otimes i)(I)s$. This leads to the first assertion; the second can be proved similarly. □

3.1.6 Corollary. $S' \cap \hat{S}' = \hat{\alpha}(N)$.

3.2 The Fourier transform

In this subsection we construct an analogue of the Fourier transform in the context of quantum groupoids.

3.2.1 Lemma. *One has* $(1 \otimes U)\sum \hat{I}I\tilde{I} = e_{\hat{\alpha},\hat{\beta}}$ *and* $\sum \hat{I}I\tilde{I}(1 \otimes U) = e_{\hat{\alpha},\beta}$.

Proof. Using Lemma 3.1.5, by the same arguments as in [BS], Proposition 6.8 b), one has $(1 \otimes U)\sum \hat{V}V\tilde{V} \in \hat{\alpha}(N) \otimes \hat{\beta}(N)$. Since for every n in N we have $\hat{\alpha}(n)e = \beta(n)e$ and $\hat{\beta}(n)e = \alpha(n)e$, $e \otimes e$ separates $\hat{\alpha}(N) \otimes \hat{\beta}(N)$. Hence, as e (resp., $\hat{e}$) is a fixed (resp., cofixed) vector for $\hat{I}$ and $\tilde{I}$, we have: $(1 \otimes U)\big(\sum \hat{I}I\tilde{I}\big)(e \otimes e) = e_{\hat{\alpha},\hat{\beta}}(e \otimes e)$, so $(1 \otimes U)\sum \hat{I}I\tilde{I} = e_{\hat{\alpha},\hat{\beta}}$; the second formula is a consequence of the first one. □

3.2.2 Lemma. *For any ξ in H we have*

$$\xi = (\omega_{\xi,\hat{e}} \otimes i)(I)g_t^{-1}e = (i \otimes \omega_{e,\xi})(I)^* g_{\hat{\beta}}^{-1}\hat{e}.$$

For any x in S and y in $\hat{S}$ respectively, we have $x = (\omega_{xe,\hat{e}} \otimes i)(I)g_t^{-1}$ and $y = (i \otimes \omega_{e,y\hat{e}})(I)^* g_{\hat{\beta}}^{-1} = \hat{\kappa}(g_s^{-1}(i \otimes \omega_{y\hat{e},e})(I))$.

Proof. Let ξ be any element of H. Due to Lemma 3.2.1 and Proposition 2.5.2 one has

$$\begin{aligned}(\omega_{\xi,\hat{e}} \otimes i)(I)e &= (i \otimes \omega_{e,\hat{e}})\Big(\sum I\Big)\xi = (i \otimes \omega_{e,\hat{e}})\Big(\sum \hat{I}I\tilde{I}(1 \otimes U)\Big)\xi \\ &= (i \otimes \omega_{e,\hat{e}})(e_{\hat{\alpha},\beta})\xi = Ug_sU\xi.\end{aligned}$$

This gives $\xi = Ug_s^{-1}U(\omega_{\xi,\hat{e}}\otimes i)(I)e$. As for any m in S one has $Ug_s^{-1}Ume = mg_t^{-1}e$, this leads to $\xi = (\omega_{\xi,\hat{e}} \otimes i)(I)g_t^{-1}e$. Applying the latter equality to $\xi = xe$ for any $x \in S$, and as e separates S, one deduces that $x = (\omega_{xe,\hat{e}} \otimes i)(I)g_t^{-1}$. On the other hand,

$$(i \otimes \omega_{e,\xi})(I)^*\hat{e} = (i \otimes \omega_{e,\hat{e}})\Big(\sum I\Big)^*\xi = Ug_sU\xi.$$

Applying the last equality to $\xi = y\hat{e}$ for any y in $\hat{S}$ and using Corollary 3.1.6 we have

$$\begin{aligned}\xi &= Ug_s^{-1}U(i \otimes \omega_{e,\xi})(I)^*\hat{e} = (i \otimes \omega_{e,y\hat{e}})(I)^*Ug_s^{-1}U\hat{e} =\\ &= (i \otimes \omega_{e,y\hat{e}})(I)^*Ug_s^{-1}pg_t^{-1}e = (i \otimes \omega_{e,y\hat{e}})(I)^*g_sg_s^{-1}pg_t^{-1}g_se\\ &= (i \otimes \omega_{e,y\hat{e}})(I)^*pg_t^{-1}g_{\hat{\beta}}^{-1}e = (i \otimes \omega_{e,y\hat{e}})(I)^*g_{\hat{\beta}}^{-1}pg_t^{-1}e\\ &= (i \otimes \omega_{e,y\hat{e}})(I)^*g_{\hat{\beta}}^{-1}\hat{e}.\end{aligned}$$

Since $\hat{e}$ separates $\hat{S}$, also the following identity is true: $y = (i \otimes \omega_{e,y\hat{e}})(I)^*g_{\hat{\beta}^{-1}} = \hat{\kappa}(g_s^{-1}(i \otimes \omega_{y\hat{e},e})(I))$.

3.2.3 Corollary. *The vector e (resp., $\hat{e}$) is cyclic for S (resp., $\hat{S}$). The space $\mathcal{F}$ of fixed vectors is $\hat{\beta}(N)e$, and the space $\hat{\mathcal{F}}$ of cofixed vectors is $\hat{\beta}(N)\hat{e}$. S and $\hat{S}$ are in standard position in H.*

Proof. The cyclicity of the two vectors follows from Lemma 3.2.2 and from the inclusion $\beta(N)e \subset \mathcal{F}$. If f is any element of $\mathcal{F}$, there exists x in S such that $f = se$; for any s' in S' and η in H one has

$$\begin{aligned}\Gamma(x)(s'e \otimes \eta) &= (s' \otimes 1)I(x \otimes 1)I^*(e \otimes \eta) = (s' \otimes 1)I(xe \otimes \eta)\\ &= (s' \otimes 1)e_{\alpha,\beta}(xe \otimes \eta) = e_{\alpha,\beta}(x \otimes 1)(s'e \otimes \eta).\end{aligned}$$

As e is cyclic for S', one has $\Gamma(x) = e_{\alpha,\beta}(x \otimes 1)$. Then x belongs to $\beta(N)$, so $\mathcal{F} = \beta(N)e$. A similar argument is valid for $\hat{e}$. The end of the proof is obvious. □

3.2.4 Notation. As S and $\hat{S}$ are in standard position in H, one can use the Tomita theory with its usual notations, so it can be easily seen that, for every $x \in S$ and $y \in \hat{S}$, one has the following: $Jxe = g_s^{\frac{1}{2}}g_t^{\frac{1}{2}}x^*g_t^{-\frac{1}{2}}g_s^{-\frac{1}{2}}e$ and $\hat{J}y\hat{e} = g_{\hat{\beta}}^{\frac{1}{2}}g_s^{\frac{1}{2}}y^*g_{\hat{\beta}}^{-\frac{1}{2}}g_s^{-\frac{1}{2}}\hat{e}$.

3.2.5 Lemma. *For any $\hat{s}$ in $\hat{S}$, we have $U\hat{s}\hat{e} = g_{\hat{\beta}}\hat{\kappa}(\hat{s})g_{\hat{\beta}}^{-1}\hat{e}$.*

Proof. Let y in $\hat{S}$, then by Lemma 3.2.2

$$\begin{aligned}\hat{\kappa}(y)\hat{e} &= \hat{\kappa}^2((i \otimes \omega_{y\hat{e},e})(I)g_s^{-1})\hat{e} = g_sg_{\hat{\beta}}^{-1}g_s^{-1}(i \otimes \omega_{y\hat{e},e})(I)g_{\hat{\beta}}g_s^{-1}\hat{e}\\ &= g_{\hat{\beta}}^{-1}(i \otimes \omega_{y\hat{e},e})(I)g_{\hat{\beta}}g_t^{-1}\hat{e} = g_{\hat{\beta}}^{-1}g_t^{-1}(i \otimes \omega_{y\hat{e},e})(I(g_{\hat{\beta}} \otimes 1))\hat{e}\end{aligned}$$

$$= g_{\hat\beta}^{-1} g_t^{-1}(i \otimes \omega_{y\hat e,e})(I(1 \otimes g_s))\hat e = g_{\hat\beta}^{-1} g_t^{-1}(i \otimes \omega_{g_s y\hat e,e})(I)\hat e$$
$$= g_{\hat\beta}^{-1} g_t^{-1}(\omega_{\hat e,e} \otimes i)\Big(\sum I\Big) g_s y\hat e = g_{\hat\beta}^{-1} g_t^{-1}(\omega_{\hat e,e} \otimes i)\Big(\sum \hat I I \tilde I\Big) g_s y\hat e$$
$$= g_{\hat\beta}^{-1} g_t^{-1}(\omega_{\hat e,e} \otimes i)(e_{\hat\alpha,\beta}(1 \otimes U)) g_s y\hat e = g_{\hat\beta}^{-1} g_t^{-1} g_t U g_s y\hat e$$
$$= U g_t^{-1} g_s y\hat e = U g_s y g_s^{-1}\hat e,$$

and the result follows. □

3.2.6 Corollary. *One has* $U\hat S U = \hat S'$.

Proof. The proof is the same as in Proposition 3.1.1. □

3.2.7 Proposition. *The Fourier transforms* $\mathcal F : S \mapsto \hat S$ *and* $\hat{\mathcal F} : \hat S \mapsto S$, *defined, for any* (x, y) *in* $S \times \hat S$, *by*

$$\mathcal F(x) = (i \otimes \omega_{g_s x g_s^{-1}e,e}(I) g_{\hat\beta}^{-1},\ \hat{\mathcal F}(y) = (\omega_{y\hat e,\hat e} \otimes i)(I) g_t^{-1},$$

are such that $\hat{\mathcal F}\mathcal F = j$ *and* $\mathcal F\hat{\mathcal F} = \hat j$, *where* $j(x) = g_s\kappa(x)g_s^{-1}$ *and* $\hat j(y) = g_{\hat\beta}^{-1}\hat\kappa(y)g_{\hat\beta}^{-1}$. *So* $\mathcal F$ *and* $\hat{\mathcal F}$ *are "of order 4".*

Proof. By Lemma 3.2.2, for any y in $\hat{\mathcal S}$ we have: $\hat{\mathcal F}(y)e = y\hat e$, then due to Lemma 3.2.5, one has, for any x in S,

$$j(x)e = g_s\kappa(x)g_s^{-1}e = (i \otimes \omega_{e,g_s\kappa(x)g_s^{-1}e}(I^*) g_{\hat\beta}^{-1}\hat e$$
$$= (i \otimes (\omega_{e,g_s\kappa(x)g_s^{-1}e} \circ \kappa))(I) g_{\hat\beta}^{-1}\hat e.$$

But for any z in S, as $\phi = \phi \circ \kappa^{-1}$ and g_s belongs to the centralizer of ϕ,

$$(\omega_{e,g_s\kappa(x)g_s^{-1}e} \circ \kappa)(z) = \phi(\kappa(z)g_s\kappa(x)g_s^{-1}) = \phi(g_t^{-1}x g_t z)$$
$$= \phi(g_t^{-1}g_s^{-1}z g_t g_s g_t^{-1} x g_t) = \phi(z g_s x g_s^{-1}).$$

Therefore $\omega_{e,g_s\kappa(x)g_s^{-1}e} \circ \kappa = \omega_{g_s x g_s^{-1}e,e}$ and $j(x)e = (i \otimes (\omega_{e,g_s x g_s^{-1}e} \circ \kappa))(I) g_{\hat\beta}^{-1}\hat e = \mathcal F(x)\hat e = \hat{\mathcal F}(\mathcal F(x))e$, which gives $\hat{\mathcal F}\mathcal F = j$.

For any y, z in $\hat S$, one has

$$(\omega_{y\hat e,\hat e} \circ \hat\kappa)(z) = \hat\phi(\hat\kappa(z)y) = \hat\phi(\hat\kappa(y)\hat\kappa^2(z)) = \hat\phi(\hat\kappa(y)g_s g_{\hat\beta}^{-1} z g_s g_{\hat\beta}^{-1})$$
$$= \hat\phi(g_s g_{\hat\beta}^{-1} z g_s^{-1} g_{\hat\beta} g_s g_{\hat\beta}\hat\kappa(y) g_s^{-1} g_{\hat\beta}^{-1}) = \hat\phi(z g_{\hat\beta}^2\hat\kappa(y) g_{\hat\beta}^{-2})$$
$$= (\omega_{g_{\hat\beta}^2\hat\kappa(y)g_{\hat\beta}^{-2}\hat e,\hat e})(z).$$

As $g_{\hat\beta}$ belongs to the centralizer of $\hat\phi$, we deduce that

$$\begin{aligned}
\mathcal{F}\hat{\mathcal{F}}(y)\hat{e} &= j(\hat{\mathcal{F}}(y))\hat{e} = g_s\kappa((\omega_{y\hat{e},\hat{e}}\otimes i)(I)g_t^{-1})g_s^{-1}\hat{e}\\
&= \kappa((\omega_{y\hat{e},\hat{e}}\otimes i)(I))g_s^{-1}\hat{e} = (\omega_{y\hat{e},\hat{e}}\otimes i)(I^*)g_s^{-1}\hat{e}\\
&= ((\omega_{y\hat{e},\hat{e}}\circ\hat{\kappa})\otimes i)(I)g_s^{-1}\hat{e}\\
&= (\omega_{g_{\hat\beta}^2\hat\kappa(y)g_{\hat\beta}^{-2}\hat{e},\hat{e}}\otimes i)(I)g_s^{-1}\hat{e} = (\omega_{g_{\hat\beta}^2\hat\kappa(y)g_{\hat\beta}^{-2}\hat{e},\hat{e}}\otimes i)(I)g_{\hat\beta}^{-1}\hat{e}\\
&= g_{\hat\beta}^{-1}(\omega_{g_{\hat\beta}^2\hat\kappa(y)g_{\hat\beta}^{-2}\hat{e},\hat{e}}\otimes i)(I)\hat{e} = g_{\hat\beta}^{-1}(\omega_{g_{\hat\beta}^2\hat\kappa(y)g_{\hat\beta}^{-2}\hat{e},\hat{e}}\otimes i)(I)g_tg_t^{-1}\hat{e}\\
&= g_{\hat\beta}^{-1}(\omega_{g_{\hat\beta}^2\hat\kappa(y)g_{\hat\beta}^{-2}\hat{e},\hat{e}}\otimes i)(I(1\otimes g_t)))g_t^{-1}\hat{e}\\
&= g_{\hat\beta}^{-1}(\omega_{g_{\hat\beta}^2\hat\kappa(y)g_{\hat\beta}^{-2}\hat{e},\hat{e}}\otimes i)((g_{\hat\beta}\otimes 1)I))g_t^{-1}\hat{e}\\
&= g_{\hat\beta}^{-1}(\omega_{g_{\hat\beta}^2\hat\kappa(y)g_{\hat\beta}^{-2}\hat{e},g_{\hat\beta}\hat{e}}\otimes i)(I))g_t^{-1}\hat{e} = g_{\hat\beta}^{-1}(\omega_{g_{\hat\beta}^2\hat\kappa(y)g_{\hat\beta}^{-1}\hat{e},e}\otimes i)(I))g_t^{-1}\hat{e}\\
&= g_{\hat\beta}^{-1}g_{\hat\beta}^2\hat\kappa(y)g_{\hat\beta}^{-1}\hat{e} = \hat{j}(y)\hat{e}.
\end{aligned}$$

□

3.3 Extensions of the Haar measures and Haar conditional expectations

3.3.1 Notations. Let I be any regular *mpi*. The basic conditional expectation is, by definition, the natural conditional expectation from $\mathcal{L}(H)$ onto $\alpha(N)$ such that $\mathrm{tr}\circ E = \mathrm{tr}$. If $\hat\delta = \frac{d\hat\phi}{d\mathrm{tr}}$, let us define special elements by the formulas $\mu = (i\otimes\mathrm{tr})(\tilde{I}^*(1\otimes\hat\delta)\tilde{I})$ and $\nu = \left(\hat\delta\alpha(n_0)\right)^{-1}\mu$. It is easy to see that μ and ν are invertible and belong to $\alpha(N)'\cap\beta(N)'\cap\hat\alpha(N)'\cap\hat\beta(N)'$, also μ belongs to S'.

3.3.2 Lemma. *If I is a regular mpi, then the source conditional expectation E^s_ϕ of S and the target conditional expectation $E^t_{\hat\phi}$ of $\hat S$ are related to the canonical conditional expectation E from $\mathcal{L}(H)$ onto $\alpha(N)$* ($\mathrm{tr}\circ E = \mathrm{tr}$) *by the following formulas:*

$$E^s_\phi(x) = \alpha(n_0)E(x\delta)$$

and

$$E^t_{\hat\phi}(y) = \alpha(n_0)E(y\hat\delta),$$

for every x in S and y in $\hat S$, where $\delta = \frac{d\phi}{d\mathrm{tr}}$ and $\hat\delta = \frac{d\hat\phi}{d\mathrm{tr}}$.

Proof. As $\dfrac{d\phi\circ\alpha}{d\tau} = n_0^{-1}$, for every n in N and x in S, one has

$$\begin{aligned}\operatorname{tr}(\alpha(n)E^s_\phi(x)) &= \phi(\alpha(n_0)\alpha(n)E^s_\phi(x)) = \phi(\alpha(n_0)\alpha(n)x)\\ &= \operatorname{tr}(\alpha(n_0)\alpha(n)x\delta) = \operatorname{tr}(\alpha(n)\alpha(n_0)E(x\delta)).\end{aligned}$$

We deduce that $E^s_\phi(x) = \alpha(n_0)E(x\delta)$, the remaining statement is proved by duality (using $\sum I^* \sum$). □

Consequently, in the case of a weak Kac algebra, (i.e., $\alpha(n_0) = \beta(n_0)$), we have $E^s_\phi = E \mid S$ and $E^t_{\hat\phi} = E \mid \hat S$.

3.3.3 Proposition. *If I is irreducible, then for every s in S and $\hat s$ in $\hat S$, one has*

$$E(s\hat s\nu) = E(\nu s\hat s) = E(s)E(\hat s).$$

Proof. By Lemma 3.1.4, for every s in S and $\hat s$ in $\hat S$, the following equality is true:

$$\tilde I(s\hat s\otimes 1)\tilde I^* = (s\otimes 1)\hat\Gamma(\hat s),$$

hence

$$\begin{aligned}\operatorname{tr}(sE^t_{\hat\phi}(\hat s)) &= \operatorname{tr}(s(i\otimes\hat\phi)\hat\Gamma(\hat s)) = (\operatorname{tr}\otimes\hat\phi))((s\otimes 1)\hat\Gamma(\hat s)) && (3.5)\\ &= (\operatorname{tr}\otimes\operatorname{tr})((1\otimes\hat\delta)\tilde I(s\hat s\otimes 1)\tilde I^*) = (\operatorname{tr}\otimes\operatorname{tr})(\tilde I^*(1\otimes\hat\delta)\tilde I(s\hat s\otimes 1)) && (3.6)\\ &= \operatorname{tr}(s\hat s\mu). && (3.7)\end{aligned}$$

Therefore, for each n in N,

$$\operatorname{tr}(\alpha(n)E(s\hat s\mu)) = \operatorname{tr}(\alpha(n)s\hat s\mu) = \operatorname{tr}(\alpha(n)sE^t_{\hat\phi}(\hat s)) = \operatorname{tr}(\alpha(n)E(s)E^t_{\hat\phi}(\hat s))$$

which gives

$$E(s\hat s\mu) = E(s)E^t_{\hat\phi}(\hat s). \tag{3.8}$$

So, due to Lemma 3.3.2, one deduces

$$E(s\hat s\nu) = E(s)E^t_{\hat\phi}(\hat s\hat\delta\alpha(n_0)) = E(s)E(\hat s).$$

As ν belongs to $\alpha(N)'$, then $E(s\hat s\nu) = E(\nu s\hat s)$. □

3.3.4 Corollary. *If $\alpha(n_0) = \beta(n_0)$ (i.e., S is a weak Kac algebra), then, for every s in S and $\hat s$ in $\hat S$, one has*

$$E(s\hat s) = E(s)E(\hat s).$$

Proof. If $\alpha(n_0) = \beta(n_0)$, then $\mu = \nu = 1$. □

3.3.5 Remark. We will prove in Chapter 4 that the formula of Proposition 3.3.3 leads to a characterization of quantum groupoids in duality.

3.3.6 Proposition. *If F is the application $\hat{\alpha}(N)'(= S\hat{S}) \mapsto S \cap \hat{S}$, defined, for every x in $\hat{\alpha}(N)'$, by $F(x) = E(x\mu)$, then F is a conditional expectation such that $F(s\hat{s}) = F(s)F(\hat{s})$ for all $s \in S$ and $\hat{s} \in \hat{S}$ (i.e., it is multiplicative).*

Proof. Let E_S be the canonical conditional expectation of $\mathcal{L}(H)$ on S (i.e., $\mathrm{tr} \circ E_s = \mathrm{tr}$). For every $\hat{s} \in \hat{S}$, from the proof of Proposition 3.3.3 one has $F(\hat{s}) = E^t_{\hat{\phi}}(\hat{s}) = E_S(\hat{s}\mu)$, which implies that F is a conditional expectation. For all $s \in S$ and $n \in N$, the following holds:

$$\begin{aligned}\mathrm{tr}(\alpha(n)F(s)F(\hat{s})) &= \mathrm{tr}(\alpha(n)E(s\mu)E(\hat{s}\mu)) = \mathrm{tr}(\alpha(n)s\mu E(\hat{s}\mu))\\ &= \mathrm{tr}(\alpha(n)sE_S(\mu)E_S(\hat{s}\mu)) = \mathrm{tr}(\alpha(n)sE_S(\hat{s}\mu))\\ &= \mathrm{tr}(\alpha(n)s\hat{s}\mu) = \mathrm{tr}(\alpha(n)E(s\hat{s}\mu))\\ &= \mathrm{tr}(\alpha(n)F(s\hat{s})).\end{aligned}$$

□

3.3.7 Remark. The element ν in Proposition 3.3.3 is unique, but F is not unique in general. One can easily prove that $\hat{\mu} = (\mathrm{tr} \otimes i)(\hat{I}^*(\delta \otimes i)\hat{I})$ satisfies the equality $E(s\hat{s}\hat{\mu}) = E^s_{\phi}(s)E(\hat{s})$. So $G : x \mapsto E(x\hat{\mu})$ has the same multiplicativity property as F. Since, by the uniqueness of ν, one has $(\hat{\delta}\alpha(n_0))^{-1}\mu = \hat{\mu}(\delta\alpha(n_0))^{-1}$ and since $\hat{\mu}$ belongs to $\hat{S}'$, then F is equal to G if and only if $\delta = \hat{\delta}$. But δ belongs to $\alpha(N)$ if and only if $\delta = \alpha(n_0)$, and we will se in the next chapter that this characterizes the weak Kac algebra case.

3.3.8 Proposition. *Both conditional expectations $E \mid S : S \to S \cap \hat{S}$ and $E^t_{\hat{\phi}} : \hat{S} \to S \cap \hat{S}$ can be extended to a multiplicative conditional expectation onto $N'_{\hat{\alpha}} = S\hat{S}$, the same is true for $E \mid \hat{S} : \hat{S} \to S \cap \hat{S}$ and $E^s_{\phi} : S \to S \cap \hat{S}$.*

Proof. From the proof of Proposition 3.3.3 one has $E_S(\mu) = 1$, so $F \mid S = E \mid S$ and by a similar argument $G \mid \hat{S} = E \mid \hat{S}$, from where the statement of the proposition follows easily. □

3.3.9 Theorem. *There is a faithful positive linear form on $S\hat{S}$ which extends both ϕ and $\hat{\phi}$, we denote this extension by ψ. There is a unique multiplicative conditional expectation $E^{\alpha}_{\psi} : S\hat{S} \to S \cap \hat{S}$, which extends both $E^t_{\hat{\phi}}$ and E^s_{ϕ} and is invariant with respect to ψ.*

Proof. As we have seen in the previous remark, one has $\mu\delta = \hat{\mu}\hat{\delta}$. This element is positive in $S\hat{S}$ and we will denote it by ρ. As $E_S(\mu) = 1$ and $E_{\hat{S}}(\hat{\mu}) = 1$, then $E_S(\rho) = \delta$ and $E_{\hat{S}}(\rho) = \hat{\delta}$, so $\psi(\cdot) = \mathrm{tr}(\rho\cdot)$ extends both ϕ and $\hat{\phi}$ to $S\hat{S}$ (even to

$\mathcal{L}(H)$). For every $n \in N$, $s \in S$ and $\hat{s} \in \hat{S}$ one has

$$\begin{aligned}\psi(\alpha(n)s\hat{s}) &= \operatorname{tr}(\alpha(n)s\hat{s}\mu\delta) = \operatorname{tr}(\alpha(n)\delta s\hat{s}\mu) = \operatorname{tr}(\alpha(n)\delta s E_S(\hat{s}\mu))\\ &= \operatorname{tr}(\alpha(n)\delta s E^t_{\hat{\phi}}(\hat{s})) = \phi(\alpha(n)sE^t_{\hat{\phi}}(\hat{s}))\\ &= \psi(\alpha(n)E^s_{\phi}(s)E^t_{\hat{\phi}}(\hat{s})).\end{aligned}$$

Using this equality and Proposition 3.1.5, one deduces that there exists a linear application $E^\alpha_\psi : S\hat{S} \to S \cap \hat{S}$ such that, for every $x \in S\hat{S}$, it holds that $\psi(\alpha(n)E^\alpha_\psi(x)) = \psi(\alpha(n)x)$ and $E^\alpha_\psi(s\hat{s}) = E^s_\phi(s)E^t_{\hat{\phi}}(\hat{s})$. Due to these two facts, one can easily see that E^α_ψ is a multiplicative faithful conditional expectation, which extends both E^s_ϕ and $E^t_{\hat{\phi}}$. The uniqueness of E^α_ψ is obvious from Proposition 3.1.5. Also the invariancy of E^α_ψ with respect to ψ is obvious. □

3.4 The uniqueness of the multiplicative partial isometry

Recall that E_S denotes the canonical conditional expectation of $\mathcal{L}(H)$ onto S.

3.4.1 Proposition. a) *For every $\hat{s}$ in $\hat{S}$ the following equality holds:*

$$E_S(\hat{s}\nu) = E(\hat{s}) = \alpha\left(\frac{d(\omega_{\hat{s}\hat{\delta}^{-1}\hat{e},\hat{e}} \circ \alpha)}{d\operatorname{tr}\circ\alpha}\right) = (i \otimes \omega_{\beta(n_0)^{-1}\hat{\delta}^{-1}\hat{e},\hat{e}})(e_{\alpha,\beta}(1 \otimes \hat{s})).$$

b) *For every x in $\hat{\alpha}(N)'$ one has*

$$E_S(x\nu) = (i \otimes \omega_{\beta(n_0)^{-1}\hat{\delta}^{-1}\hat{e},\hat{e}})(\hat{I}^*(1 \otimes x)\hat{I}).$$

Proof. a) For every $\hat{s}$ in $\hat{S}$, the fact that $E_S(\hat{s}\nu) = E(\hat{s})$ follows from the proof of Proposition 3.3.3. Then, for all n in N, we have

$$\begin{aligned}\operatorname{tr}(\alpha(n)\hat{s}) &= \hat{\phi}(\alpha(n)\hat{s}\hat{\delta}^{-1}) = \omega_{\hat{e}}(\alpha(n)\hat{s}\hat{\delta}^{-1}) = (\omega_{\hat{s}\hat{\delta}^{-1}\hat{e},\hat{e}} \circ \alpha)(n)\\ &= (\operatorname{tr}\circ\alpha)\left(\frac{d\omega_{\hat{s}\hat{\delta}^{-1}\hat{e},\hat{e}} \circ \alpha}{d\operatorname{tr}\circ\alpha} n\right),\end{aligned}$$

which gives the two first equalities of assertion a). Since for every n in N one has $\alpha(n)\hat{e} = \beta(n)\hat{e}$, then, for any η, η' in H,

$$\begin{aligned}\left(\alpha\left(\frac{d\omega_{\hat{s}\hat{\delta}^{-1}\hat{e},\hat{e}} \circ \alpha)}{d\operatorname{tr}\circ\alpha}\right)\eta, \eta'\right) &= \left(\alpha\left(\frac{d\omega_{\hat{s}\hat{\delta}^{-1}\hat{e},\hat{e}} \circ \beta)}{d\operatorname{tr}\circ\alpha}\right)\eta, \eta'\right)\\ &= (e_{\alpha,\beta}(\eta \otimes \beta(n_0)^{-\frac{1}{2}}\hat{s}\hat{\delta}^{-1}\hat{e}), e_{\alpha,\beta}(\eta' \otimes \beta(n_0)^{-\frac{1}{2}}\hat{e}))\\ &= ((e_{\alpha,\beta}(\eta \otimes \beta(n_0)^{-1}\hat{s}\hat{\delta}^{-1}\hat{e}), \eta' \otimes \hat{e})\\ &= (i \otimes \omega_{\beta(n_0)^{-1}\hat{\delta}^{-1}\hat{e},\hat{e}})(e_{\alpha,\beta}(1 \otimes \hat{s}))\end{aligned}$$

which is the third equality.

b) Due to Proposition 3.1.5, it is sufficient to verify the formula for $x = s\hat{s}$, where s is in S and $\hat{s}$ is in $\hat{S}$. Due to Proposition 3.1.4, one has

$$\begin{aligned}(i \otimes \omega_{\beta(n_0)^{-1}\hat{\delta}^{-1}\hat{e},\hat{e}})(\hat{I}^*(1 \otimes s\hat{s})\hat{I}) &= (i \otimes \omega_{\hat{s}\beta(n_0)^{-1}\hat{\delta}^{-1}\hat{e},\hat{e}})(\hat{I}^*(1 \otimes s)\hat{I}) \\ &= (i \otimes \omega_{\beta(n_0)^{-1}\hat{\delta}^{-1}\hat{s}\hat{e},\hat{e}})(\Gamma(s)) \\ &= (i \otimes \omega_{\beta(n_0)^{-1}\hat{\delta}^{-1}\hat{s}\hat{e},\hat{e}})(I(s \otimes 1)I^*) \\ &= (i \otimes \omega_{\beta(n_0)^{-1}\hat{\delta}^{-1}\hat{s}\hat{e},\hat{e}})((s \otimes 1)e_{\alpha,\beta}) \\ &= s(i \otimes \omega_{\beta(n_0)^{-1}\hat{\delta}^{-1}\hat{e},\hat{e}})(e_{\alpha,\beta}(1 \otimes \hat{s})) \\ &= sE_S(\hat{s}\nu).\end{aligned}$$

□

3.4.2 Lemma. *For every x, x' in S and y, y' in $\hat{S}$ one has*

$$(I(xe \otimes y\hat{e}), x'e \otimes y'\hat{e}) = \psi(x'^* y'^* xy).$$

Proof. For every x, x' in S and y, y' in $\hat{S}$ it holds that

$$\begin{aligned}(I(xe \otimes y\beta(n_0)^{-1}\hat{\delta}^{-1}\hat{e}), x'e \otimes y'\hat{e}) &= (\Gamma(x)(e \otimes y\beta(n_0)^{-1}\hat{\delta}^{-1}\hat{e}), x'e \otimes y'\hat{e}) \\ &= ((i \otimes \omega_{\beta(n_0)^{-1}\hat{\delta}^{-1}\hat{e},\hat{e}})((1 \otimes y'^*)\Gamma(x)(1 \otimes y))e, x'e) \\ &= ((i \otimes \omega_{\beta(n_0)^{-1}\hat{\delta}^{-1}\hat{e},\hat{e}})(\hat{I}^*(1 \otimes y'^* xy)\hat{I})e, x'e) \\ &= (E_S(y'^* xy\nu)e, x'e) = \phi(E_S(x'^* y'^* xy\nu)) = \mathrm{tr}(x'^* y'^* xy\nu\delta).\end{aligned}$$

Applying this last formula to $y' = y\alpha(n_0)\hat{\delta}$ and using the facts that $\alpha(n_0)$ commutes with $\hat{\delta}$ and $\alpha(n_0)\hat{e} = \beta(n_0)\hat{e}$, one proves the formula. □

3.4.3 Proposition. *A multiplicative partial isometry I in the irreducible situation is completely determined by the involutive algebras S and $\hat{S}$ and the spaces $\mathcal{F}$ and $co\mathcal{F}$.*

Proof. A binormalized fixed vector e is an element in $\mathcal{F}$ such that $\omega_e \mid S \cap \hat{S}$ is the canonical trace of the algebra $S \cap \hat{S}$ and $\omega_e \mid S \cap \hat{S}'$ is the one of $S \cap \hat{S}'$; an equivalent property holds for a cofixed binormalized vector f. As $\phi = \omega_e \mid S$ and $\hat{\phi} = \omega_f \mid \hat{S}$, then $E^t_{\hat{\phi}}$, E^s_{ϕ}, g_t depend only on S, $\hat{S}$, $\mathcal{F}$ and $co\mathcal{F}$. But, for any $s \in S$ and $\hat{s} \in \hat{S}$,

$$\psi(s\hat{s}) = \psi(E^{\alpha}_{\psi}(s\hat{s})) = \psi(E^{\alpha}_{\psi}(s)E^{\alpha}_{\psi}(\hat{s})) = \psi(E^s_{\phi}(s)E^t_{\hat{\phi}}(\hat{s})) = \mathrm{tr}_{\alpha}(E^s_{\phi}(s)E^t_{\hat{\phi}}(\hat{s})),$$

where tr_{α} is the canonical trace of the algebra $\alpha(N)$. By Proposition 3.1.5, one deduces also that ψ depends only on S, $\hat{S}$, $\mathcal{F}$ and $co\mathcal{F}$; if e is any binormalized fixed vector, then one has $\hat{e} = pg_t^{-1}e$, which also depends only on S, $\hat{S}$, $\mathcal{F}$ and $co\mathcal{F}$, and Lemma 3.4.2 determines a unique operator I, as e (resp., $\hat{e}$) is cyclic for S (resp., $\hat{S}$).

□

4 A characterisation of quantum groupoids in duality operating in the same Hilbert space

4.1 Notations

In what follows let us fix a finite dimensional Hilbert space H and two involutive subalgebras A and B of $\mathcal{L}(H)$. Let us define four corner bases $N_\alpha = A \cap B$, $N_\beta = A \cap B'$, $N_{\hat{\beta}} = A' \cap B$ and $N_{\hat{\alpha}} = A' \cap B'$. Each of the involutive algebras N_i is isomorphic to a product $\oplus M_{n_b^i}(\mathbb{C})$. We will denote the canonical conditional expectation from $\mathcal{L}(H)$ to N_i such that $\mathrm{tr} \circ E_i = \mathrm{tr}$, by E_i. The (flipper) projection associated to N_i ($i \in \{\alpha, \beta, \hat{\alpha}, \hat{\beta}\}$) by Lemma 2.0.1, will be denoted by f_i, and we will denote the element $(\mathrm{tr}^o \otimes i)(f_i)$ by n_i, hence n_i belongs to $Z(N_i)$. Let us define, for each i, a linear form on N_i: $\phi_i(\cdot) = \mathrm{tr}(n_i^{-1} \cdot)$ (i.e., *the canonical trace of* N_i), and a N_i - "scalar product" by the formula: $\langle \xi, \eta \rangle_i = \dfrac{d\omega_{\xi,\eta} \mid N_i}{d\phi_i \mid N_i}$. We will denote, for any ξ, η in H and $i \in \{\alpha, \beta, \hat{\alpha}, \hat{\beta}\}$, the element of $\mathcal{L}(H)$ defined by the equality: $T^i_{\xi,\eta}(\theta) = \langle \theta, \xi \rangle_i \eta$, by $T^i_{\xi,\eta}$. Let us also recall that $N_i'^o$, the opposite algebra of N_i', acts on $\overline{H}$, the conjugate Hilbert space of H, by the usual formula $x^o \overline{\xi} = \overline{x^* \xi}$.

4.1.1 Lemma. *For any $i \in \{\alpha, \beta, \hat{\alpha}, \hat{\beta}\}$ the vector space generated by the operators $T^i_{\xi,\eta}$ is equal to N_i'. There is a natural linear isomorphism $f_i(\overline{H} \otimes H) \to N_i'$ such that $f_i(\overline{\xi} \otimes \eta) \mapsto T^i_{\xi,\eta}$.*

Proof. For any $i \in \{\alpha, \beta, \hat{\alpha}, \hat{\beta}\}$, the set $\{T^i_{\xi,\eta}\}$ is stable with respect to the multiplication; using the same arguments as in Lemma 2.6.5 of [Val1], one can conclude that this set generates N_i' as a vector space. Hence, the natural homomorphism $\overline{H} \otimes H \mapsto N_i'$, such that $\overline{\xi} \otimes \eta \mapsto T^i_{\xi,\eta}$, is surjective and it factorizes through the projection f_i. The injectivity of the factorization is due to [EV]. □

4.1.2 Corollary. $\dim N_i' = \mathrm{tr}(n_i)$.

4.2 Normalized and separating vectors

4.2.1 Definitions. For any vector ξ in H, let us define the element Θ^i_ξ in $\mathcal{L}(H)$ by the formula: $\Theta^i_\xi(\eta) == \langle \eta, \xi \rangle_i \xi$; we say that ξ is i-normalized if $\langle \xi, \xi \rangle_i = 1$. One can easily check that this means that $\omega_\xi \mid N_i$ is the *canonical trace* of N_i, and we say that ξ is completely normalized if it is i-normalized for every i.

4.2.2 Lemma. *Let ξ be an i-normalized vector of H and $e^{n^i_b}_{p,q}$ a matrix unit of N_i, then the family $\eta^{n^i_b}_{p,q} = \frac{1}{\sqrt{n^i_b}} e^{n^i_b}_{p,q}\xi$ is an orthogonal base of $N_i\xi$ such that*

$$\langle \eta^{n^i_b}_{p,q}, \eta^{n^i_{b'}}_{p',q'} \rangle_i = \delta_{b,b'}\delta_{q,q''}\frac{1}{n^i_b} e^{n^i_b}_{p,p'}.$$

Proof. As ξ is i-normalized, it is cyclic and separating for N_i in $K = N_\beta\xi$. Obviously, the family $\eta^{n^i_b}_{p,q} = \frac{1}{\sqrt{n^i_b}} e^{n^i_b}_{p,q}\xi$ is a base of the vector space K. One has

$$\begin{aligned}\langle \eta^{n^i_b}_{p,q}, \eta^{n^i_{b'}}_{p',q'} \rangle_i &= \left\langle \frac{1}{\sqrt{n^i_b}} e^{n^i_b}_{p,q}\xi, \frac{1}{\sqrt{n^i_{b'}}} e^{n^i_{b'}}_{p,q}\xi \right\rangle_i = \frac{1}{\sqrt{n^i_b}} e^{n^i_b}_{p,q} \langle \xi, \xi\rangle_i \frac{1}{\sqrt{n^i_{b'}}} e^{n^i_{b'}}_{q',p'} \\ &= \delta_{b,b'}\delta_{p,p'}\frac{1}{n^i_b} e^{n^i_b}_{p,p'}.\end{aligned}$$

Let $m^i_b = \mathrm{tr}(e^{n^i_b}_{1,1})$ and q^b_i be the family of maximal projections on N_β. Then one has

$$\begin{aligned}\langle \eta^{n^i_b}_{p,q}, \eta^{n^i_{b'}}_{p',q'} \rangle &= \mathrm{tr}\left(n_i^{-1} \left\langle \frac{1}{\sqrt{n^i_b}} e^{n^i_b}_{p,q}\xi, \frac{1}{\sqrt{n^i_{b'}}} e^{n^i_{b'}}_{p,q}\xi \right\rangle_i \right) \\ &= \delta_{b,b'}\delta_{q,q'}\frac{1}{n^i_b}\,\mathrm{tr}(n_i^{-1} e^{n^i_b}_{p,p'}) = \delta_{b,b'}\delta_{q,q'}\frac{1}{n^i_b}\,\mathrm{tr}\left(\sum_{b'} \frac{n^i_{b'}}{m^i_{b'}} q^{b'}_i e^{n^i_b}_{p,p'} \right) \\ &= \delta_{b,b'}\delta_{q,q'}\delta_{p,p'}.\end{aligned}$$

□

4.2.3 Lemma. *If ξ is an i-normalized vector, then Θ^i_ξ is the orthogonal projection on $N_i\xi$ and $\mathrm{tr}(x\Theta^i_\xi) = \omega_\xi(x)$ for every x in N'_i. Then $\langle x^*\xi, \xi\rangle_j = \langle x\xi, \xi\rangle^*_j$, for all x in N'_i and all j such that Θ^i_ξ belongs to N'_j.*

Proof. Let $\eta^{n^i_b}_{p,q}$ be the base of $N_i\xi$ given by Lemma 4.2.2, then every vector η in H can be written in the form $\eta = \sum \lambda^{n^i_b}_{p,q}\eta^{n^i_b}_{p,q} + \eta'$, where η' is orthogonal to $N_i\xi$. As the orthogonal complement to $N_i\xi$ is stable under N_i, then $\langle \eta', \xi\rangle_i = 0$; so, for every b' and p', q', one has

$$\begin{aligned}\langle \eta - \Theta^i_\xi(\eta), e^{n^i_{b'}}_{p',q'}\xi \rangle &= \langle \eta - \langle \eta, \xi\rangle_i \xi, e^{n^i_{b'}}_{p',q'}\xi \rangle \\ &= \left\langle \sum \lambda^{n^i_b}_{p,q}\eta^{n^i_b}_{p,q} - \left\langle \sum \lambda^{n^i_b}_{p,q}\eta^{n^i_b}_{p,q}, \xi \right\rangle_i \xi, e^{n^i_{b'}}_{p',q'}\xi \right\rangle\end{aligned}$$

$$= \sum \lambda_{p,q}^{n_b^i} \langle \eta_{p,q}^{n_b^i} - \langle \eta_{p,q}^{n_b^i}, \xi \rangle_i \xi, e_{p',q'}^{n_{b'}^i} \xi \rangle$$

$$= \sum \lambda_{p,q}^{n_b^i} \langle \eta_{p,q}^{n_b^i} - \frac{1}{\sqrt{n_b^i}} e_{p,q}^{n_b^i} \langle \xi, \xi \rangle_i \xi, e_{p',q'}^{n_{b'}^i} \xi \rangle = 0.$$

Thus, Θ_ξ^i is the orthogonal projection on $N_i H$, and for any x in N_i', one has

$$\mathrm{tr}(x\Theta_\xi^i) = \sum_{b,p,q} \frac{1}{n_b^i} \langle x\Theta_\xi^i e_{p,q}^{n_b^i} \xi, e_{p,q}^{n_b^i} \xi \rangle = \sum_{b,p,q} \frac{1}{n_b^i} \langle x e_{p,q}^{n_b^i} \Theta_\xi^i e_{p,q}^{n_b^i} \xi, \xi \rangle$$

$$= \sum_{b,q} \langle x e_{q,q}^{n_b^i} \xi, \xi \rangle = \omega_\xi(x).$$

We have, for any j such that Θ_ξ^i belongs to N_j', and for all z_j in N_j and x in N_i':

$$\begin{aligned}\mathrm{tr}(n_j^{-1} z_j \langle x^* \xi, \xi \rangle_j) &= \omega_\xi(z_j x^*) = \mathrm{tr}(z_j x^* \Theta_\xi^i) = \mathrm{tr}(x^* z_j \Theta_\xi^i) = \omega_\xi(x^* z_j) \\ &= \overline{\omega_\xi(z_j^* x)} = \overline{\mathrm{tr}(n_j^{-1} z_j^* \langle x\xi, \xi \rangle_j)} \\ &= \mathrm{tr}(\langle x\xi, \xi \rangle_j^* z_j n_j^{-1}) = \mathrm{tr}(n_j^{-1} z_j \langle x\xi, \xi \rangle_j^*),\end{aligned}$$

which gives the result. □

4.2.4 Proposition. *In the notations of* 4.1, *let H, A, B be such that there exists an element ν in $\bigcap_i N_i'$ for which $E_\alpha(ab\nu) = E_\alpha(a)E_\alpha(b)$ for all (a, b) in $A \times B$. Then for any completely normalized vector ξ in H such that $\Theta_\xi^{\hat{\alpha}}$ belongs to B (resp., A), and for every a in A (resp., b in B), one has $\phi_\alpha(a) = \omega_\xi(\nu a)$ (resp. $\phi_\alpha(b) = \omega_\xi(b\nu)$).*

Proof. Let us suppose that there exists an element ν in $N_{\hat{\alpha}}'$ such that $E(ab\nu) = E(a)E(b)$ for all (a, b) in $A \times B$ and let ξ be an α and $\hat{\alpha}$-normalized vector in H such that $\Theta_\xi^{\hat{\alpha}}$ belongs to B. For any n in N_α we have: $\mathrm{tr}(n\Theta_\xi^{\hat{\alpha}}) = \omega_\xi(n) = \mathrm{tr}(n_\alpha^{-1} n)$, from which one deduces that

$$E_\alpha(\Theta_\xi^{\hat{\alpha}}) = n_\alpha^{-1}.$$

This leads to the fact that, for any a in A,

$$\begin{aligned}\mathrm{tr}(a n_\alpha^{-1}) &= \mathrm{tr}(a E_\alpha(\Theta_\xi^{\hat{\alpha}})) = \mathrm{tr}(E_\alpha(a) E_\alpha(\Theta_\xi^{\hat{\alpha}})) = \mathrm{tr}(a \Theta_\xi^{\hat{\alpha}} \nu) \\ &= \mathrm{tr}(\nu a \Theta_\xi^{\hat{\alpha}}) = \omega_\xi(\nu a).\end{aligned}$$

The proof of the second assertion is similar. □

4.2.5 Corollary. *In the conditions of Proposition* 4.2.4, *ξ is separating for A (resp., for B).*

Proof. Let $a \in A$ be such that $a\xi = 0$, then $a^* a \xi = 0$, so $\mathrm{tr}(n_\alpha a^* a) = \omega_\xi(\nu a^* a) = 0$, hence $a = 0$. For B one uses a similar reasoning. □

4.2.6 Corollary. *In the notations of Section 3, the following assertions are equivalent:*

i) *I is related to a weak Kac algebra (i.e., $\alpha(n_0) = \beta(n_0)$).*

ii) *For every s in S and $\hat{s}$ in $\hat{S}$, one has $E(s\hat{s}) = E(s)E(\hat{s})$.*

iii) *$\delta = \alpha(n_0)$.*

Proof. Due to Corollary 3.3.4, i) implies ii). Let us suppose that ii) is true, then by Lemma 3.1.3, one has $\Theta^{\hat{\alpha}}_{e} = \Theta^{\beta}_{e}$, which is the Haar projection of $\hat{S}$; also $\Theta^{\hat{\alpha}}_{\hat{e}} = \Theta^{\hat{\beta}}_{\hat{e}}$, which is the Haar projection of $\hat{S}$. Applying Proposition 4.2.4, one has $\phi(s) = \mathrm{tr}(s\alpha(n_0)^{-1})$, that is iii). If iii) is true and F is the conditional expectation from $\mathcal{L}(H)$ to $\beta(N)$ such that $\mathrm{tr} \circ F = F$, then due to [Val1], Proposition 3.1.8. (4): $F(\alpha(n_0)^{-1}) = F(\delta) = \beta(n_0)^{-1}$; so by the Schwarz inequality, one has

$$\begin{aligned}\mathrm{tr}(\beta(n_0)^{-2}) = \mathrm{tr}(\beta(n_0)^{-1}\alpha(n_0)^{-1}) &\leq (\mathrm{tr}(\beta(n_0)^{-2})^{\frac{1}{2}}(\mathrm{tr}(\alpha(n_0)^{-2})^{\frac{1}{2}}\\ &\leq \mathrm{tr}(\beta(n_0)^{-2}).\end{aligned}$$

Hence there exists a λ in $\mathbb{C}$ such that $\beta(n_0)^{-1} = \lambda\alpha(n_0)^{-1}$; but these operators have the same trace, so $\lambda = 1$ and i) is true. □

4.2.7 Proposition. *In the notations of* 4.1, *let us suppose that A, B, H have the same finite dimension and that there exists v invertible in $\bigcap_i N_i'$ such that $E_\alpha(abv) = E_\alpha(a)E_\alpha(b)$ for all (a, b) in $A \times B$. Let us also suppose the existence of two vectors e and $\hat{e}$, completely normalized in H and such that $\Theta^{\hat{\alpha}}_{e}$ belongs to B (resp., $\Theta^{\hat{\alpha}}_{\hat{e}}$ belongs to A). Then e (resp., $\hat{e}$) is cyclic and separating for A (resp., B), and for all (a, b) in $A \times B$, we have*

$$a\Theta^{\hat{\alpha}}_{\hat{e}} = \langle a\hat{e}, \hat{e}\rangle_\alpha \Theta^{\hat{\alpha}}_{\hat{e}} \quad and \quad \Theta^{\hat{\alpha}}_{\hat{e}} a = \Theta^{\hat{\alpha}}_{\hat{e}} \langle a^*\hat{e}, \hat{e}\rangle^*_\alpha,$$
$$b\Theta^{\hat{\alpha}}_{e} = \langle be, e\rangle_\alpha \Theta^{\hat{\alpha}}_{e} \quad and \quad \Theta^{\hat{\alpha}}_{e} b = \Theta^{\hat{\alpha}}_{e} \langle b^*e, e\rangle^*_\alpha.$$

Proof. Due to Corollary 4.2.5, e (resp., $\hat{e}$) is separating for A (resp., B); as $\dim A = \dim B = \dim \mathcal{L}(H)$, then these vectors are also cyclic.

For every n in N_α and a in A, Lemma 4.2.3 shows that

$$\mathrm{tr}(nE_\alpha(a\Theta^{\hat{\alpha}}_{\hat{e}})) = \mathrm{tr}(na\Theta^{\hat{\alpha}}_{\hat{e}}) = \omega_{\hat{e}}(na) = \mathrm{tr}(n_\alpha^{-1}\langle na\hat{e}, \hat{e}\rangle_\alpha).$$

Then one deduces that $E_\alpha(a\Theta^{\hat{\alpha}}_{\hat{e}}) = \langle a\hat{e}, \hat{e}\rangle_\alpha n_\alpha^{-1}$. So, if we denote $\hat{\delta} = \dfrac{d\omega_{\hat{e}} \mid B}{d\mathrm{tr} \mid B}$, then we have the following for every b in B:

$$\begin{aligned}(va\hat{e}, b^*\hat{e}) &= \omega_{\hat{e}}(bva) = \mathrm{tr}(bva\Theta^{\hat{\alpha}}_{\hat{e}}) = \mathrm{tr}(a\Theta^{\hat{\alpha}}_{\hat{e}}bv) = \mathrm{tr}(E_\alpha(a\Theta^{\hat{\alpha}}_{\hat{e}})E_\alpha(b))\\ &= \mathrm{tr}(E_\alpha(a\Theta^{\hat{\alpha}}_{\hat{e}})b) = \mathrm{tr}(\langle a\hat{e}, \hat{e}\rangle_\alpha n_\alpha^{-1}b) = \mathrm{tr}(b\langle a\hat{e}, \hat{e}\rangle_\alpha n_\alpha^{-1})\\ &= \omega_{\hat{e}}(b\langle a\hat{e}, \hat{e}\rangle_\alpha n_\alpha^{-1}\hat{\delta}^{-1}) = (\langle a\hat{e}, \hat{e}\rangle_\alpha n_\alpha^{-1}\hat{\delta}^{-1}\hat{e}, b^*\hat{e}).\end{aligned}$$

As $\hat{e}$ is separating for B, this proves that

$$a\hat{e} = \nu^{-1}\langle a\hat{e}, \hat{e}\rangle_\alpha n_\alpha^{-1}\hat{\delta}^{-1}\hat{e}.$$

Since ν belongs to N'_α,

$$\begin{aligned} a\Theta^{\hat{\alpha}}_{\hat{e}}(\eta) &= a\langle \eta, \hat{e}\rangle_{\hat{\alpha}}\hat{e} = \langle \eta, \hat{e}\rangle_{\hat{\alpha}} a\hat{e} = \langle \eta, \hat{e}\rangle_{\hat{\alpha}}\nu^{-1}\langle a\hat{e}, \hat{e}\rangle_\alpha n_\alpha^{-1}\hat{\delta}^{-1}\hat{e} \\ &= \langle a\hat{e}, \hat{e}\rangle_\alpha \nu^{-1} n_\alpha^{-1}\hat{\delta}^{-1}\langle \eta, \hat{e}\rangle_{\hat{\alpha}}\hat{e} = \langle a\hat{e}, \hat{e}\rangle_\alpha \nu^{-1} n_\alpha^{-1}\hat{\delta}^{-1}\Theta^{\hat{\alpha}}_{\hat{e}}(\eta) \end{aligned}$$

holds for every η in H. For $a = 1$ we have

$$\Theta^{\hat{\alpha}}_{\hat{e}} = \nu^{-1} n_\alpha^{-1}\hat{\delta}^{-1}\Theta^{\hat{\alpha}}_{\hat{e}}, \tag{4.1}$$

so

$$a\Theta^{\hat{\alpha}}_{\hat{e}} = \langle a\hat{e}, \hat{e}\rangle_\alpha \Theta^{\hat{\alpha}}_{\hat{e}}.$$

Applying this formula to a^* and using Lemma 4.2.3, one has $a^*\Theta^{\hat{\alpha}}_{\hat{e}} = \langle a^*\hat{e}, \hat{e}\rangle_\alpha \Theta^{\hat{\alpha}}_{\hat{e}}$. The remaining statements follow similarly. □

4.2.8 Corollary. *In the conditions of Proposition* 4.2.7, $\Theta^{\alpha}_{\hat{e}} \in A'$ *and* $\Theta^{\alpha}_{e} \in B'$.

Proof. For every a in A one has $a\hat{e} = \langle a\hat{e}, \hat{e}\rangle_\alpha \hat{e}$, so $N_\alpha \hat{e}$ is stable under action of A and $\Theta^{\alpha}_{\hat{e}} \in A'$. A similar reasoning is applicable for Θ^{α}_{e}. □

4.2.9 Corollary. *If* $\mu = \dfrac{d\omega_{\hat{e}} \mid B}{d\mathrm{tr} \mid B} n_\alpha \nu$ *and* $\hat{\mu} = \nu n_\alpha \dfrac{d\omega_e \mid A}{d\,\mathrm{tr} \mid A}$, *then* $\mu\hat{e} = \hat{e}$ *and* $\hat{\mu}^* e = e$.

Proof. The first equality follows from the proof of the latter proposition, and the second equality can be proved by a similar calculation. □

4.3 The construction of a multiplicative partial isometry

In what follows we will assume that the conditions of Subsection 4.2 hold and *also that* μ *and* $\hat{\mu}$ *are positive and belong to* A' *and* B' *respectively.* (If $\nu = 1$, then, by Proposition 4.2.4, $\mu = \hat{\mu} = 1$, and this condition holds automatically.)

For the sake of simplification, we will denote $\delta = \dfrac{d\omega_e \mid A}{d\,\mathrm{tr} \mid A}$ and $\hat{\delta} = \dfrac{d\omega_{\hat{e}} \mid B}{d\mathrm{tr} \mid B}$; as $\delta = n_\alpha^{-1}\nu^{-1}\hat{\mu}$ and $\hat{\delta} = n_\alpha^{-1}\mu\nu^{-1}$, then N_α is included into the centralizator of $\omega_e \mid A$ and into that of $\omega_{\hat{e}} \mid B$. As $\hat{\delta} n_\alpha \nu \delta = \mu\delta = \hat{\delta}\hat{\mu}$, this defines a strictly positive element ρ of $N'_{\hat{\alpha}}$ and one can define on $\mathcal{L}(H)$ a faithful positive form $\psi(x) = \mathrm{tr}(\rho x)$. One can easily verify that ψ extends both $\omega_e \mid A$ and $\omega_{\hat{e}} \mid B$.

4.3.1 Notations. Let us denote by $\kappa : N_\beta \to N_\alpha$ (resp., $\hat{\kappa} : N_{\hat{\beta}} \to N_\alpha$) the application defined for every x_β in N_β by $\kappa(x_\beta) = \langle x_\beta \hat{e}, \hat{e}\rangle_\alpha$ (resp., for every $x_{\hat{\beta}}$ in $N_{\hat{\beta}}$ by $\hat{\kappa}(x_{\hat{\beta}}) = \langle x_{\hat{\beta}} e, e\rangle_\alpha$).

4.3.2 Lemma. *The applications κ and $\hat{\kappa}$ are C^*-anti-isomorphisms.*

Proof. Due to Lemma 4.2.3, κ is a $*$-morphism. By Proposition 4.2.7, for any x_β in N_β one has $x_\beta \hat{e} = \kappa(x_\beta)\hat{e}$. As $\hat{e}$ is completely normalized, κ is injective, and also $\dim N_\beta = \mathrm{tr}(n_\beta^{-1}) = \mathrm{tr}(n_\alpha^{-1}) = \dim N_\alpha$, so κ is a linear isomorphism which preserves the involution by Lemma 4.2.3. Let y_β be another element of N_β. Then

$$\begin{aligned}\kappa(x_\beta y_\beta)\hat{e} &= \langle x_\beta y_\beta \hat{e}, \hat{e}\rangle_\alpha \hat{e} = \langle x_\beta \kappa(y_\beta)\hat{e}, \hat{e}\rangle_\alpha \hat{e} = \langle \kappa(y_\beta) x_\beta \hat{e}, \hat{e}\rangle_\alpha \hat{e} \\ &= \kappa(y_\beta)\langle x_\beta \hat{e}, \hat{e}\rangle_\alpha \hat{e} = \kappa(y_\beta)\kappa(x_\beta)\hat{e}.\end{aligned}$$

Hence, κ is an involutive C^*-anti-isomorphism. The proof for $\hat{\kappa}$ is similar. □

4.3.3 Definition and Remark. Let us denote $N = N_\alpha$. We define two representations, $\alpha, \hat{\alpha}$, and two antirepresentations $\beta, \hat{\beta}$ by the formulas $\alpha(n) = n$, $\beta(n) = J_B \alpha(n^*) J_B$, $\hat{\beta}(n) = J_A \alpha(n^*) J_A$, $\hat{\alpha}(n) = J_A J_B \alpha(n) J_B J_A$, for every n in N. Then one has $\mathrm{tr} \circ \alpha = \mathrm{tr} \circ \beta = \mathrm{tr} \circ \hat{\beta} = \mathrm{tr} \circ \hat{\alpha}$.

4.3.4 Lemma. *For every n in N we have $\alpha(n)e = \hat{\beta}(n)e$, $\alpha(n)\hat{e} = \beta(n)\hat{e}$, $\beta(n)e = \hat{\alpha}(n)e$, $\alpha(n)\theta_{\hat{e}}^{\hat{\alpha}} = \beta(n)\theta_{\hat{e}}^{\hat{\alpha}}$.*

Proof. As $\alpha(N)$ is in the centralizer of $\omega_{\hat{e}}$, hence, using Lemma 4.3.2, $\beta(n)\hat{e} = J_B \alpha(n^*) J_B \hat{e} = \alpha(n)\hat{e}$. The end of the proof is easy. □

4.3.5 Corollary. *One has $N_\alpha = \alpha(N)$, $N_\beta = \beta(N)$, $N_{\hat{\alpha}} = \hat{\alpha}(N)$, $N_{\hat{\beta}} = \hat{\beta}(N)$, and there exists a unique element n_0 in the center of N such that for all $i \in \{\alpha, \beta, \hat{\alpha}, \hat{\beta}\}$ one has $n_i = i(n_0)$ and $k^{-1}(\langle \xi, \eta\rangle_k) = j^{-1}(\langle \xi, \eta\rangle_j)$ for every $\xi, \eta \in H$ and $k, j \in \{\alpha, \beta, \hat{\alpha}, \hat{\beta}\}$.*

4.3.6 Lemma. *The element $\langle \hat{e}, e\rangle_{\hat{\alpha}}$ is invertible in $\hat{\alpha}(N)$ and one can suppose that $\langle \hat{e}, e\rangle_{\hat{\alpha}} = \langle \delta \hat{e}, \hat{e}\rangle_{\hat{\alpha}}^{\frac{1}{2}} = \langle \hat{\delta} e, e\rangle_{\hat{\alpha}}^{\frac{1}{2}}$.*

Proof. If $\langle \hat{e}, e\rangle_{\hat{\alpha}}$ is not invertible in $\hat{\alpha}(N)$, then so is $\langle e, \hat{e}\rangle_{\hat{\alpha}}$. Then there must exist a positive element $n \neq 0$ of N such that $\hat{\alpha}(n)\langle e, \hat{e}\rangle_{\hat{\alpha}} = 0$. Applying this to $\hat{e}$ one has $\theta_{\hat{e}}^{\hat{\alpha}} \beta(n) e = \theta_{\hat{e}}^{\hat{\alpha}} \hat{\alpha}(n) e = \hat{\alpha}(n)\theta_{\hat{e}}^{\hat{\alpha}} e = \hat{\alpha}(n)\langle e, \hat{e}\rangle_{\hat{\alpha}} \hat{e} = 0$. Now e separates A, hence $\theta_{\hat{e}}^{\hat{\alpha}} \beta(n) = 0$, and also $\beta(n)\theta_{\hat{e}}^{\hat{\alpha}} = 0$. Applying the latter equality to $\hat{e}$ one has $\beta(n)\hat{e} = 0$, which leads to a contradiction as $\hat{e}$ separates $\hat{S}'$.

For every $n \in N$ one has

$$\begin{aligned}\operatorname{tr}(\hat{\alpha}(n)\langle\hat{e}, e\rangle_{\hat{\alpha}}{}^*\langle\hat{e}, e\rangle_{\hat{\alpha}}) &= \operatorname{tr}(\hat{\alpha}(n)\langle\langle e, \hat{e}\rangle_{\hat{\alpha}}\hat{e}, e\rangle_{\hat{\alpha}}) \\ &= \operatorname{tr}(\hat{\alpha}(n)\langle\theta^{\hat{\alpha}}_{\hat{e}} e, e\rangle_{\hat{\alpha}}) = \operatorname{tr}(\langle\hat{\alpha}(n)\theta^{\hat{\alpha}}_{\hat{e}} e, e\rangle_{\hat{\alpha}}) \\ &= \langle n_{\hat{\alpha}}\hat{\alpha}(n)\theta^{\hat{\alpha}}_{\hat{e}} e, e\rangle = \langle\theta^{\hat{\alpha}}_{\hat{e}} e, n_\beta\beta(n^*)e\rangle \\ &= \langle n_\beta\beta(n)\theta^{\hat{\alpha}}_{\hat{e}} e, e\rangle = \langle n_\alpha\alpha(n)\theta^{\hat{\alpha}}_{\hat{e}} e, e\rangle.\end{aligned}$$

So, on the one hand,

$$\begin{aligned}&= \operatorname{tr}(\delta n_\alpha\alpha(n)\theta^{\hat{\alpha}}_{\hat{e}}) = \langle\delta n_\alpha\alpha(n)\hat{e}, \hat{e}\rangle = \langle n_\alpha\alpha(n)\delta\hat{e}, \hat{e}\rangle \\ &= \langle n_\alpha\alpha(n)\langle\delta\hat{e}, \hat{e}\rangle_\alpha\hat{e}, \hat{e}\rangle = \operatorname{tr}(\alpha(n)\langle\delta\hat{e}, \hat{e}\rangle_\alpha) \\ &= \operatorname{tr}(\hat{\alpha}(n)\langle\delta\hat{e}, \hat{e}\rangle_{\hat{\alpha}}),\end{aligned}$$

and on the other hand

$$\begin{aligned}&= \operatorname{tr}(n_\alpha\alpha(n)\theta^{\hat{\alpha}}_{\hat{e}}\theta^{\hat{\alpha}}_{e}) = \operatorname{tr}(\theta^{\hat{\alpha}}_{e} n_\alpha\alpha(n)\theta^{\hat{\alpha}}_{\hat{e}}) = \langle\vartheta^{\hat{\alpha}}_{e} n_\alpha\alpha(n)\hat{e}, \hat{e}\rangle \\ &= \operatorname{tr}(\hat{\delta}\theta^{\hat{\alpha}}_{e} n_\alpha\alpha(n)) = \operatorname{tr}(\langle\hat{\delta}e, e\rangle_\alpha\theta^{\hat{\alpha}}_{e} n_\alpha\alpha(n)) \\ &= \operatorname{tr}(\langle\hat{\delta}e, e\rangle_\alpha\theta^{\hat{\alpha}}_{e} n_\alpha\alpha(n)) = \langle n_\alpha\alpha(n)\langle\hat{\delta}e, e\rangle_{\alpha} e, e\rangle \\ &= \operatorname{tr}(\alpha(n)\langle\hat{\delta}e, e\rangle_\alpha) \\ &= \operatorname{tr}(\hat{\alpha}(n)\langle\hat{\delta}e, e\rangle_{\hat{\alpha}}).\end{aligned}$$

This yields $\langle\hat{e}, e\rangle^*_{\hat{\alpha}}\langle\hat{e}, e\rangle_{\hat{\alpha}} = \langle\delta\hat{e}, \hat{e}\rangle_{\hat{\alpha}} = \langle\hat{\delta}e, e\rangle_{\hat{\alpha}}$. Therefore there exists a unitary $u \in N_{\hat{\alpha}}$ such that $\langle\hat{e}, e\rangle_{\hat{\alpha}} = u\langle\delta\hat{e}, \hat{e}\rangle_{\hat{\alpha}}{}^{\frac{1}{2}} = u\langle\hat{\delta}e, e\rangle_{\hat{\alpha}}{}^{\frac{1}{2}}$, but $u^*\hat{e}$ is also completely normalized and $\theta^{\hat{\alpha}}_{u^*\hat{e}} = \theta^{\hat{\alpha}}_{\hat{e}}$. Now the statement of the lemma follows. □

4.3.7 Notations. Let us denote $g_{\hat{\alpha}} = \langle\hat{e}, e\rangle_{\hat{\alpha}}$, $g_s = (\alpha\circ\hat{\alpha}^{-1})(g_{\hat{\alpha}})$, $g_t = (\beta\circ\hat{\alpha}^{-1})(g_{\hat{\alpha}})$, $g_{\hat{\beta}} = (\hat{\beta}\circ\hat{\alpha}^{-1})(g_{\hat{\alpha}})$. In particular one has $\hat{e} = g_{\hat{\alpha}}^{-1}\theta^{\hat{\alpha}}_{e} e = \theta^{\hat{\alpha}}_{\hat{e}} g_{\hat{\alpha}}^{-1} e = \theta^{\hat{\alpha}}_{\hat{e}} g_t^{-1} e$.

4.3.8 Lemma. *One has $J_A\hat{e} = \hat{e}$, $J_B e = e$ and, for every $n \in N$, $\hat{\alpha}(n)\hat{e} = \hat{\beta}(n)\hat{e}$.*

Proof. For every $\eta \in H$ one has

$$\begin{aligned}\delta^{\frac{1}{2}}\theta^{\hat{\alpha}}_{\hat{e}}(\eta) &= \delta^{\frac{1}{2}}\langle\eta, \hat{e}\rangle_{\hat{\alpha}}\hat{e} = \langle\eta, \hat{e}\rangle_{\hat{\alpha}}\delta^{\frac{1}{2}}\hat{e} = \langle\eta, \hat{e}\rangle_{\hat{\alpha}}\langle\delta^{\frac{1}{2}}\hat{e}, \hat{e}\rangle_\alpha\hat{e} \\ &= \langle\eta, \hat{e}\rangle_{\hat{\alpha}}\langle\delta^{\frac{1}{2}}\hat{e}, \hat{e}\rangle_\alpha\hat{e} = \langle\delta^{\frac{1}{2}}\hat{e}, \hat{e}\rangle_\alpha\langle\eta, \hat{e}\rangle_{\hat{\alpha}}\hat{e} \\ &= \langle\delta^{\frac{1}{2}}\hat{e}, \hat{e}\rangle_\alpha\theta^{\hat{\alpha}}_{\hat{e}}(\eta).\end{aligned}$$

Therefore $\delta^{\frac{1}{2}}\theta^{\hat{\alpha}}_{\hat{e}} = \langle\delta^{\frac{1}{2}}\hat{e}, \hat{e}\rangle_\alpha\theta^{\hat{\alpha}}_{\hat{e}}$ and, by a similar argument, $\delta\theta^{\hat{\alpha}}_{\hat{e}} = \langle\delta\hat{e}, \hat{e}\rangle_\alpha\theta^{\hat{\alpha}}_{\hat{e}} = g_\alpha^2\theta^{\hat{\alpha}}_{\hat{e}}$, hence $g_s^2\theta^{\hat{\alpha}}_{\hat{e}} = \langle\delta\hat{e}, \hat{e}\rangle_\alpha\theta^{\hat{\alpha}}_{\hat{e}} = \delta\theta^{\hat{\alpha}}_{\hat{e}} = \delta^{\frac{1}{2}}\delta^{\frac{1}{2}}\theta^{\hat{\alpha}}_{\hat{e}} = \langle\delta^{\frac{1}{2}}\hat{e}, \hat{e}\rangle_\alpha^2\theta^{\hat{\alpha}}_{\hat{e}}$. Applying this to $\hat{e}$ one deduces that $g_s^2\hat{e} = \langle\delta^{\frac{1}{2}}\hat{e}, \hat{e}\rangle_\alpha^2\hat{e}$. So, as $\hat{e}$ separates B, one has $g_s^2 = \langle\delta^{\frac{1}{2}}\hat{e}, \hat{e}\rangle_\alpha^2$ and $g_s = \langle\delta^{\frac{1}{2}}\hat{e}, \hat{e}\rangle_\alpha$. Hence $\delta^{\frac{1}{2}}\theta^{\hat{\alpha}}_{\hat{e}} = g_s\theta^{\hat{\alpha}}_{\hat{e}}$ and, passing to the adjoints, $\theta^{\hat{\alpha}}_{\hat{e}}\delta^{\frac{1}{2}} = \theta^{\hat{\alpha}}_{\hat{e}} g_s$.

Thus, we have $\theta_{\hat{e}}^{\hat{\alpha}} = \theta_{\hat{e}}^{\hat{\alpha}} g_s \delta^{-\frac{1}{2}} = \theta_{\hat{e}}^{\hat{\alpha}} \delta^{-\frac{1}{2}} g_s$, from which we deduce that $\theta_{\hat{e}}^{\hat{\alpha}} \delta^{-\frac{1}{2}} = \theta_{\hat{e}}^{\hat{\alpha}} g_s^{-1}$. Then, using Lemma 4.3.4, we have

$$\delta^{\frac{1}{2}} \theta_{\hat{e}}^{\hat{\alpha}} \delta^{-\frac{1}{2}} = g_s \theta_{\hat{e}}^{\hat{\alpha}} g_s^{-1}.$$

One deduces that

$$\begin{aligned} J_A \hat{e} &= J_A \theta_{\hat{e}}^{\hat{\alpha}} g_t^{-1} e = \delta^{\frac{1}{2}} g_t^{-1} \theta_{\hat{e}}^{\hat{\alpha}} \delta^{-\frac{1}{2}} e \\ &= \delta^{\frac{1}{2}} g_s^{-1} \theta_{\hat{e}}^{\hat{\alpha}} \delta^{-\frac{1}{2}} e = g_s^{-1} \delta^{\frac{1}{2}} \theta_{\hat{e}}^{\hat{\alpha}} \delta^{-\frac{1}{2}} e = g_s^{-1} g_s \theta_{\hat{e}}^{\hat{\alpha}} g_s^{-1} e \\ &= \hat{e}. \end{aligned}$$

A similar argument is valid for the equality $J_B e = e$. This implies that, for any $n \in N$, $\hat{\alpha}(n)\hat{e} = J_A J_B \alpha(n) J_B J_A \hat{e} = J_A \beta(n) J_A \hat{e} = J_A \beta(n) \hat{e} = J_A \alpha(n) \hat{e} = J_A \alpha(n) J_A \hat{e} = \hat{\beta}(n)\hat{e}$. □

4.3.9 Lemma. *Let f be the (flipper) projection associated to N by Lemma 2.0.1. Let $\overline{f}$ be the same operator viewed as an element of $N \otimes N$ and let us denote $(\alpha \otimes \alpha)(\overline{f})$ by $e_{\alpha,\alpha}$, then, for every $n \in N$: $e_{\alpha,\alpha}(1 \otimes \alpha(n)) = (\alpha(n) \otimes 1)e_{\alpha,\alpha}$.*

Proof. For every $n \in N$, the equality $f(1 \otimes n) = f(n^o \otimes 1)$ gives $\overline{f}(1 \otimes n) = (n \otimes 1)\overline{f}$, and the statement of the lemma follows. □

4.3.10 Proposition. *The two applications $Z, Z' : H \otimes H \to \Lambda_\psi \mathcal{L}(H)$, defined for every $x \in A$ and $y \in B$ by $Z(xe \otimes y\hat{e}) = \Lambda_\psi(xy)$ and $Z'(xe \otimes y\hat{e}) = \Lambda_\psi(yx)$, are partial isometries and $I = Z'^* Z$ is also a partial isometry whose initial (resp., final) support is $e_{\hat{\beta},\alpha}$ (resp., $e_{\alpha,\beta}$). We also have the following:*

i) *For every n, n' in N: $I(\beta(n) \otimes \hat{\beta}(n')) = (\beta(n) \otimes \hat{\beta}(n'))I$ and $I(\alpha(n) \otimes \beta(n')) = (\hat{\beta}(n') \otimes \alpha(n))I$.*

ii) *For all a, x in A and b, y in B: $(I(ae \otimes b\hat{e}), xe \otimes y\hat{e}) = \psi(x^* y^* ab)$.*

iii) $(J_A \otimes J_B) I (J_A \otimes J_B) = I^*$.

iv) *For all (a, b) in $A \times B$, one has $(\omega_{ae,\hat{e}} \otimes i)(I) = a g_t$ and $(i \otimes \omega_{e,b\hat{e}})(I) = g_{\hat{\beta}} b^*$.*

Proof. For every $x \in A$ and $y \in B$, as $\hat{\mu} \in B'$, using Proposition 3.3.3 and Lemma 4.3.9, one has

$$\begin{aligned} (Z^* Z(xe \otimes y\hat{e}), x'e \otimes y'\hat{e}) &= (\Lambda_\psi(xy), \Lambda_\psi(x'y')) = \operatorname{tr}(\hat{\delta}\hat{\mu} y'^* x'^* xy) \\ &= \operatorname{tr}(x'^* xy\hat{\delta}\hat{\mu} y'^*) = \operatorname{tr}(x'^* xy\hat{\delta} y'^* \hat{\mu}) = \operatorname{tr}(x'^* xy\hat{\delta} y'^* \nu\alpha(n_0)\delta) \\ &= \operatorname{tr}(\alpha(n_0)\delta x'^* xy\hat{\delta} y'^* \nu) = \operatorname{tr}(\alpha(n_0) E(\delta x'^* x) E(y\hat{\delta} y'^*)) \\ &= \operatorname{tr}((\operatorname{tr} \otimes i)(e_{\alpha,\alpha})(E(\delta x'^* x) E(y\hat{\delta} y'^*)) \\ &= (\operatorname{tr} \otimes \operatorname{tr})(e_{\alpha,\alpha}(1 \otimes E(\delta x'^* x) E(y\hat{\delta} y'^*)) \end{aligned}$$

$$
\begin{aligned}
&= (\mathrm{tr}\otimes\mathrm{tr})((E(\delta x'^{*}x)\otimes 1)e_{\alpha,\alpha}(1\otimes E(y\hat{\delta}y'^{*})) \\
&= (\mathrm{tr}\otimes\mathrm{tr})(e_{\alpha,\alpha}(E(\delta x'^{*}x)\otimes E(y\hat{\delta}y'^{*})) \\
&= (\mathrm{tr}\otimes\mathrm{tr})(e_{\alpha,\alpha}(\delta x'^{*}x\otimes y\hat{\delta}y'^{*})) \\
&= (\mathrm{tr}\otimes\mathrm{tr})((\delta\otimes\hat{\delta}y'^{*})e_{\alpha,\alpha}(x'^{*}x\otimes y)) \\
&= (\phi\otimes\hat{\phi})((1\otimes y'^{*})e_{\alpha,\alpha}(x'^{*}x\otimes y)) \\
&= (\omega_e\otimes\omega_{\hat{e}})((1\otimes y'^{*})e_{\alpha,\alpha}(x'^{*}x\otimes y)) \\
&= (\omega_e\otimes\omega_{\hat{e}})((1\otimes y'^{*})e_{\hat{\beta},\alpha}(x'^{*}x\otimes y)) \\
&= (\omega_e\otimes\omega_{\hat{e}})((x'^{*}\otimes y'^{*})e_{\hat{\beta},\alpha}(x\otimes y)) \\
&= (e_{\hat{\beta},\alpha}(xe\otimes y\hat{e}), x'e\otimes y'\hat{e}).
\end{aligned}
$$

So Z is a partial isometry with initial support equal to $e_{\hat{\beta},\alpha}$. On the other hand,

$$
\begin{aligned}
(Z'^{*}Z'(xe\otimes y\hat{e}), x'e\otimes y'\hat{e}) &= (\Lambda_\psi(yx), \Lambda_\psi(y'x')) = \mathrm{tr}(\mu\delta x'^{*}y'^{*}yx) \\
&= \mathrm{tr}(x\mu\delta x'^{*}y'^{*}y) = \mathrm{tr}(\mu x\delta x'^{*}y'^{*}y) = \mathrm{tr}(x\delta x'^{*}y'^{*}y\mu) \\
&= \mathrm{tr}(x\delta x'^{*}y'^{*}y\hat{\delta}n_\alpha \nu) = \mathrm{tr}(E(x\delta x'^{*})E(y'^{*}y\hat{\delta})n_\alpha) \\
&= \mathrm{tr}((\mathrm{tr}\otimes i)(e_{\alpha,\alpha})(E(x\delta x'^{*})E(y'^{*}y\hat{\delta})) \\
&= (\mathrm{tr}\otimes\mathrm{tr})(e_{\alpha,\alpha}(1\otimes E(x\delta x'^{*})E(y'^{*}y\hat{\delta})) \\
&= (\mathrm{tr}\otimes\mathrm{tr})((E(x\delta x'^{*})\otimes 1)e_{\alpha,\alpha}(1\otimes E(y'^{*}y\hat{\delta})) \\
&= (\mathrm{tr}\otimes\mathrm{tr})(e_{\alpha,\alpha}(E(x\delta x'^{*})\otimes E(y'^{*}y\hat{\delta})) \\
&= (\mathrm{tr}\otimes\mathrm{tr})(e_{\alpha,\alpha}(x\delta x'^{*}\otimes y'^{*}y\hat{\delta})) \\
&= (\mathrm{tr}\otimes\mathrm{tr})((x'^{*}\otimes 1)e_{\alpha,\alpha}(x\delta\otimes y'^{*}y\hat{\delta})) \\
&= (\phi\otimes\hat{\phi})((x'^{*}\otimes 1)e_{\alpha,\alpha}(x\otimes y'^{*}y)) \\
&= (\omega_e\otimes\omega_{\hat{e}})((x'^{*}\otimes 1)e_{\alpha,\alpha}(x\otimes y'^{*}y)) \\
&= (\omega_e\otimes\omega_{\hat{e}})((x'^{*}\otimes 1)e_{\alpha,\beta}(x\otimes y'^{*}y)) \\
&= (\omega_e\otimes\omega_{\hat{e}})((x'^{*}\otimes y'^{*})e_{\alpha,\beta}(x\otimes y)) \\
&= (e_{\alpha,\beta}(xe\otimes y\hat{e}), x'e\otimes y'\hat{e}).
\end{aligned}
$$

Thus Z' is a partial isometry with initial support equal to $e_{\alpha,\beta}$. Then the images of Z and Z' have the same dimension equal to $\mathrm{tr}(n_{\hat{\alpha}})$ and belong to $\Lambda_\psi N'_{\hat{\alpha}}$, so they are both equal to $\Lambda_\psi N'_{\hat{\alpha}}$ by Corollary 4.1.2. Hence $I = Z'^{*}Z$ is a partial isometry. Due to Proposition 4.3.10 ii), for any n, n' in N, a, x in A and b, y in B we have

$$
\begin{aligned}
(I(\beta(n)\otimes\hat{\beta}(n')(ae\otimes b\hat{e}), xe\otimes y\hat{e}) &= \mathrm{tr}(\mu\delta x^{*}y^{*}\beta(n)a\hat{\beta}(n')b) \\
&= \mathrm{tr}(\mu\delta x^{*}\beta(n)y^{*}\hat{\beta}(n')ab) = \mathrm{tr}(\mu\delta(\beta(n^{*})x)^{*}(\hat{\beta}(n'^{*})y)^{*}ab) =
\end{aligned}
$$

$$= (I(ae \otimes b\hat{e}), (\beta(n) \otimes \hat{\beta}(n'))^*(xe \otimes y\hat{e}))$$
$$= ((\beta(n) \otimes \hat{\beta}(n'))I(ae \otimes b\hat{e}), xe \otimes y\hat{e}).$$

By the same reasons and by Lemma 4.3.4, one has

$$\begin{aligned}(I(\alpha(n) \otimes \beta(n')(ae \otimes b\hat{e}), xe \otimes b\hat{e}) &= (I(\alpha(n)ae \otimes \beta(n')y\hat{e}), xe \otimes y\hat{e}) \\ &= (I(\alpha(n)ae \otimes y\beta(n')\hat{e}), xe \otimes y\hat{e}) \\ &= (I(\alpha(n)ae \otimes y\alpha(n')\hat{e}), xe \otimes y\hat{e}) \\ &= \mathrm{tr}(\mu\delta x^* y^* \alpha(n)ab\alpha(n')) = \mathrm{tr}(\mu\delta\alpha(n')x^*y^*\alpha(n)ab) \\ &= (I(ae \otimes b\hat{e}), (x\alpha(n'^*)e \otimes \alpha(n^*)y\hat{e})) \\ &= (I(ae \otimes b\hat{e}), (x\hat{\beta}(n'^*)e \otimes \alpha(n^*)y\hat{e})) \\ &= (I(ae \otimes b\hat{e}), (\hat{\beta}(n'^*)xe \otimes \alpha(n^*)y\hat{e})) \\ &= ((\hat{\beta}(n') \otimes \alpha(n))I(ae \otimes y\hat{e}), xe \otimes b\hat{e}).\end{aligned}$$

The computation proving ii) is obvious. Under the conditions of 4.3 one has $\delta^{\frac{1}{2}}\hat{\delta}^{-\frac{1}{2}} = \mu^{-\frac{1}{2}}\hat{\mu}^{\frac{1}{2}}$, $\delta^{-\frac{1}{2}}\hat{\delta}^{\frac{1}{2}} = \mu^{\frac{1}{2}}\hat{\mu}^{-\frac{1}{2}}$, $\mu \in S'$, $\hat{\mu} \in \hat{S}'$. So for every a, x in A and b, y in B we obtain

$$\begin{aligned}&(xe \otimes y\hat{e}, (J_A \otimes J_B)I(J_A \otimes J_B)(ae \otimes b\hat{e})) \\ &\quad= (I(J_A \otimes J_B)(ae \otimes b\hat{e}), (J_A \otimes J_B)(xe \otimes y\hat{e})) \\ &\quad= (I(\delta^{\frac{1}{2}}a^*\delta^{-\frac{1}{2}}e \otimes \hat{\delta}^{\frac{1}{2}}b^*\hat{\delta}^{-\frac{1}{2}}\hat{e}), (\delta^{\frac{1}{2}}x^*\delta^{-\frac{1}{2}}e \otimes \hat{\delta}^{\frac{1}{2}}y^*\hat{\delta}^{-\frac{1}{2}}\hat{e})) \\ &\quad= \mathrm{tr}(\delta^{-\frac{1}{2}}x\delta^{\frac{1}{2}}\hat{\delta}^{-\frac{1}{2}}y\hat{\delta}^{\frac{1}{2}}\delta^{\frac{1}{2}}a^*\delta^{-\frac{1}{2}}\hat{\delta}^{\frac{1}{2}}b^*\hat{\delta}^{-\frac{1}{2}}\mu\delta) \\ &\quad= \mathrm{tr}(\delta^{-\frac{1}{2}}x\mu^{-\frac{1}{2}}\hat{\mu}^{\frac{1}{2}}y\hat{\delta}^{\frac{1}{2}}\delta^{\frac{1}{2}}a^*\mu^{\frac{1}{2}}\hat{\mu}^{-\frac{1}{2}}b^*\hat{\delta}^{-\frac{1}{2}}\mu\delta) \\ &\quad= \mathrm{tr}(\delta^{\frac{1}{2}}\mu^{\frac{1}{2}}xy\hat{\mu}^{\frac{1}{2}}\hat{\delta}^{\frac{1}{2}}\delta^{\frac{1}{2}}\mu^{\frac{1}{2}}a^*b^*\hat{\mu}^{-\frac{1}{2}}\hat{\delta}^{-\frac{1}{2}}) \\ &\quad= \mathrm{tr}(xy\delta\mu a^*b^*) = \psi(a^*b^*xy) \\ &\quad= (xe \otimes y\hat{e}, I(xe \otimes y\hat{e})).\end{aligned}$$

Hence iii) is true.

Using Lemma 4.2.3, then 4.3.7 and Corollary 4.2.8, one has, for all c, b in B,

$$\begin{aligned}((\omega_{ae,\hat{e}} \otimes i)(I)b\hat{e}, c\hat{e}) &= (I(ae \otimes b\hat{e}), \theta_{\hat{e}}^{\hat{\alpha}} g_t^{-1} e \otimes c\hat{e}) \\ &= \psi(g_t^{-1}\theta_{\hat{e}}^{\hat{\alpha}} c^*ab) \\ &= \mathrm{tr}(c^*ab\mu\delta g_t^{-1}\theta_{\hat{e}}^{\hat{\alpha}}) = \omega_{\hat{e}}(c^*ab\mu\delta g_t^{-1}) = (ab\mu\delta g_t^{-1}\hat{e}, c\hat{e}) \\ &= (ag_t^{-1}b\mu\delta\hat{e}, c\hat{e}) = (ag_t^{-1}b\delta\mu\hat{e}, c\hat{e}) = (ag_t^{-1}b\delta\hat{e}, c\hat{e}) \\ &= (ag_t^{-1}b\langle\delta\hat{e}, \hat{e}\rangle_\alpha\hat{e}, c\hat{e}) = (ag_t^{-1}b\beta(\alpha^{-1}(\langle\delta\hat{e}, \hat{e}\rangle_\alpha))\hat{e}, c\hat{e}) \\ &= (ag_t^{-1}\beta(\alpha^{-1}(\langle\delta\hat{e}, \hat{e}\rangle_\alpha))b\hat{e}, c\hat{e}).\end{aligned}$$

As $\hat{e}$ is cyclic for B, this gives the first identity of iv); the second can be proved by a similar calculation. □

4.3.11 Lemma. *The following assertions are equivalent:*

i) I *is in* $B \otimes \mathcal{L}(H)$,

ii) I *is in* $\mathcal{L}(H) \otimes A$,

iii) I *is in* $B \otimes A$.

Proof. If i) is true, then, due to Proposition 4.3.10 iv), $(\omega \otimes i)(I)$ belongs to A for every linear form ω in B^*, so iii) is true. A similar argument shows that ii) implies iii). Then the statement of the lemma follows. □

4.4 A characterization of quantum groupoids in duality

In what follows, we suppose that I satisfies the conditions of Subsection 4.3 and, in particular, Lemma 4.3.11.

4.4.1 Lemma. *If I belongs to $\mathcal{L}(H) \otimes A$, then $J_A B J_A = B$, $J_B A J_B = A$ and $J_A J_B = J_B J_A$.*

Proof. Using the assertion iv) of Proposition 4.3.10, the first part follows from Lemma 2.3 b) and c) of [BBS]. As $J_B e = e$, then, for any $x \in A$: $\omega_e(J_B x^* J_B) = \omega_e(x) = \mathrm{tr}(\delta x) = \mathrm{tr}(\delta J_B x^* J_B)$, from which one can deduce that $J_B \delta J_B = \delta$. Hence, for every a in A, $J_A J_B a e = J_A (J_B a J_B) J_B e = J_A (J_B a J_B) e = \delta^{\frac{1}{2}} (J_B a^* J_B) \delta^{-\frac{1}{2}} e = (J_B \delta^{\frac{1}{2}} a^* \delta^{-\frac{1}{2}} J_B) e = J_B J_A a e$ and $J_A J_B = J_B J_A$. □

4.4.2 Lemma. *The pairs of C^*-algebras (B', A), (A, B') satisfy the conditions of Proposition* 4.2.7 *and* 4.3.

Proof. Let E_β be the natural conditional expectation from $\mathcal{L}(H)$ onto $\beta(N)$, then, for any (a, b) in $A \times B$ and n in N, one has

$$\begin{aligned}
\mathrm{tr}(\beta(n) E_\beta(J_B v^* J_B J_B b J_B a)) &= \mathrm{tr}(\beta(n) J_B v^* b J_B a) \\
&= \mathrm{tr}(J_B a^* J_B b^* v J_B \beta(n^*) J_B) \\
&= \mathrm{tr}(J_B a^* J_B b^* v \alpha(n)) = \mathrm{tr}(\alpha(n) J_B a^* J_B b^* v) \\
&= \mathrm{tr}(\alpha(n) E_\alpha(J_B a^* J_B) E_\alpha(b^*)) \\
&= \mathrm{tr}(\beta(n) (\beta \circ \alpha^{-1})(E_\alpha(J_B a^* J_B) E_\alpha(b^*))) \\
&= \mathrm{tr}(\beta(n) (\beta \circ \alpha^{-1})(E_\alpha(b^*)) (\beta \circ \alpha^{-1})(E_\alpha(J_B a^* J_B))),
\end{aligned}$$

so $J_B\nu^* J_B E_\beta(J_B b J_B a) = (\beta\circ\alpha^{-1})(E_\alpha(b^*))(\beta\circ\alpha^{-1})(E_\alpha(J_B a^* J_B))$. But obviously, for any $x \in \mathcal{L}(H)$ one has $\beta^{-1}(E_\beta(J_B b J_B)) = \alpha^{-1}(E_\alpha(b^*))$. Thus,

$$E_\beta(J_B\nu^* J_B J_B b J_B a) = E_\beta(J_B b J_B)E_\beta(a).$$

Next one has that dim A = dim B' = dim H, and by Corollary 4.2.8, $\theta_e^{\hat\beta}(=\theta_e^{\alpha})$ is in B' and $\theta_{\hat e}^{\hat\beta}(=\theta_{\hat e}^{\hat\alpha})$ is in A. Hence, (B', A) satisfies all the conditions of Proposition 4.2.7 with $J_B\nu^* J_B$, and an easy computation gives that $\mu(B', A) = J_B\mu^* J_B = J_B\mu J_B$ and $\hat\mu(B', A) = J_A\hat\mu^* J_A = J_A\mu J_A$, so (B', A) verifies all the conditions of 4.3. A similar argument works also for (A, B'). □

4.4.3 Notation. We will denote by U the orthogonal symmetry $J_A J_B$ and $\hat\rho$ the equivalent of ρ for (B', A).

4.4.4 Lemma. *The applications I and $\hat I = \sum(U\otimes 1)I(U\otimes 1)\sum$ satisfy the following conditions:*

i) $(\hat I(b'\hat e\otimes ae), y'\hat e\otimes xe) = \mathrm{tr}(\hat\rho y'^* x^* b' a)$, *for all x, a in A and b', y' in B';*

iii) $(I(\theta_{\hat e}^{\hat\alpha}\otimes 1)I^*(ae\otimes b\hat e), xe\otimes y\hat e) = \omega_{\hat e}(ba\rho xy)$, *for all x, a in A and b, y in B;*

iii) $(\hat I^*(1\otimes\theta_{\hat e}^{\hat\alpha})\hat I(y'\hat e\otimes ae), b'\hat e\otimes xe) = \omega_{\hat e}(y' a\hat\rho x^* b'^*)$, *for all x, a in A and b', y' in B';*

iv) $I(\theta_{\hat e}^{\hat\alpha}\otimes 1)I^* = \hat I^*(1\otimes\theta_{\hat e}^{\hat\alpha})\hat I$.

Proof. Using Lemma 4.2.3, one can see that the first two assertions coincide with Lemma 2.4 and Lemma 2.5 b) of [BBS] respectively. The assertion i) proves that $\hat I$ is the application $I(B', A)$ which belongs to $A\otimes\mathcal{L}(H)$, so $\sum\hat I^*\sum$ is associated to the pair (A, B'). Then assertion iii) is just assertion ii) for the pair (A, B'). The assertion iv) is similar to Lemma 2.5 d) of [BBS]. □

4.4.5 Lemma. *The application $\pi : N'_{\hat\alpha} \mapsto \mathcal{L}(H\otimes H)$, defined by $\pi(x) = Z'^*\pi_\psi(s)Z'$ is a representation and*

i) *for all (a, b) in $A\times B$ one has $\pi(a) = I(a\otimes 1)I^*$ and $\pi(b) = e_{\alpha,\beta}(1\otimes b)$,*

ii) *for all x in $N'_{\hat\alpha}$ one has $\pi(x) = \hat I^*('1\otimes x)\hat I$.*

Proof. As the image of the partial isometry Z' is $\Lambda_\psi N'_{\hat\alpha}$, π is a representation. The proof of assertion i) is similar to that of Lemma 2.6 a) of [BBS]. For any b, b' in B, one has $b'\theta_{\hat e}^{\hat\alpha}b = T^{\hat\alpha}_{b'^*\hat e, b\hat e}$, so $N'_{\hat\alpha}$ is generated by B and $\theta_{\hat e}^{\hat\alpha}$. But the final support of $\hat I$ is equal to $e_{\beta,\hat\alpha}$. Therefore the applications $x\mapsto\hat I^*(1\otimes x)\hat I$ and $x\mapsto\pi(x)$ are $*$-morphisms for x in $N'_{\hat\alpha}$ and equal for $x\in B$ and $x = \theta_{\hat e}^{\hat\alpha}$. This proves ii). □

4.4.6 Proposition. *For any (a, b) in $A \times B$ one has $I(a \otimes 1)I^* \in A \otimes A$ and $I^*(1 \otimes b)I \in B \otimes B$.*

Proof. See Proposition 2.7. of [BBS]. □

4.4.7 Lemma. *The set $\{\omega_{xe,\hat{e}} \otimes \omega_{\alpha(n)e,\hat{e}} \otimes \omega_{\eta,e} \mid B \otimes N \otimes A; x \in A, n \in N, \eta \in H\}$ generates $(B \otimes N \otimes A)^*$.*

Proof. The applications $x \mapsto \omega_{xe,\hat{e}} \mid B$, $\eta \mapsto \omega_{\eta,e} \mid A$ from A to B^* and from H to A^* are injective homomorphisms of vector spaces of the same dimension, so they are isomorphisms. The application $n \mapsto \omega_{\alpha(n)e,\hat{e}} \mid N$ is a homomorphism of vector spaces of the same dimension and, for any n in N, one has

$$\begin{aligned}\omega_{\alpha(n)e,\hat{e}}(\alpha(n^*)) &= (\alpha(n^*n)e, \theta_{\hat{e}}^{\hat{\alpha}} g_t^{-1} e) = \omega_e(g_t^{-1}\theta_{\hat{e}}^{\hat{\alpha}}\alpha(n^*n)) = \operatorname{tr}(\delta g_t^{-1}\theta_{\hat{e}}^{\hat{\alpha}}\alpha(n^*n)) \\ &= \operatorname{tr}(\delta g_s^{-1}\theta_{\hat{e}}^{\hat{\alpha}}\alpha(n^*n)) = \operatorname{tr}(g_s^{-1}\delta\theta_{\hat{e}}^{\hat{\alpha}}\alpha(n^*n)) \\ &= \operatorname{tr}(g_s^{-1}g_s^2\theta_{\hat{e}}^{\hat{\alpha}}\alpha(n^*n)) = \omega_{\hat{e}}(\alpha(n^*n)g_s).\end{aligned}$$

So $n \mapsto \omega_{\alpha(n)e,\hat{e}}$ is an isomorphism and the statement of the lemma follows. □

4.4.8 Theorem. *Let A, B be two involutive subalgebras of $\mathcal{L}(H)$ of the same finite dimension as the Hilbert space H, and let e, $\hat{e}$ be two completely normalized vectors for the pair (A, B). We suppose that*

i) *there exists $\nu \in \bigcap_i N_i'$, invertible and such that $E_\alpha(ab\nu) = E_\alpha(a)E_\alpha(b)$ for all (a, b) in $A \times B$,*

ii) *the orthogonal projection onto $(A' \cap B')\hat{e}$ belongs to A and the one onto $(A' \cap B')e$ belongs to B,*

iii) *$\mu = \dfrac{d\omega_{\hat{e}} \mid B}{d\mathrm{tr} \mid B} n_\alpha \nu$ and $\hat{\mu} = \nu n_\alpha \dfrac{d\omega_e \mid A}{d\mathrm{tr} \mid A}$ are positive and belong to A' and B' respectively,*

iv) *the application $I \in \mathcal{L}(H \otimes H)$, defined for all $a, x \in A$ and $b, y \in B$ by $(I(ae \otimes b\hat{e}), xe \otimes y\hat{e}) = \operatorname{tr}\left(\mu \dfrac{d\omega_e \mid A}{d\mathrm{tr} \mid A} x^*y^*ab\right)$, belongs to $B \otimes \mathcal{L}(H)$.*

Then I is a regular multiplicative partial isometry, its right (resp., left) leg generates A (resp., B): the pair (A, B) is in a natural way a pair of quantum groupoids in duality.

Proof. As $I_{12}^*(I_{23}I_{12}I_{23}^*) = (I_{12}^*I_{23}I_{12})I_{23}^*$, so $I_{12}^*I_{23}I_{12}I_{23}^*$ belongs to $B \otimes N_\alpha \otimes A$ by Proposition 4.4.6, also $I_{13}(1 \otimes e_{\alpha,\beta})$ belongs to $B \otimes N_\alpha \otimes A$. As Z and Z' are partial isometries with initial support respectively equal to $e_{\hat{\beta},\alpha}$ and $e_{\alpha,\beta}$, then for every η in H one has $I(e \otimes \eta) = e_{\alpha,\beta}(e \otimes \eta)$ and $I(\eta \otimes \hat{e}) = e_{\hat{\beta},\alpha}(\eta \otimes \hat{e})$. Hence, for any x

in A, n in N and η in H, one has

$$\begin{aligned}
&(I_{12}^*I_{23}I_{12})I_{23}^*(xe\otimes\alpha(n)e\otimes\eta),\hat{e}\otimes\hat{e}\otimes e)\\
&\quad=(I_{23}I_{12}I_{23}^*(xe\otimes e\otimes\beta(n)\eta),I_{12}(\hat{e}\otimes\hat{e}\otimes e))\\
&\quad=(I_{23}I_{12}(1\otimes e_{\hat{\beta},\alpha})(xe\otimes e\otimes\beta(n)\eta),(e_{\alpha,\beta}\otimes 1)(\hat{e}\otimes\hat{e}\otimes e))\\
&\quad=((e_{\alpha,\beta}\otimes 1)I_{23}(1\otimes e_{\hat{\beta},\alpha})I_{12}(xe\otimes e\otimes\beta(n)\eta),(\hat{e}\otimes\hat{e}\otimes e))\\
&\quad=(I_{23}I_{12}(xe\otimes e\otimes\beta(n)\eta),(\hat{e}\otimes\hat{e}\otimes e))\\
&\quad=(I((\omega_{xe,\hat{e}}\otimes i)(I)\otimes 1)(e\otimes\beta(n)\eta),\hat{e}\otimes e)\\
&\quad=(I(xg_te\otimes\beta(n)\eta),\hat{e}\otimes e)\\
&\quad=((\omega_{xg_te,\hat{e}}\otimes i)(I)\beta(n)\eta,e)=(xg_t^2\beta(n)\eta,e)\\
&\quad=(g_t\beta(n)\eta,g_tx^*e)\\
&\quad=((\omega_{e,\hat{e}}\otimes i)(I)\beta(n)\eta,g_tx^*e)\\
&\quad=(I(e\otimes\beta(n)\eta),\hat{e}\otimes g_tx^*e)\\
&\quad=((1\otimes xg_t)e_{\alpha,\beta}(e\otimes\beta(n)\eta),\hat{e}\otimes e)\\
&\quad=((1\otimes(\omega_{xe,\hat{e}}\otimes i)(I))e_{\alpha,\beta}(e\otimes\beta(n)\eta),\hat{e}\otimes e)\\
&\quad=(I_{13}(xe\otimes e_{\alpha,\beta}(e\otimes\beta(n)\eta),\hat{e}\otimes e)\\
&\quad=(I_{13}(1\otimes e_{\alpha,\beta})(xe\otimes\alpha(n)e\otimes\eta),\hat{e}\otimes\hat{e}\otimes e).
\end{aligned}$$

Due to Lemma 4.4.7, one deduces $I_{12}^*I_{23}I_{12}I_{23}^* = I_{13}(1\otimes e_{\alpha,\beta})$. Now, multiplying on the left by I_{12} and on the right by I_{23}, one gets the pentagonal equality $I_{23}I_{12} = I_{12}I_{13}I_{23}$, and the theorem follows immediately. □

4.4.9 Corollary. *Let A, B be two involutive subalgebras of $\mathcal{L}(H)$ of the same finite dimension as the Hilbert space H, and let e, $\hat{e}$ be two completely normalized vectors for the pair (A, B). We suppose that*

i) $E_\alpha(ab) = E_\alpha(a)E_\alpha(b)$ *for all (a, b) in $A\times B$,*

ii) *the orthogonal projection onto $(A'\cap B')\hat{e}$ belongs to A and the one onto $(A'\cap B')e$ belongs to B,*

iii) *the application $I\in\mathcal{L}(H\otimes H)$ defined, for all a, x in A and b, y in B, by*

$$(I(ae\otimes b\hat{e}),xe\otimes y\hat{e}) = \mathrm{tr}(n_\alpha^{-1}x^*y^*ab),$$

belongs to $B\otimes\mathcal{L}(H)$.

Then I is a regular multiplicative partial isometry, its right (resp., left) leg generates A (resp. B); the pair (A, B) is in a natural way a pair of weak Kac algebras in duality.

Proof. Let A, B satisfy conditions i), ii) and iii); then, by 4.3, condition iii) of Theorem 4.4.8 holds and this theorem is applicable. Now $\delta = n_\alpha$, so A and B are weak Kac algebras by Corollary 4.2.6. □

References

[BS] S. Baaj and G. Skandalis, Unitaires multiplicatifs et dualité pour les produits croisés de C*-algèbres, *Ann. Sci. École Norm. Sup.* (4) 26 (1993), 425–488.

[BBS] S. Baaj , E. Blanchard and G. Skandalis, Unitaires multiplicatifs en dimension finie et leurs sous-objets, *Ann. Inst. Fourier* 49 (1999), 1305–1344.

[BoSz] G. Böhm and K. Szlachányi, Weak C*-Hopf algebras: the coassociative symmetry of non integral dimensions, in *Quantum groups and quantum spaces* (R. Budzynski et al., eds.), Banach Center Publ. 40 Warsaw 1997, 9–19.

[BoSzNi] G. Böhm, K. Szlachányi and F. Nill, Weak Hopf Algebras I, Integral Theory and C^*-structure, *J. Algebra* 221 (1999), 385–438.

[EV] M. Enock and J. M. Vallin, Inclusions of von Neumann algebras and quantum groupoids, *J. Funct. Anal.* 172 (2000), 249–300.

[NV1] D. Nikshych and L. Vainerman, Algebraic versions of a finite-dimensional quantum groupoid, in *Hopf algebras and quantum groups* (S. Caenepeel et al., eds.), Lecture Notes in Pure and Appl. Math. 209, Marcel Dekker, New York 2000, 189–221.

[NV2] D. Nikshych and L. Vainerman, A characterization of depth 2 subfactors of II_1 factors, *J. Funct. Anal.* 171 (2000), 278–307 (2000).

[NV3] D. Nikshych and L. Vainerman, Finite Quantum Groupoids and Their Applications, preprint `math.QA/0006057` (2000).

[NV4] D. Nikshych and L. Vainerman, A Galois correspondence for II_1-factors and quantum groupoids, *J. Funct. Anal.* 178 (2000), 113–142.

[Val0] J. M. Vallin, Unitaire pseudo-multiplicatif associé à un groupoïde. Applications à la moyennabilité, *J. Operator Theory* 44, No.2 (2000), 347-368.

[Val1] J. M. Vallin, Groupoïdes quantiques finis. *J. Algebra* 26 (2001), 425–488.

Multiplier Hopf *-algebras with positive integrals: A laboratory for locally compact quantum groups

Alfons Van Daele

Department of Mathematics
K.U. Leuven, Celestijnenlaan 200B, 3001 Heverlee, Belgium
email: `Alfons.VanDaele@wis.kuleuven.ac.be`

Abstract. Any multiplier Hopf *-algebra with positive integrals gives rise to a locally compact quantum group (in the sense of Kustermans and Vaes). As a special case of such a situation, we have the compact quantum groups (in the sense of Woronowicz) and the discrete quantum groups (as introduced by Effros and Ruan). In fact, the class of locally compact quantum groups arising from such multiplier Hopf *-algebras is self-dual.

The most important features of these objects are (1) that they are of a purely algebraic nature and (2) that they have already a great complexity, very similar to the general locally compact quantum groups. This means that they can serve as a good model for the general objects, at least from the purely algebraic point of view. They can therefore be used to study various aspects of the general case, without going into the more difficult technical aspects, due to the complicated analytic structure of a general locally compact quantum group.

In this paper, we will first recall the notion of a multiplier Hopf *-algebra with positive integrals. Then we will illustrate how these *algebraic quantum groups* can be used to gain a deeper understanding of the general theory. An important tool will be the Fourier transform. We will also concentrate on certain actions and how they behave with respect to this Fourier transform. On the one hand, we will study this in a purely algebraic context while on the other hand, we will also pass to the Hilbert space framework.

1 Introduction

A locally compact quantum group is a pair (A, Δ) of a C^*-algebra A and a comultiplication Δ on A, satisfying certain properties. If A is an abelian C^*-algebra, it then has the form $C_0(G)$, the C^*-algebra of all continuous complex functions, tending to 0 at infinity on a locally compact group G and the comultiplication Δ is given by the formula $(\Delta(f))(p, q) = f(pq)$ where $f \in C_0(G)$ and pq is the product in G of the elements p, q. Observe that in this case, $\Delta(f)$ is a bounded continuous complex function on $G \times G$ that in general, will not belong to $C_0(G \times G)$. Indeed, also in the general case, the comultiplication Δ is a *-homomorphism on A with values in

$M(A \otimes A)$, the multiplier algebra of the spatial C^*-tensor product $A \otimes A$ of A with itself.

Any locally compact group G carries a left and a right Haar measure. This is also true for a locally compact quantum group. In this case, these are (nice) invariant weights on the C^*-algebra. An important fact is that, in the quantum case, the existence of these weights is part of the axioms, whereas in the classical theory, it is possible to prove the existence of the Haar measures. Such existence theorems exist only in special cases for locally compact quantum groups however (and it seems that a general existence theorem is still out of sight).

The structure of a locally compact quantum group is very rich, but also technically difficult to work with. It requires not only the standard results on operator algebras (like the Tomita–Takesaki theory), but also fundamental skills with weights on C^*-algebras, unbounded operators on Hilbert spaces, ... And all of this comes on top of a highly non-trivial algebraic structure, involving a lot of objects. This makes it rather hard to work with locally compact quantum groups. Moreover, the present non-trivial examples are very complicated (although something is changing here, thanks to work done by Vainerman and Vaes, see e.g. [20] and also [21]). All of this makes it difficult, and perhaps not very attractive, to try to learn the theory and start working in it. Nevertheless, all the people familiar with the theory know that the structure is very rich and that this is a nice piece of mathematics.

Fortunately, there are the 'algebraic quantum groups'. These are multiplier Hopf (*)-algebras with (positive) integrals (see Section 2 where we start by recalling this notion). As we mentioned already in the abstract, any multiplier Hopf *-algebra with positive integrals gives rise (in a straightforward and easy way) to a locally compact quantum group. However, not all locally compact quantum groups are of this form. The compact and discrete quantum groups belong to this class and some combinations of those two (like the Drinfel'd double of a compact quantum group). The class is also self-dual. Among the locally compact groups it seems to be possible to characterize those coming from a multiplier Hopf algebra ([15]). Such a result is not yet available for locally compact quantum groups.

Algebraic quantum groups are of a purely algebraic nature and it is possible to work with them without going into deep analysis. Nevertheless, the structure is very rich and from the algebraic point of view, contains all features of the general locally compact quantum groups. All of the relevant data are present and essentially no extra relations are imposed by the restriction to these algebraic quantum groups. For completeness, we have to mention however that we still don't know of examples of algebraic quantum groups where the scaling group is not leaving the integrals invariant – this is still open. On the other hand, in the non *-case, such examples are known (and are in fact not so complicated), see e.g. [24] or [27].

It seems fair to say that the development of the general theory of locally compact quantum groups (by Kustermans and Vaes, see [9] and [10]) has been possible, among other reasons, because of the work done before by Kustermans and myself on algebraic quantum groups (see [13]). And, as we indicated already above, it is a common practice to verify general results about locally compact quantum groups first for algebraic quantum groups (where only the algebraic aspects have to be considered). Extending these results to the general case later is usually fairly complicated, but there is always a good chance that it can be done. Indeed, algebraic quantum groups are *a good model* for general locally compact quantum groups.

And there is more. Not only to obtain new results, but also for understanding the old ones, it is important to get first some familiarity with the framework of algebraic quantum groups (as probably, the authors themselves have done before obtaining their general result; of course, not publishing this intermediate step).

This is precisely what this note is all about: After recalling some of the basics of multiplier Hopf *-algebras with positive integrals (in Section 2), we illustrate the above strategy in the two following sections. In Section 3, we take a certain point of view, starting from a dual pair of multiplier Hopf *-algebras. The *-structure however does not play an essential role here. On the other hand, in Section 4, we pass to the Hilbert space level and there the *-operation and positivity of the integrals becomes essential.

The key to our approach here is the Fourier transform. In the general theory, the Fourier transform is not very explicit. The main reason is that the Hilbert spaces, $L^2(G)$ and $L^2(\hat{G})$ in the classical case of an abelian locally compact group, are identified through this Fourier transform in the general quantum case. This common practice has clear advantages, but it also makes some features less transparent.

In the present note, very few proofs are given. In the first part of this paper we recall some of the basic notions and known results. Details can be found elsewhere and references will be given. On the other hand, many other results that we present later, are not yet found (in this form) in other papers and it is our intention to publish details together with J. Kustermans in [14]. However, we must say that essentially most of the results are, in some form, already present in one of the papers [10], [11] and [13]. The main difference is the explicit use of the Fourier transform. Recall that after all, this paper is meant to serve mainly as a tool for learning and understanding the subject.

For the standard notions and results on Hopf algebras, we refer to the basic works of Abe [1] and Sweedler [18]. For some information about dual pairs of Hopf algebras, we refer to [22]. For the theory of multiplier Hopf algebras, the reference is [23] while algebraic quantum groups (multiplier Hopf algebras with integrals) are studied in [24]. Dual pairs of multiplier Hopf algebras are treated in [3] and actions of multiplier Hopf algebras in [4]. A survey on the theory of multiplier Hopf algebras is given in [27].

Then, as part of this work also takes place in Hilbert spaces, we need to give some references about operator algebras also. Much of this can be found in [6] but also a good reference is [16]. For the theory of weights and the Tomita–Takesaki theory, we refer to [17]. The theory of Kac algebras is to be found in [5].

Finally, we would like to say something about conventions. The algebras we deal with are algebras over the complex numbers and may or may not have an identity. If there is no identity however, the product is assumed to be non-degenerate (as a bilinear form). We are mainly interested in *-algebras. These are algebras with an involution $a \mapsto a^*$ satisfying the usual properties. Essentially, these *-algebra structures are always of a certain type because we assume that there is a faithful positive linear functional. For the comultiplications, we use the symbol Δ. This comes from Hopf algebra theory. However, this choice is not completely obvious here as the same symbol is also commonly used for the modular function of a non-unimodular locally compact group and (related) for the modular operator in the Tomita–Takesaki theory. We will use other symbols for these objects.

In fact, the difficulty arises from the fact that this material is relating two completely different fields in mathematics. The first one is the theory of Hopf algebras and the second one is the theory of operator algebras. Different customs are usual in these two areas. Since we are mainly interested in the theory of locally compact quantum groups, that is formulated in the operator algebra framework, we will follow what is common there. We will use however the Sweedler notation as it is justified to do so and of course it makes many formulas and arguments much more transparent. Indeed, we would like this paper also to be readable for the Hopf algebra people and we hope that our third section (where we do not emphasize on the involutive structure) will serve as a bridge between the two areas.

Acknowledgements. First, I would like to thank my colleagues (and friends) at the Institute of Mathematics in Oslo, where part of this work was done, for their hospitality during my visit in November 2001. Secondly, I am grateful to the organizers of the meeting in Strasbourg, in particular L. Vainerman, for giving me the opportunity to talk about my work. I also like to thank my coworkers J. Kustermans and S. Vaes for many fruitful discussions on this subject. Finally, I like to thank A. Jacobs for some LaTeX-help.

2 Algebraic quantum groups

We will first briefly recall the notion of a multiplier Hopf *-algebra. For details, we refer to [23], see also [27].

Definition 2.1. A *multiplier Hopf*-algebra* is a pair (A, Δ) of a *-algebra A (with a non-degenerate product) and a comultiplication Δ on A such that the linear maps T_1 and T_2 defined on $A \otimes A$ by

$$T_1(a \otimes a') = \Delta(a)(1 \otimes a')$$
$$T_2(a \otimes a') = (a \otimes 1)\Delta(a')$$

are one-to-one and have range equal to $A \otimes A$.

We have to give some more explanation.

The *-algebra may or may not have an identity. However, the product, as a bilinear map, must be *non-degenerate*. This is automatic if an identity exists. For an algebra A with a non-degenerate product, one can define the so-called *multiplier algebra*. It contains A as an essential ideal and it has an identity. In fact, it is the largest algebra with these properties. Because A is a *-algebra, the multiplier algebra $M(A)$ is also a *-algebra. The tensor product $A \otimes A$ is again a *-algebra with a non-degenerate product and also the multiplier algebra $M(A \otimes A)$ can be constructed. Elements of the form $1 \otimes a$ and $a \otimes 1$ exist in $M(A \otimes A)$ for all $a \in A$.

A *comultiplication* on A is a *-homomorphism $\Delta : A \to M(A \otimes A)$ which is non-degenerate and coassociative. To be non-degenerate here means that $\Delta(A)(A \otimes A) = A \otimes A$. This property is automatic when A has an identity 1 and when Δ is unital, i.e. $\Delta(1) = 1 \otimes 1$. Because of the non-degeneracy of Δ, it is possible to extend the obvious maps $\Delta \otimes \iota$ and $\iota \otimes \Delta$ (where ι is the identity map) on $A \otimes A$ to maps from $M(A \otimes A)$ to $M(A \otimes A \otimes A)$. This is why coassociativity makes sense in the form $(\Delta \otimes \iota)\Delta = (\iota \otimes \Delta)\Delta$. Finally, the linear maps T_1 and T_2, as defined in the definition, will be maps from $A \otimes A$ to $M(A \otimes A)$. The requirement is that they are injective, have range in $A \otimes A$ and that all of $A \otimes A$ is in the range of these maps.

The following is the motivating example for this notion.

Example 2.2. Let G be a group and let A be the algebra $K(G)$ of complex functions with finite support in G. Then $A \otimes A$ is identified with $K(G \times G)$ while $M(A \otimes A)$ is the algebra of all complex functions on $G \times G$. The map Δ, defined by $\Delta(f)(p, q) = f(pq)$ whenever $p, q \in G$ and $f \in K(G)$, will be a comultiplication on A. Coassociativity is a consequence of the associativity of the group multiplication in G.

Here is the relation with the notion of a Hopf *-algebra (see [23]).

Proposition 2.3. *If (A, Δ) is a Hopf *-algebra, then it is a multiplier Hopf *-algebra. Conversely, if (A, Δ) is a multiplier Hopf *-algebra and if A has an identity, then it is a Hopf *-algebra.*

Proof (sketch). i) If (A, Δ) is a Hopf *-algebra, the inverses of the maps T_1 and T_2 in Definition 2.1 are given in terms of the antipode S:

$$T_1^{-1}(a \otimes a') = (\iota \otimes S)(\Delta(a))(1 \otimes a')$$
$$T_2^{-1}(a \otimes a') = (a \otimes 1)(S \otimes \iota)(\Delta(a')).$$

ii) On the other hand, if (A, Δ) is any multiplier Hopf *-algebra, the above formulas can be used to construct an antipode (and a counit). The counit is a *-homomorphism $\varepsilon : A \to \mathbb{C}$ such that $(\varepsilon \otimes \iota)\Delta(a) = a$ and $(\iota \otimes \varepsilon)\Delta(a) = a$ for all $a \in A$. The antipode is a anti-homomorphism $S : A \to A$ satisfying $S(S(a)^*)^* = a$ and

$$m(S \otimes \iota)\Delta(a) = \varepsilon(a)1$$
$$m(\iota \otimes S)\Delta(a) = \varepsilon(a)1$$

for all $a \in A$ (where m is multiplication as a linear map from $A \otimes A$ to A). These formulas are given a meaning in $M(A)$.

iii) So, if (A, Δ) is a multiplier Hopf *-algebra with an identity, then it is automatically a Hopf *-algebra. □

It is obvious that the antipode S and the counit ε in the case of the Example 2.2 are given by $S(f)(p) = f(p^{-1})$ and $\varepsilon(f) = f(e)$ where e is the identity in the group and where p^{-1} is the inverse of p.

Next we recall the notion of an integral on a multiplier Hopf *-algebra (see [24]).

Definition 2.4. A linear functional φ on A satisfying $(\iota \otimes \varphi)\Delta(a) = \varphi(a)1$ for all $a \in A$ is called left invariant. A linear functional ψ on A is called right invariant if $(\psi \otimes \iota)\Delta(a) = \psi(a)1$ for all $a \in A$. A non-zero left invariant functional is called a *left integral* while a non-zero right invariant functional is called a *right integral.*

Observe that the above formulas, expressing invariance, again must be considered in $M(A)$. Left invariance of φ should e.g. be written in the form $(\iota\otimes\varphi)((a'\otimes 1)\Delta(a)) = \varphi(a)a'$ for all $a, a' \in A$.

We have the following results on integrals on a multiplier Hopf *-algebra (see [24]).

Proposition 2.5. *Let (A, Δ) be a multiplier Hopf *-algebra and assume that a left integral φ exists. Then we have:*

i) *There is also a right integral ψ.*
ii) *The left and right integrals are unique, up to a scalar.*
iii) *The integrals are faithful.*
iv) *There exists an invertible multiplier δ in $M(A)$ such that $(\varphi \otimes \iota)\Delta(a) = \varphi(a)\delta$ and $(\iota \otimes \psi)\Delta(a) = \psi(a)\delta^{-1}$.*
v) *There exists automorphisms σ and σ' of A such that $\varphi(aa') = \varphi(a'\sigma(a))$ and $\psi(aa') = \psi(a'\sigma'(a)))$ for all $a, a' \in A$.*
vi) *There exists a scalar $\nu \in \mathbb{C}$ such that $\varphi(S^2(a)) = \nu\varphi(a)$ for all $a \in A$.*

In fact, this result is true for any regular multiplier Hopf algebra (again see [24]).

In the case of a multiplier Hopf *-algebra, it is natural to assume that the left integral φ is positive, i.e. that $\varphi(a^*a) \geq 0$ for all $a \in A$. In that case, we have some more consequences for the above data. It can be shown e.g. that automatically a positive right integral exists. This is not obvious however and the only argument available now is given in [13]. Also, we have that $|\nu| = 1$. Furthermore, some very nice analytic properties can be proven about the multiplier δ and about the automorphisms σ and σ' (see [8]).

There are many extra properties and relations among the different data that appear in Proposition 2.5. We refer to [14] for a collection of them.

From now on, we will assume that (A, Δ) is a multiplier Hopf *-algebra with positive integrals. We will also use the term (*-)*algebraic quantum group.* Observe that the adjective '*algebraic*' is not referring to the concept of an algebraic group, but rather to the purely algebraic framework that one can use in the study of this type of locally compact quantum groups.

We will use φ to denote a positive left integral and we use ψ for a positive right integral. It is possible to give a standard relative normalization of φ and ψ (see e.g. [14]), but we will not need it.

We now turn our attention to *the dual.* We have the following result (again see [24]).

Theorem 2.6. *Let (A, Δ) be a multiplier Hopf*-algebra with positive integrals. Let $\hat{A}$ be the subspace of the dual space of A, consisting of linear functionals on A of the form $\varphi(\cdot\, a)$ where φ is a left integral and $a \in A$. Then $\hat{A}$ can be made into a multiplier Hopf*-algebra and again it has positive integrals.*

First observe that the elements of $\hat{A}$ are also the ones of the form $\varphi(a\,\cdot)$, $\psi(\cdot\, a)$ or $\psi(a\,\cdot)$ where always a runs through A. The product in $\hat{A}$ is obtained by dualizing the coproduct and the involution is given by $\omega^*(a) = \omega(S(a)^*)^-$ where S is the antipode. The coproduct $\hat{\Delta}$ on $\hat{A}$ is dual to the product in A. A right integral $\hat{\psi}$ on A is obtained by letting $\hat{\psi}(\omega) = \varepsilon(a)$ when $\omega = \varphi(\cdot\, a)$. With this definition, we get $\hat{\psi}(\omega^*\omega) = \varphi(a^*a)$ when as before $\omega = \varphi(\cdot\, a)$. So indeed, $\hat{\psi}$ is again positive. Applying the procedure once more takes us back to the original multiplier Hopf *-algebra. For details here see [24].

There are also many formulas relating the data for (A, Δ), namely ε, S, φ, ψ, σ, σ', ν with the corresponding data for the dual algebra $(\hat{A}, \hat{\Delta})$. See e.g. [8] and also [14].

In the *remaining of this paper*, we will let B be the dual algebra $\hat{A}$, but we will consider it with the *opposite comultiplication* $\hat{\Delta}^{\mathrm{op}}$. We will systematically use letters $a, a', \ldots$ (and sometimes x, x') to denote elements in A and letters $b, b', \ldots$ (or y, y') for elements in B. We will use $\langle a, b\rangle$ for the evaluation of the element b in the element a. We will (in general) also drop the symbol ^ on the objects related with B. In other words, we will be working with a (modified) dual pair (A, B) of multiplier Hopf *-algebras with positive integrals (as introduced in [3]). We call it '*modified*' because of the fact that the coproduct in B has been reversed (contrary to the original definition in [3]). This has e.g. as a consequence that $S = \hat{S}^{-1}$ on B and so

$$\langle S(a), b\rangle = \langle a, S^{-1}(b)\rangle$$

for all $a \in A$ and $b \in B$. Also $\varphi = \hat{\psi}$ and $\psi = \hat{\varphi}$ on B.

The reason for this modification is to get the theory here in accordance with the general theory of locally compact quantum groups.

A last remark for this section: we will use the Sweedler notation. It is justified as long as we have the right *coverings*: e.g. $\Delta(a)(1 \otimes a')$ is written as $a_{(1)} \otimes a_{(2)}a'$ and we say that $a_{(2)}$ is covered by a'. See [3] for a detailed discussion on the use of the Sweedler notation for multiplier Hopf algebras. Observe that sometimes, the covering is through the pairing. The reason is that given an element b in B e.g. there exists an element e in A such that $\langle a, b\rangle = \langle ea, b\rangle$ for all $a \in A$ (cf. [4]) so that the element b will, in a sense, cover the element a, through the pairing, by means of this element e.

3 Actions and the Fourier transform

As in the previous section, also here (A, B) will be a dual pair of multiplier Hopf *-algebras with positive integrals. Again we will assume that $B = \hat{A}$, and that B is endowed with the opposite coproduct $\Delta = \hat{\Delta}^{\mathrm{op}}$.

We will now define an action of A and an action of B on A.

Definition 3.1. We define linear maps $\pi(a)$ for $a \in A$ and $\lambda(b)$ for $b \in B$ on the space A by

$$\pi(a)x = ax$$
$$\lambda(b)x = \langle S^{-1}(x_{(1)}), b\rangle x_{(2)}$$

where $x \in A$.

Observe that $x_{(1)}$ is covered, through the pairing, by b.

One can verify that these are indeed actions of A and of B.

Remark that we have used x for an element in A. We will do this systematically for elements in A when we treat A as the space on which the algebras act. We will use the letter y for elements in B when we have actions on the space B.

Throughout this paper, the reader should have in mind that $\pi(\,\cdot\,)$ will be used for '*multiplication*' operators and that $\lambda(\,\cdot\,)$ will be used for '*convolution*' operators.

We have the following commutation rules.

Lemma 3.2. *For $a \in A$ and $b \in B$ we have*

$$\pi(a)\lambda(b) = \langle a_{(1)}, b_{(1)}\rangle\, \lambda(b_{(2)})\pi(a_{(2)}).$$

These relations are called the *Heisenberg commutation relations*. One can show that the linear span of the operators $\lambda(b)\pi(a)$ is precisely the linear span of the rank one maps on A of the form $x \mapsto \langle x, b'\rangle a'$ where $a' \in A$ and $b' \in B$.

We have already considered the bijection $a \mapsto \varphi(\,\cdot\, a)$ from A to B. This is one possible *Fourier transform*. We will also consider another one:

Definition 3.3. We define two maps $F_1, F_2 : A \to B$ by

$$F_1(a) = \varphi(\,\cdot\, a)$$
$$F_2(a) = \psi(S(\,\cdot\,)a)$$

Observe that these maps depend on the normalization of φ and ψ. They are related by a simple formula: F_2 is a scalar multiple of $F_1 \circ \sigma \circ S^{-1}$. The two maps can also be regarded in their dual forms. Then, we get the following:

Proposition 3.4. *For all $a \in A$ and $b \in B$ we have*

$$\text{i)}\quad b = \varphi(\,\cdot\, a) \quad \Leftrightarrow \quad a = \varphi(S^{-1}(\,\cdot\,)b)$$
$$\text{ii)}\quad b = \psi(S(\,\cdot\,)a) \quad \Leftrightarrow \quad a = \psi(\,\cdot\, b).$$

In this case, we need the relative normalization of the left integrals on A and B together with the relative normalization of the right integrals on A and B w.r.t. each other (cf. [23] and also [14]).

Later in this section, we will give another form of these formulas.

As we mentioned already, these maps play the role of the Fourier transforms. Observe that $\varphi(b^*b) = \varphi(a^*a)$ when $b = F_1(a) = \varphi(\,\cdot\, a)$ (see further). This is a result that we have mentioned before already, taken into account that $\varphi = \hat{\psi}$ on B. For the other transform, we have $\psi(b^*b) = \psi(a^*a)$ when $b = F_2(a) = \psi(S(\,\cdot\,)a)$. As we have chosen to work basically with the left integrals, we will be mainly using the Fourier transform F_1 and we will drop the index and simply write F for F_1.

What happens with the actions under this Fourier transform? The Fourier transform does convert the multiplication operators to convolution operators and the other way around. More precisely, we get the following:

Proposition 3.5. *For all $a, x \in A$ and $b \in B$ we have*

$$F(\pi(a)x) = \lambda(a)F(x)$$
$$F(\lambda(b)x) = \pi(b)F(x)$$

where $\lambda(a)$ and $\pi(b)$ are defined on B by

$$\lambda(a)y = \langle a, y_{(1)}\rangle y_{(2)}$$
$$\pi(b)y = by$$

with $y \in B$.

One can verify e.g. that, for $a, a' \in A$ and $y \in B$,

$$\begin{aligned}\lambda(aa')y &= \langle aa', y_{(1)}\rangle y_{(2)} \\ &= \langle a, y_{(2)}\rangle \langle a', y_{(1)}\rangle y_{(3)} \\ &= \lambda(a)\langle a', y_{(1)}\rangle y_{(2)} \\ &= \lambda(a)\lambda(a')y.\end{aligned}$$

Observe the difference with the action λ of B on A where the antipode was involved. This is not so here, for the action λ of A on B, a fact that is related with taking the opposite comultiplication on the dual (as we can notice in the above argument).

We have a similar set of formulas when we look at F_2 in stead of F_1.

Now, we relate all of this with the so-called *left regular representation.*

Definition 3.6. Consider the map V from $A \otimes A$ to itself defined by

$$V(x \otimes x') = \Delta(x')(x \otimes 1).$$

This map is bijective and the inverse is given by

$$V^{-1}(x \otimes x') = S^{-1}(x'_{(1)})x \otimes x'_{(2)}$$

(using the Sweedler notation). We will denote V^{-1} by W. Then we get the following:

Lemma 3.7. *The map W verifies the Pentagon equation*

$$W_{12}W_{13}W_{23} = W_{23}W_{12}$$

(where we use the leg-numbering notation).

This equation is to be considered, in the first place, as an equality of linear maps from $A \otimes A \otimes A$ to itself.

In the present context, however, we can say more.

Proposition 3.8. *We have $W \in M(A \otimes B)$ and $\langle W, b \otimes a\rangle = \langle a, b\rangle$ for all $a \in A$ and $b \in B$. Also $W^{-1} \in M(A \otimes B)$ and $\langle W^{-1}, b \otimes a\rangle = \langle S(a), b\rangle = \langle a, S^{-1}(b)\rangle$.*

These formulas need some explanation. The algebra $A \otimes B$ acts on $A \otimes A$ in the obvious way (where we use the action λ of B and π of A as defined in Definition 3.1). This action of $A \otimes B$ is non-degenerate and therefore extends to an action of the multiplier algebra $M(A \otimes B)$ in a unique way (see e.g. [4]). The first statement in the above proposition says that the linear operator W is the action of some multiplier in $M(A \otimes B)$. For the second result, we observe that we have a natural pairing of $A \otimes B$ with $B \otimes A$ (coming from the pairing of A with B) and that this pairing can be extended to a bilinear map pairing $M(A \otimes B)$ with $B \otimes A$.

The fact that $W \in M(A \otimes B)$ makes it possible to interpret certain (well-known) formulas in another, nicer way. One can now show that the formula

$$(\Delta \otimes \iota)W = W_{13}W_{23}$$

makes sense in the algebra $M(A \otimes A \otimes B)$. For this we have to observe that $\Delta \otimes \iota$ is a non-degenerate *-homomorphism from $A \otimes B$ to $M(A \otimes A) \otimes B$, which is naturally imbedded in $M(A \otimes A \otimes B)$, and hence has a unique extension to a unital *-homomorphism from $M(A \otimes B)$ to $M(A \otimes A \otimes B)$.

The other equation, namely $\Delta(a) = W^{-1}(1 \otimes a)W$ is somewhat more tricky. This equation is clear when considered as an equation of linear maps from $A \otimes A$ to itself. But in these circumstances, it can also be viewed as an equation in $M(A \otimes C)$ where C is the algebra generated by A and B, taking into account the commutation relations of Lemma 3.2. This algebra C is called the *Heisenberg algebra* (or sometimes the Heisenberg double) but, as we mentioned before, its structure only depends on the pairing of the linear spaces A and B. Therefore we tend to think of the Heisenberg algebra C as an algebra with two special subalgebras A and B, sitting in the multiplier algebra $M(C)$, rather then of the algebra itself.

Finally, there is the equation

$$W^{-1} = (S \otimes \iota)W = (\iota \otimes S^{-1})W.$$

This equation can be given a meaning using not only that $W \in M(A \otimes B)$ but also that $(a \otimes 1)W(1 \otimes b)$ and $(1 \otimes b)W(a \otimes 1)$ are in fact elements in $A \otimes B$. If e.g. we apply $S \otimes \iota$ to the first of these two elements, we get formally, because S is an anti-homomorphism $((S \otimes \iota)W)(S(a) \otimes b)$ and this is then $W^{-1}(S(a) \otimes b)$.

Using the fact that W is (in a way) the duality, we can rewrite the first formula of Proposition 3.4 as

$$b = F(a) = (\varphi \otimes \iota)(W(a \otimes 1))$$
$$a = F^{-1}(b) = (\iota \otimes \varphi)(W^{-1}(1 \otimes b)).$$

Using these expressions, we have an easy way to obtain the Plancherel formula. Indeed, using the same notations as above,

$$\begin{aligned}\varphi(b^*b) &= \varphi(b^*((\varphi \otimes \iota)W(a \otimes 1)))\\ &= (\varphi \otimes \varphi)((1 \otimes b^*)W(a \otimes 1))\\ &= \varphi((\iota \otimes \varphi)((1 \otimes b^*)W)a) = \varphi(a^*a)\end{aligned}$$

We have used that W is a unitary in the *-algebra case and so $W^* = W^{-1}$.

The Pentagon equation for W (Lemma 3.7) is in fact equivalent with the Heisenberg commutation relations (Lemma 3.2) given that W is the duality (cf. Proposition 3.8). The relation with the Heisenberg commutation rules becomes also more apparent in the following formula.

Proposition 3.9. *If we transform the map W on $A \otimes A$ to a map on $B \otimes A$ using the Fourier transform F (i.e. F_1 in Definition 3.3) on the first leg, we get the map*

$$y \otimes x \mapsto \langle S^{-1}(x_{(1)}), y_{(1)}\rangle y_{(2)} \otimes x_{(2)}.$$

Indeed, we had $W(x' \otimes x) = S^{-1}(x_{(1)})x' \otimes x_{(2)}$ and we know from Proposition 3.5 that $F(ax') = \lambda(a)F(x')$ where $\lambda(a)y = \langle a, y_{(1)}\rangle y_{(2)}$.

It should be observed that the inverse of the above map is

$$y \otimes x \mapsto \langle x_{(1)}, y_{(1)}\rangle y_{(2)} \otimes x_{(2)}$$

and that these two maps are indeed well-defined maps from $B \otimes A$ to itself (which is not completely obvious but follows from the previous considerations, see also [3]). Also remark that these two maps precisely govern the Heisenberg commutation relations (see Lemma 3.2).

We finish this section by another application of the above expressions.

Proposition 3.10. *Let A be finite-dimensional and denote by* tr *the trace on the algebra C. Then, for some scalar $k \in \mathbb{C}$, we have*

$$\varphi(a) = k\,\mathrm{tr}(\pi(a))$$

for all $a \in A$.

Proof. We will show that the right hand side is left invariant and then we get the formula by uniqueness of left invariant functionals. Observe that $\mathrm{tr}(\pi(1)) = \mathrm{tr}(1) \neq 0$.

So let $a \in B$. We will write $\sum u_i \otimes v_i$ for W. We will also drop π and simply write a for $\pi(a)$. Then we have

$$\begin{aligned}(\iota \otimes \mathrm{tr})\Delta(a) &= (\iota \otimes \mathrm{tr})(W^{-1}(1 \otimes a)W) \\ &= (\iota \otimes \mathrm{tr})((S \otimes \iota)W(1 \otimes a)W) \\ &= \sum (\iota \otimes \mathrm{tr})((1 \otimes a)(S(u_i) \otimes 1)W(1 \otimes v_i)) \\ &= \sum S((\iota \otimes \mathrm{tr})((1 \otimes a)W^{-1}(u_i \otimes v_i))) \\ &= \mathrm{tr}(a)S(1) = \mathrm{tr}(a).\end{aligned}$$

□

Observe that we have used the trace property and also the fact that in this case $S^2 = \iota$. By composing with the antipode, we see that here the trace is also left invariant. The above argument can be used to prove the existence of integrals on Hopf *-algebras with a nice underlying *-algebra structure.

In fact, the argument can be modified so that it also works for any finite-dimensional Hopf algebra. In this case, one has to make one modification because possibly $S^2 \neq \iota$ and another one because it might happen that the above functional is trivially 0. We refer to [14].

4 Actions on Hilbert spaces

In the previous section, we essentially did not use the *-structure. In this section, we will see the consequences of the positivity of the integrals. We refer to [10], [11] and [13] where more details can be found; see also [14].

A basic construction is the so-called *G.N.S.-construction*:

Definition 4.1. Let $\mathcal{H}$ denote the Hilbert space obtained by completing A for the scalar product given by $(x', x) \mapsto \varphi(x^*x')$. Denote by $x \mapsto \eta(x)$ the canonical imbedding of A in $\mathcal{H}$ so that $\langle \eta(x'), \eta(x) \rangle = \varphi(x^*x')$ for all $x, x' \in A$. Similarly, we will use $\hat{\mathcal{H}}$ and $\hat{\eta}$ for these objects for the integral φ on B.

We now have the following results concerning the actions π and λ as defined in the previous section.

Proposition 4.2. *The maps $\eta(x) \mapsto \eta(ax)$ and $\eta(x) \mapsto \langle S^{-1}(x_{(1)}), b\rangle \eta(x_{(2)})$ extend to bounded linear maps on $\mathcal{H}$. We will use $\pi(a)$ for the first one and $\lambda(b)$ for the second one (in agreement with the notations used in the previous section). We have $\pi(a^*) = \pi(a)^*$ and $\lambda(b^*) = \lambda(b)^*$ whenever $a \in A$ and $b \in B$.*

These last formulas are relatively easy to verify, using the left invariance of φ. However, the boundedness of the maps $\pi(a)$ and $\lambda(b)$ is not so obvious. We will give an argument later.

We have similar results for B. That is, we have bounded maps $\pi(b)$ and $\lambda(a)$ on $\hat{\mathcal{H}}$ given by $\pi(b)\hat{\eta}(y) = \hat{\eta}(by)$ and $\lambda(a)\hat{\eta}(y) = \langle a, y_{(1)}\rangle\hat{\eta}(y_{(2)})$ whenever $a \in A$ and $b, y \in B$. Also here $\pi(b^*) = \pi(b)^*$ and $\lambda(a^*) = \lambda(a)^*$.

Because of the 'Plancherel formula' (cf. a remark after Proposition 3.4), we have the following:

Proposition 4.3. *The Fourier transform F is an isometry of $\mathcal{H}$ onto $\hat{\mathcal{H}}$ and it transforms the operators $\pi(a)$ and $\lambda(b)$ on $\mathcal{H}$ to respectively $\lambda(a)$ and $\pi(b)$ on $\hat{\mathcal{H}}$.*

This clarifies in a way the boundedness of the operators $\lambda(a)$ and $\lambda(b)$, provided we know the boundedness of the operators $\pi(a)$ and $\pi(b)$ already, but that is more obvious.

Now, we will see what we can say about the map W.

Proposition 4.4. *The linear operator W as defined from $A \otimes A$ to itself, 'extends' to a unitary operator on $\mathcal{H} \otimes \mathcal{H}$, still denoted by W.*

So we have

$$W(\eta(x) \otimes \eta(x')) = \eta(S^{-1}(x'_{(1)})x) \otimes \eta(x'_{(2)})$$
$$W^*(\eta(x) \otimes \eta(x')) = \eta(x'_{(1)}x) \otimes \eta(x'_{(2)}).$$

The unitary W is called the *left regular representation* of (A, Δ). If we take any other $x'' \in A$, and if we define a normal linear functional ω on $\mathcal{B}(\mathcal{H})$ (the algebra of all bounded linear operators on $\mathcal{H}$) by $\omega(z) = \langle z\eta(x'), \eta(x'')\rangle$, we see that $(\iota \otimes \omega)W^* = \varphi(x''^* x'_{(2)})\pi(x'_{(1)})$. It is this type of formula that can be used to argue that $\pi(a)$ is a bounded operator on $\mathcal{H}$. And similarly for $\lambda(b)$.

Again, we can apply the Fourier transform F. If we only apply it on the first leg, we get a unitary map U from $\hat{\mathcal{H}} \otimes \mathcal{H}$ to itself given by

$$U(\hat{\eta}(y) \otimes \eta(x)) = \langle S^{-1}(x_{(1)}), y_{(1)}\rangle\, \hat{\eta}(y_{(2)}) \otimes \eta(x_{(2)}).$$

This is the Hilbert space version of the map given in Proposition 3.9.

We can look at this formula in two ways. First, the right hand side is

$$\lambda(S^{-1}(x_{(1)}))\hat{\eta}(y) \otimes \eta(x_{(2)}).$$

This is the form we get when we start from the formula with W and apply the Fourier transform F on the first leg in the tensor product.

We can also look at the right hand side as

$$\hat{\eta}(y_{(2)}) \otimes \lambda(y_{(1)})\eta(x)$$

and when we apply the Fourier transform once more, now on the second leg, we get the operator on $\hat{\mathcal{H}} \otimes \hat{\mathcal{H}}$ given by

$$\hat{\eta}(y) \otimes \hat{\eta}(y') \mapsto \hat{\eta}(y_{(2)}) \otimes \hat{\eta}(y_{(1)}y').$$

When we flip the two components in the tensor product $\hat{\mathcal{H}} \otimes \hat{\mathcal{H}}$, we precisely get the adjoint of the unitary operator $\hat{W}$ defined by

$$\hat{W}(\hat{\eta}(y) \otimes \hat{\eta}(y')) = \hat{\eta}(S^{-1}(y'_{(1)})y') \otimes \hat{\eta}(y_{(2)}).$$

This is the left regular representation of (B, Δ). So we find that the Fourier transform carries W into $\Sigma \hat{W}^* \Sigma$ where Σ is used here to denote the flip on $\hat{\mathcal{H}} \otimes \hat{\mathcal{H}}$.

We will come back to this operator later. Now, we want to apply the Tomita–Takesaki theory. According to this theory, we get the following (see e.g. [17]). We denote by M the weak operator closure of $\pi(A)$ and by M' the commutant of M.

Proposition 4.5. *There exists a positive, self-adjoint, non-singular operator ∇ and a unitary, conjugate linear involution J on $\mathcal{H}$ such that $\eta(x) \in \mathcal{D}(\nabla^{\frac{1}{2}})$ and $\eta(x^*) = J\nabla^{\frac{1}{2}}\eta(x)$ for all $x \in A$. The space $\eta(A)$ is a core for the domain of $\nabla^{\frac{1}{2}}$. We also have that $JMJ = M'$ and $\nabla^{it}M\nabla^{-it} = M$ for all $t \in \mathbb{R}$.*

The maps σ_t, defined by $\sigma_t(x) = \nabla^{it}x\nabla^{-it}$ are called the *modular automorphisms*. One can show that $\pi(a)$ for $a \in A$ is analytic for this one-parameter group of automorphisms and that $\sigma_{-i}(\pi(a)) = \pi(\sigma(a))$ (see [13]). Do not confuse the automorphism σ on A with the one-parameter group of automorphisms (σ_t) on M.

Similarly, we have operators $\hat{\nabla}$ and $\hat{J}$ on $\hat{\mathcal{H}}$ satisfying $\hat{\eta}(y^*) = \hat{J}\hat{\nabla}^{\frac{1}{2}}\hat{\eta}(y)$ whenever $y \in B$. They have the same properties w.r.t. the von Neumann algebra $\hat{M}$, obtained by taking the weak closure of $\pi(B)$ on $\hat{\mathcal{H}}$.

Let us now consider a very *special functional* on the Heisenberg algebra C, introduced in the previous section. For a treatment of this functional in the case of a general locally compact quantum group, see [19].

Definition 4.6. Define a linear functional f on the Heisenberg algebra C by $f(ba) = \varphi(b)\varphi(a)$ for $a \in A$ and $b \in B$.

Lemma 4.7. *This functional is positive and $f(a^*ba) = \varphi(b)\varphi(a^*a)$ for all $a \in A$ and $b \in B$.*

This is easy to verify. Indeed

$$aba' = \langle a_{(1)}, b_{(1)}\rangle b_{(2)}a_{(2)}a'$$

so that

$$\begin{aligned} f(aba') &= \langle a_{(1)}, b_{(1)}\rangle \varphi(b_{(2)})\varphi(a_{(2)}a') \\ &= \langle a_{(1)}, 1\rangle \varphi(b)\varphi(a_{(2)}a') \\ &= \varphi(b)\varphi(aa'). \end{aligned}$$

Now, we want to apply the G.N.S.-construction for f and relate the Tomita–Takesaki data for f with those for φ on A and φ on B as introduced above.

We first have the following.

Proposition 4.8. *We can identify the Hilbert space $\mathcal{H}_f$ with $\hat{\mathcal{H}} \otimes \mathcal{H}$ and we get for the canonical imbedding*

$$\eta_f(yx) = \hat{\eta}(y) \otimes \eta(x)$$

whenever $x \in A$ and $y \in B$. The G.N.S.-representation π_f is given by

$$\pi_f(a) = \lambda(a_{(1)}) \otimes \pi(a_{(2)})$$
$$\pi_f(b) = \pi(b) \otimes 1$$

when $a \in A$ and $b \in B$.

Proof. The first statement is obvious as

$$f((yx)^*(yx)) = f(x^*y^*yx) = \varphi(y^*y)\varphi(x^*x)$$

for all $x \in A$ and $y \in B$ by Lemma 4.7. As $\pi_f(b)\eta_f(yx) = \eta_f(byx)$, we see immediately that $\pi_f(b) = \pi(b) \otimes 1$ when $b \in B$. Finally, using the commutation rules of Lemma 3.2, we get

$$\begin{aligned}\pi_f(a)\eta_f(yx) = \eta_f(ayx) &= \langle a_{(1)}, y_{(1)}\rangle\ \eta_f(y_{(2)}a_{(2)}x)\\ &= \langle a_{(1)}, y_{(1)}\rangle\ \hat{\eta}(y_{(2)}) \otimes \pi(a_{(2)})\eta(x)\\ &= \lambda(a_{(1)})\hat{\eta}(y) \otimes \pi(a_{(2)})\eta(x)\end{aligned}$$

whenever $a, x \in A$ and $b \in B$, giving the last formula. □

We know from the previous observations that the comultiplication Δ on A is implemented by W in the sense that $\Delta(a) = W^*(1 \otimes a)W$. Consider this formula on $\mathcal{H} \otimes \mathcal{H}$ and apply the Fourier transform on the first leg. Then, we find

$$\pi_f(a) = U^*(1 \otimes \pi(a))U$$

where U is, as before, given by

$$\begin{aligned}U(\hat{\eta}(y) \otimes \eta(x)) &= \langle S^{-1}(x_{(1)}), y_{(1)}\rangle\ \hat{\eta}(y_{(2)}) \otimes \eta(x_{(2)})\\ &= \hat{\eta}(y_{(2)}) \otimes \lambda(y_{(1)})\eta(x)\\ &= \lambda(S^{-1}(x_{(1)}))\hat{\eta}(y) \otimes \eta(x_{(2)}).\end{aligned}$$

Next, we consider the polar decomposition $J_f\nabla_f^{\frac{1}{2}}$ of the operator T_f defined as the closure of the map $\eta_f(z) \mapsto \eta_f(z^*)$ with z in the Heisenberg algebra C.

Proposition 4.9. *We have*

$$J_f = (\hat{J} \otimes J)U = U^*(\hat{J} \otimes J)$$
$$\nabla_f = \hat{\nabla} \otimes \nabla$$

where U and also J, $\hat{J}$, ∇ and $\hat{\nabla}$ are as above. Moreover, U commutes with ∇_f.

Proof. For $x \in A$ and $y \in B$ we have

$$\begin{aligned} T_f \eta_f(yx) &= \eta_f(x^* y^*) \\ &= \langle x^*_{(1)}, y^*_{(1)} \rangle \eta_f(y^*_{(2)} x^*_{(2)}) \\ &= U^* \hat{\eta}(y^*) \otimes \eta(x^*). \end{aligned}$$

It follows, by the uniqueness of the polar decomposition, that

$$\begin{aligned} J_f &= U^*(\hat{J} \otimes J) \\ \nabla_f &= \hat{\nabla} \otimes \nabla. \end{aligned}$$

Using the properties of the operators involved, we find easily that also $J_f = (\hat{J} \otimes J)U$ and that U commutes with ∇_f. □

This result for general locally compact quantum groups can be found e.g. in [11], Corollary 2.2. And as in that same paper, some nice conclusions can be drawn from the previous result.

For this we need to introduce some notations. We have used M and $\hat{M}$ to denote the von Neumann algebras on $\mathcal{H}$ and $\hat{\mathcal{H}}$ generated by $\pi(a)$ with $a \in A$ and $\pi(b)$ with $b \in B$ respectively. We will use $\hat{N}$ and N to denote the von Neumann algebras on $\mathcal{H}$ and $\hat{\mathcal{H}}$ generated by $\lambda(b)$ with $b \in B$ and $\lambda(a)$ with $a \in A$ respectively.

Observe that the pair $(M, \hat{N})$ acts on $\mathcal{H}$ and that the pair $(N, \hat{M})$ acts on $\hat{\mathcal{H}}$. The Fourier transform sends M to N and $\hat{N}$ tot $\hat{M}$ (because of Proposition 3.5).

The operator U has its first leg in N and its second leg in $\hat{N}$ (see e.g. the formulas for U given before Proposition 4.9). Because the two legs of U generate these von Neumann algebras, we get, as a consequence of Proposition 4.9, the following (see e.g. Proposition 2.1 in [11]).

Proposition 4.10.

$$\begin{aligned} \hat{\nabla}^{it} N \hat{\nabla}^{-it} &= N, & \hat{J} N \hat{J} &= N, \\ \nabla^{it} \hat{N} \nabla^{-it} &= \hat{N}, & J \hat{N} J &= \hat{N}. \end{aligned}$$

We can be more specific. Define linear maps R and $\hat{R}$ on N and $\hat{N}$ respectively by $R(x) = \hat{J} x^* \hat{J}$ and $\hat{R}(y) = J y^* J$. Also define one-parameter groups of automorphisms (τ_t) and $\hat{\tau}_t$ on N and $\hat{N}$ respectively by $\tau_t(x) = \hat{\nabla}^{it} x \hat{\nabla}^{-it}$ and $\hat{\tau}_t(y) = \nabla^{it} y \nabla^{-it}$. Then we have the following.

Proposition 4.11. *We have $\lambda(a) \in \mathcal{D}(\tau_{-\frac{i}{2}})$ and $\lambda(S(a)) = R\tau_{-\frac{i}{2}}(\lambda(a))$ for all $a \in A$. Similarly, $\lambda(b) \in \mathcal{D}(\hat{\tau}_{-\frac{i}{2}})$ and $\lambda(S(b)) = \hat{R}\hat{\tau}_{-\frac{i}{2}}(\lambda(b))$ for all $b \in B$.*

We will not give a rigorous proof, just indicating why these formulas are true.

Take first $a \in A$ and $y \in B$. Then

$$\begin{aligned}
\hat{J}\hat{\nabla}^{\frac{1}{2}}\lambda(a)\hat{\eta}(y) &= \hat{J}\hat{\nabla}^{\frac{1}{2}}\langle a, y_{(1)}\rangle\hat{\eta}(y_{(2)}) \\
&= \langle a, y_{(1)}\rangle^{-}\hat{\eta}(y^*_{(2)}) \\
&= \langle S(a)^*, y^*_{(1)}\rangle\hat{\eta}(y^*_{(2)}) \\
&= \lambda(S(a)^*)\hat{\eta}(y^*) \\
&= \lambda(S(a)^*)\hat{J}\hat{\nabla}^{\frac{1}{2}}\hat{\eta}(y).
\end{aligned}$$

We get formally

$$\lambda(S(a)^*) = \hat{J}\hat{\nabla}^{\frac{1}{2}}\lambda(a)\hat{\nabla}^{-\frac{1}{2}}\hat{J}$$

and so $\lambda(S(a)) = R\tau_{-\frac{i}{2}}(\lambda(a))$.

Similarly, take $b \in B$ and $x \in A$. Then

$$\begin{aligned}
J\nabla^{\frac{1}{2}}\lambda(b)\eta(x) &= J\nabla^{\frac{1}{2}}\langle S^{-1}(x_{(1)}), b\rangle\eta(x_{(2)}) \\
&= \langle x_{(1)}, S(b)\rangle^{-}\eta(x^*_{(2)}) \\
&= \langle x^*_{(1)}, b^*\rangle\eta(x^*_{(2)}) \\
&= \langle S^{-1}(x^*_{(1)}), S^{-1}(b^*)\rangle\eta(x^*_{(2)}) \\
&= \lambda(S^{-1}(b^*)\eta(x^*) \\
&= \lambda(S(b)^*)J\nabla^{\frac{1}{2}}\eta(x).
\end{aligned}$$

Again formally

$$\lambda(S(b)^*) = J\nabla^{\frac{1}{2}}\lambda(b)\nabla^{-\frac{1}{2}}J$$

and so $\lambda(S(b)) = \hat{R}\hat{\tau}_{-\frac{i}{2}}(\lambda(b))$.

As we see from the above, the use of the functional f, as defined in Definition 4.6 (and studied in more detail in [19]), in combination with the use of the Fourier transform, admits a certain way of understanding the very basic formulas obtained by Kustermans and Vaes in [10] and [11]. The remaining step to arrive really to their framework is to identify the spaces $\mathcal{H}$ and $\hat{\mathcal{H}}$ so that the Fourier transform F becomes the identity map. Then the pairs $(M, \hat{N})$ and $(N, \hat{M})$ will become the same (namely $(M, \hat{M})$ in their terminology). We believe that it has some advantage to wait for this identification until some level of understanding has been obtained (as we did in this paper).

References

[1] E. Abe, *Hopf algebras,* Cambridge University Press, Cambridge 1977.

[2] S. Baaj and G. Skandalis, Unitaires multiplicatifs et dualité pour les produits croisés de C^*-algèbres, *Ann. Sci. École Norm. Sup.* (4) 26 (1993), 425–488.

[3] B. Drabant and A. Van Daele, Pairing and the quantum double of multiplier Hopf algebras, *Algebr. Represent. Theory* 4 (2001), 109–132.

[4] B. Drabant, A. Van Daele and Y. Zhang, Actions of Multiplier Hopf Algebras, *Comm. Algebra* 27 (1999), 4117–4127.

[5] M. Enock and J.-M. Schwartz, *Kac algebras and duality for locally compact groups,* Springer-Verlag, Berlin 1992.

[6] R. V. Kadison and J. R. Ringrose, *Fundamentals of the theory of operator algebras,* Grad. Stud. in Math. 16, Amer. Math. Soc., Providence, RI, 1997.

[7] J. Kustermans, *C^*-algebraic quantum groups arising from algebraic quantum groups,* Ph.D. thesis, K.U. Leuven 1997.

[8] J. Kustermans, The analytic structure of algebraic quantum groups, to appear in *J. Algebra,* preprint, K.U. Leuven 2000, a rewritten form of `#funct-An/970710` with some results added from `#funct-An/9704004`.

[9] J. Kustermans and S. Vaes, A simple definition for locally compact quantum groups, *C. R. Acad. Sci. Paris Sér.* I 328 (1999), 871–876.

[10] J. Kustermans and S. Vaes, Locally compact quantum groups, *Ann. Sci. École Norm. Sup.* (4) 33 (2000), 837–934.

[11] J. Kustermans and S. Vaes, Locally compact quantum groups in the von Neumann algebra setting, to appear in *Math. Scand.*

[12] J. Kustermans, S. Vaes, A. Van Daele, L. Vainerman and S. L. Woronowicz, Locally compact quantum groups, in *Lecture Notes for the School on Noncommutative Geometry and Quantum Groups in Warsaw (September* 17–29, 2001), Banach Center Publ., Warsaw, to appear.

[13] J. Kustermans and A. Van Daele, C^*-algebraic quantum groups arising from algebraic quantum groups, *Internat. J. Math.* 8 (1997), 1067–1139.

[14] J. Kustermans and A. Van Daele, The Heisenberg commutation relations for an algebraic quantum group I, II, in preparation.

[15] M. Landstad, private communication, 2001.

[16] S. Stratila and L. Zsidó, *Lectures on von Neumann algebras*, Editura Academiei/Abacus Press, Bucuresti/Tunbridge Wells, Kent, 1979.

[17] S. Stratila, *Modular theory in operator algebras.* Editura Academiei/Abacus Press, Bucuresti/Tunbridge Wells, Kent, 1981.

[18] M. E. Sweedler, *Hopf algebras,* Math. Lecture Note Ser., W. A. Benjamin, New York 1969.

[19] S. Vaes and A. Van Daele, The Heisenberg commutation relations, commuting squares and the Haar measure on locally compact quantum groups, preprint K.U. Leuven/Paris VI, 2002, 26pp., to appear in *Proceedings of the OAMP conference in Constantza, Roumania*, 2001.

[20] S. Vaes and L. Vainerman, Extensions of locally compact quantum groups and the bicrossed product construction, to appear in *Adv. Math.*

[21] S. Vaes and L. Vainerman, On low-dimensional locally compact quantum groups, in *Locally compact quantum groups and groupoids* (L. Vainerman, ed.), IRMA Lectures in Math. Theoret. Phys. 2, Walter de Gruyter, Berlin–New York 2003, 127–187.

[22] A. Van Daele, Dual pairs of Hopf *-algebras, *Bull. London Math. Soc.* 25 (1993), 209–230.

[23] A. Van Daele, Multiplier Hopf algebras, *Trans. Amer. Math. Soc.* 342 (1994), 917–932.

[24] A. Van Daele, An algebraic framework for group duality, *Adv. Math.* 140 (1998), 323–366.

[25] A. Van Daele, *The Haar measure on some locally compact quantum groups*, preprint K.U. Leuven 2001, 51pp., preprint `math.OA/0109004`.

[26] A. Van Daele, The Haar measure on finite quantum groups, *Proc. Amer. Math. Soc.* 125 (1997), 3489–3500.

[27] A. Van Daele and Y. Zhang, A survey on multiplier Hopf algebras, in *Hopf Algebras and Quantum Groups* (S. Caenepeel and F. Van Oystaeyen, eds.), Lecture Notes in Pure and Appl. Math. 209, Marcel Dekker, New York 2000, 269–309.

Bei Fragen zur Produktsicherheit wenden Sie sich bitte an:
If you have any questions regarding product safety,
please contact:

Walter de Gruyter GmbH
Genthiner Straße 13
10785 Berlin
productsafety@degruyterbrill.com